THE COMMERCIAL AIRCRAFT FINANCE HANDBOOK

T0173907

'This handbook is the perfect match for each practitioner wishing to obtain condensed and focussed ad-hoc information at a highly sophisticated level within a short timeframe. The concept of this publication is unique among the textbooks on Aircraft Finance and will become an indispensable element of the practitioners' daily work.'
— **Matthias Reuleaux** LLM, *Managing Director & Legal Counsel,*
Aircraft Finance Department, NORD/LB

'I have known Ron for most of my professional career as a friend and trusted advisor, he is a brilliant lawyer with a keen business sense and great attention to detail. *The Commercial Aircraft Finance Handbook* is a needed, informative and refreshing reference filling a void in the complicated and complex Aviation Finance industry.'
— **Scott Paige**, *Portfolio Manager,*
AAG Capital Markets LLC, Miami

The Commercial Aircraft Finance Handbook is a resource for every type of aircraft finance practitioner – seasoned and starter alike. The handbook offers a comprehensive overview of the multifaceted matters that arise in the process of financing commercial aircraft. The book clearly reviews the different topics on a high-level basis, and then explains the terminology used for each particular area of specialization. It can be used as both a learning aid and reference resource.

The area of commercial aircraft finance is a multidisciplinary one, touching professionals across law, finance, insurance and leasing (to name a few) and this book arms these diverse practitioners with a framework for knowing the questions and issues that should be considered in an aircraft financing transaction.

This book will also provide practitioners just starting out in this field with an introduction to the myriad of topics in aircraft finance while providing more seasoned professionals with explanations of matters outside their normal area of expertise. As well, all practitioners will benefit from the resources provided in the appendices.

Ronald Scheinberg is a shareholder in the New York office of the international law firm of Vedder Price P.C. He has been working in the transportation equipment finance area since 1985. He received his law degree at Harvard Law School (*cum laude*) and his Bachelor of Arts degree at Brown University (*magna cum laude*, honors, *phi beta kappa*). Mr Scheinberg often lectures on equipment finance matters at industry conferences and has written widely on the subject.

THE COMMERCIAL AIRCRAFT FINANCE HANDBOOK

RONALD SCHEINBERG

Routledge
Taylor & Francis Group

LONDON AND NEW YORK

First published 2014 by Euromoney
Second edition published 2018
by Routledge

2 Park Square, Milton Park, Abingdon, Oxon, OX14 4RN
605 Third Avenue, New York, NY 10017

Routledge is an imprint of the Taylor & Francis Group, an informa business

First issued in paperback 2020

British Library Cataloguing-in-Publication Data
A catalogue record for this book is available from the British Library

Library of Congress Cataloging-in-Publication Data
A catalog record has been requested for this book

ISBN: 978-1-138-55899-1 (hbk)
ISBN: 978-0-367-73548-7 (pbk)

Typeset in Plantin
by Florence Production Ltd, Stoodleigh, Devon, UK

This book is dedicated to the memory of my father, Sidney Scheinberg (z'l). My dad was a 'lead' navigator of the 96th Bomb Group of the Eighth Air Force during World War II. He flew 35 missions in a B-17 bomber fighting for his country and earned the Distinguished Flying Cross and multiple other commendations. My dad's plane, the Sans Souci, was the lead plane in the largest air armada ever assembled (for the Battle of the Bulge). Much as he led dozens of aircraft to their targets, he was so instrumental in helping me navigate my way in life, instilling in me love of family, love of country, love of learning and love of ma'a'sim tovim (good deeds).

CONTENTS

PREFACE

It was an early Spring day in 1986. On that fateful day, I was a third year banking associate at Milbank, Tweed, Hadley & McCloy LLP, and I received a phone call from a Milbank partner whom I did not know (and who was not even in the banking department). This partner told me that so-and-so said he could call me with an assignment. 'Okay', I said (not that I had any choice in the matter). He directed me to appear at a midtown conference room[1] of (the now defunct law firm) Lord Day & Lord (counsel to US Air) the next morning to assist with the closing of a Japanese Leveraged Lease[2] on behalf of one his bank clients.

By this time of my nascent career, I was certainly adept at closing deals on behalf of the firm's banking clients. *But what is a Japanese Leveraged Lease, anyway? I must have thought to myself. Ahh. How complicated could it be?*

So, I showed up at the appointed hour, document checklist in hand, and reviewed the closing table to verify that everything on the list was in the right folder, with the right signatures and otherwise copacetic. All seemed to be in good order. We were about to start the closing call when that same Milbank partner came into the conference room and started poking around the closing table, summarily reviewing the items in the respective folders. I watched him rummaging through the papers as I ate a bagel. As he flipped open yet another of the folders, I saw him grimace.

'Rob', he said looking over at me, 'get over here'. For that moment, anyway, I was 'Rob'.

Uh oh, what did I do? I thought.

'Look at this!' he says to me, pointing at an attachment to the UCC-1 financing statement relating to the Japanese lease that listed the collateral securing the financing. I looked. It looked good to me, since the collateral description was properly lifted from the Japanese lease. I shrugged.

'Here!' he said, pointing at the language at the end of the schedule. It read: 'This is a precautionary UCC filing . . .'.[3]

'Isn't this a lease?' I asked.

'It's a Finance Lease! [You idiot]!' He didn't say 'You idiot', but he may just as well have done so.

'Okay', I replied, 'so it doesn't need that precautionary language since the Japanese Lease is properly construed as a security agreement?'

'Right', he said. So we fixed the UCC-1, and closed the deal. That partner never failed to remind me of that 'mistake' on my part during my periodic performance reviews.[4]

But how the heck was I supposed to know that a Japanese Leveraged Lease was not an Operating Lease? No one had explained that to me.

Anyhow, thinking back on that incident, and plenty of others in my career like it, I recently came to the realization that it would be useful to have a handbook to which I can turn that would give me a (very) basic understanding of terms of art that were (and are) thrown my way. My friend and client, Claudia Ziemer from NordLB similarly expressed the same sort of frustration of there not being a handy book to turn to for guidance. So, I decided to write this book to provide Aircraft Finance practitioners with such a resource. As I put together this Handbook, I found it extraordinary how many such terms there are in Aircraft Finance, and how diverse and complicated our language is.

This Handbook is geared not just for the novice Aircraft Finance banker, investment banker, lawyer, appraiser, hedge fund analyst or airline/lessor treasury officer, but also the seasoned professional who might have the need for quick access to any of the resources I supply here, whether insurance forms, aviation treaties, appraisal definitions or what have you.

The Commercial Aircraft Finance Handbook, as broad as it may seem, is not an all-encompassing guide to every area of aircraft finance. It is largely limited to *commercial* aircraft (as compared with corporate aircraft) and, while not specified in the title, *jet* aircraft (sorry, ATR). So, I do not delve into, for example, the fractional ownership issues relating to business aircraft, helicopter propeller perfection or blimp financing issues. The Handbook is not (my lawyers tell me to say) the provision of legal advice and, importantly, is limited to facts and circumstances as of the date of submission for publication. Revised and updated new editions may be produced as conditions (and demand!) warrant.[5]

I would like to thank my friends and clients Bertrand Grabowski, Eelco van de Stadt of DVB Bank, Michael Lypka of Seabury, Matthias Reuleaux of NordLB, JoAnne Chambers of KfW-IPEX Bank, Robert Agnew of MBA and Kostya Zolotusky of Boeing Capital for their valuable insights, suggestions and editorial comments on the earlier edition of this Handbook.[6] I would also like to thank Shiro Kambara who, unsolicited(!), offered me useful feedback on certain of the Japanese financing techniques. In addition, I would like to thank my Vedder Price colleagues Cameron Gee, Mehtap Cevher Conti, Gavin Hill, Geoffrey Kass and especially, Dean Gerber for spending the time to provide me with all manner of suggestions as to content and scope for this book. Also, thanks go to my Vedder Price associate Andrew Ceppos for his thorough read of an early draft that resulted in helpful edits and photo selections. Since I write in long-hand, I must extend a hearty thank you to my assistant, Elsa Batista, for typing up my drafts of this book, as well as to the Vedder Price Document Center folks here in New York who also assisted with the typing, formatting and chart production: Jeff Janus, Ermena Vinluan and especially, David Reichley who had an uncanny ability to decipher my chicken scratch. Also, I would like to thank Maziel Abrego from Vedder Price's marketing department for spearheading the effort for publication with Routledge Press.

Thank you to Routledge Press, especially my editor, Guy Loft and my copy editor, Florence Production Ltd, for making the second edition a reality.

Finally, this book would not have been possible without the support and encouragement of my family; my wife, Stacy has been so supportive throughout my career and was instrumental in having me 'go for it' in the writing of this book. She was a wonderful sounding board for so many matters pertaining to editorial decisions and was way smarter than I was in respect of the silly photos I had wanted to include.

Ronald Scheinberg

Notes

1 Yes, we had to all assemble in person at a central location since there was no Internet; no ability to shoot PDFs around the world to obviate the need for a physical in-person closing. Faxing on that awful slippery paper used at the that time was simply too difficult.
2 Capitalized terms used in this Handbook are defined terms explained in the text.
3 Language that is used only in Operating Leases because that type of lease is not a security agreement. A UCC-1 is filed in any case in the event that lease is somehow construed as a security agreement.
4 Notwithstanding that event, I was brought into, and became a member of, Milbank's aircraft-finance group and have since remained in that practice area; continuing to do so when I joined Vedder Price in 1992.
5 I invite readers to let me know of any mistakes they have found in the book, and to suggest new areas that can be covered and covered areas that can be expounded upon.
6 My friend and client, Jim Palen from Jeffries would have been cited here, but he never gave me comments. His excuse for not reviewing this book was that he was too busy generating business that would end up being business for me; so that excuse was happily accepted.

ABOUT THE AUTHOR

Ronald Scheinberg is a shareholder in the New York office of the international law firm of Vedder Price P.C. He specializes in the financing of transportation equipment (primarily aircraft), representing the financiers (banks, investment banks, insurance companies and hedge funds), lessors, airlines/lessees, manufacturers and underwriters in financings that are structured as leveraged leases, secured and unsecured loans, off-balance sheet leases, operating leases, cross-border leases, US Eximbank/ATSB-guaranteed facilities and securitizations utilizing EETC and other structures. He has been working in the transportation equipment finance area since 1985. Mr Scheinberg has been recognized as one of the leading practitioners in the area of equipment finance in numerous publications.

Mr Scheinberg received his law degree at Harvard Law School (*cum laude*) and his Bachelor of Arts degree at Brown University (*magna cum laude*, honors, *phi beta kappa*). Mr Scheinberg often lectures on equipment finance matters at industry conferences and has written widely on the subject.

Mr Scheinberg lives in New York City with his wife, three daughters and two dogs (Oreo and Millie).

PART 1

INTRODUCTION

'Aircraft Finance' is the art of financing aircraft. The Aircraft Finance market is in excess of U.S.$100 billion per year for the financing of new deliveries.[1] Insofar as commercial aircraft (and related assets) are highly expensive (the smallest in-production Boeing aircraft, the Boeing 737–700, has a list price of some U.S.$75 million; the largest Airbus aircraft, the A380 has a list price of almost U.S.$390 million; and the GE-90 engine that powers the Boeing 777 aircraft list at approximately U.S.$24 million *each*),[2] they often need to be financed. With inherent demand for financing on the operator side, there is similarly a pool of financiers ready, willing and able to supply financing to meet this demand.

Aircraft are perceived (especially by aircraft lenders and operating lessors – those who are putting the money in ('Aircraft Financiers' or 'Finance Parties')) to be assets that are highly desirable to finance: they hold their value, earn a decent return, are relatively easy to find and, in many jurisdictions, are subject to beneficial legal regimes that allow timely recovery in default situations. The world of Aircraft Finance is dominated by three principal actors: the aircraft operator/airline, the aircraft lessor and the aircraft lender.[3] Aircraft Finance transactions may have all three of these actors involved in a single transaction, such as one would find in a Back-leveraged Operating Lease, or any sub-combination:

- airline/operator-lessor, in a leasing transaction;
- airline/operator-lender, in a mortgage financing (reliant on the operator's credit); and
- lessor-lender, in a mortgage financing (reliant on the lessor's credit).

Each of these three parties brings a very different perspective to an Aircraft Finance transaction.

- The operator wants the cheapest possible cost (rent and return conditions or fees and interest charges).
- The lessor (vis-à-vis the operator) wants the maximum return (rent and return condition) and least amount of risk to its Aircraft Asset.
- The lessor (vis-à-vis the lender) wants the cheapest possible cost and maximum flexibility for use of the Aircraft Asset.
- The lender (vis-à-vis the operator and lessor) similarly wants the maximum return (fees and interest income) and least amount of risk to the Aircraft Asset.

Those different perspectives can be summarized, then, by the following relationships.

- Return versus Cost.
- Risk versus Flexibility.

Of course, these two categories are themselves linked: the greater the risk, the greater the required return. Risk tolerance and required levels of return, therefore, necessarily dictate different approaches between lessor and lenders, and between lessors and lenders, on the one hand, and the airline/operator on the other.

Exhibit 1.1 summarizes these primary risk/reward categories.

A further, and critical, overlay to this dynamic between the parties is the debtor-credit relationship among them. One of the very essential elements of the risk component is that obligations are largely uni-directional: the operator has the ongoing obligation to pay the rent/debt service and, in the case of a lease, return the Aircraft Asset,[4] while the Aircraft Financiers satisfy their obligations on day one by delivering the Aircraft and/or financing.[5]

Efforts to balance all these various rights and risks in the case of any given financing require a broad skill set. Participants are required to have, or have access

Exhibit 1.1 Primary risk/reward categories

	Primary risk vs flexibility issues	*Primary reward/return vs cost [economic] issues*
Operator (vs lessor or lender)	• Manner in which maintenance (for example, deferrals) is conducted • (Sub)-leasing rights • Re-registration rights • Insurance deductibles/ self-insurance	*Lessor:* • Rent • Term • Term renewal options • Purchase options • Security deposits • Maintenance reserves • Return conditions *Lender:* • Amortization profile (average life) • Maturity • Security deposits • Maintenance reserves • Interest rate • Prepayment premium/make whole • Breakage costs

Lessor (vs lender)	• Limit restrictions on re-leasing:
	  jurisdiction
	  economic terms
	  contractual terms
	• Cure rights
	• Remarketing rights
	• Buy-out rights
	• Equity squeeze protection
	• Recourse versus non-recourse
		• Amortization profile
		• LTV coverage
		• Maturity
		• Security deposits
		• Maturity extension options
		• Interest rate
		• Prepayment premium/make whole
		• Breakage costs

Source: Author's own.

to, investment banking structuring skills, legal analysis, technical (metal) comprehension, insurance expertise, tax and accounting expertise and loan pricing know-how, to name but a few of the skill sets in which Aircraft Finance practitioners of *all* professions must have some level of working knowledge proficiency. Take an airline bankruptcy, for example. Finance parties holding lease or debt financing of an aircraft operated by the bankrupt airline must be able to analyze their contractual rights under the transaction documents (with the airline and the other finance parties), their rights under law, their rights in respect of the aircraft, the value and operational condition of the aircraft, their sale/lease remarketing options if they were to repossess the aircraft, their financing and refinancing (or exit) options and their risks of ownership and insurance needs on a repossession (among others!). That is quite an interdisciplinary understanding. While no professional is expected to be an expert in all these areas, any Aircraft Finance professional worth his salt will at least know the questions that need to be asked outside his personal area of expertise so as to be able to build a sufficient knowledge base to respond to developments and opportunities.

This Handbook is an attempt to provide Aircraft Finance professionals – bankers, investment bankers, appraisers, lawyers, insurance brokers, airline and lessor treasury-types and hedge fund, private equity and rating agency analysts – with resources, not to make them experts in an area outside of their particular expertise, but to provide a framework for knowing what questions and issues may be coming into play in a given situation.

Accordingly, this Handbook touches on the various disciplines with which professionals in this area must deal. In Part 2, the type of assets encountered in Aircraft Finance are examined. In addition to the usual suspects, aircraft and engines, we touch on other assets encountered in Aircraft Finance transactions, including assets that may serve as collateral. In Part 3 we explore the potpourri of financing structures frequently employed and devices that are utilized to enhance different structures. Part 4 gets into the weeds with the multiplicity of, principally debt, pricing-related provisions. Part 5 touches on the different technical terms used with aircraft and engines. Part 6 reviews terminology employed in matters relating to the maintenance and return of aircraft and engines. Part 7 covers Aircraft Finance legal terms, including specialized terms used in contracts and terms that pertain to laws governing, among other things, the granting of security interests in aircraft and related assets and the perfection of those interests therein. Part 8 lists and explains terms relevant in the bankruptcy and work-out contexts. Part 9 details the terms and concepts frequently encountered addressing intercreditor issues among the parties involved in Aircraft Finance transactions. Part 10 explores the nomenclature pertinent to the hull and liability (and other) insurances pertinent to Aircraft Financiers. Part 11 explains the terminology utilized by appraisers in their efforts to appraise aircraft. Part 12 covers the disparate contracts and legal instruments that are frequently utilized in Aircraft Finance transactions. Part 13 reviews a number of the terms that are often used as short-hand references to certain Aircraft Finance-oriented contractual provisions. Part 14 goes through a number of other terms that are 'terms of art' commonly used in Aircraft Finance. Finally, Part 15 provides an overview of the risk factors associated with Aircraft Finance transactions. Each of these parts has, as appropriate, an introduction that seeks to supply some context to the relevant practice area.

Any Aircraft Finance transaction is built on five legs: (i) the credit of the debtor; (ii) the value of the aircraft; (iii) the soundness of the transaction structure; (iv) the economics of the transaction; and (v) the legal framework of the contracts and governing law. This Handbook touches on certain fundamentals of all five of these items.

Notes

1 The Aircraft Finance market for *used* Aircraft is yet another multi-billion market.
2 Of course, no one pays list price. A better guide to this price level might, at a minimum, be appraised 'new' prices. But even at these appraised prices, and even with significant discounts offered by Original Equipment Manufacturers (OEMs), these are still very expensive assets to purchase and collateral to maintain.
3 It might be suggested that there is yet a fourth principal actor: the private equity investor investing in airlines and operating lessors. Such investors typically take a long-term view on the aircraft/airline sector as the basis for their investments but seek at some point to flip their investments at a profit.
4 Or, in the case of financing by a lender of a lessor, the lessor has to pay the debt service.
5 Of course, until such performance, the operator is taking credit risk on the lessor and/or lender to perform its obligations. This reverse credit scrutiny has taken on a life of its own in recent years coincident with periods of illiquidity and the credit crunches experienced by many lessors and lenders.

PART 2

AIRCRAFT AND OTHER AIRCRAFT FINANCE ASSETS/COLLATERAL

Introduction

The assets listed below include those that are both typical (and obvious), such as aircraft and engines and their related leases, but also others that are: (i) targeted to a standard aircraft financing, such as insurance and security deposits; and (ii) from time to time financed in association with this space. Aircraft, airframes and engines hardly need further explication as defined terms. Everyone knows what they are. However, there may be certain subtleties for these terms worth bringing to the reader's attention (as done here). As well, these assets, and the other types of assets listed here, each have particular ways to perfect interests therein, which are summarized below.

The Perfection summaries, unless otherwise noted, speak to Perfection in the U.S. only. Aircraft Finance professionals should consult with local counsel on matters of Security Interests and Perfection in all relevant jurisdictions (including the U.S.).

The assets in Aircraft Finance are at the crux of the business. For lessors, Aircraft assets *are* their business: owning, leasing and managing them is what they do. For lenders, the Aircraft assets serve as the collateral assuring repayment; only with these assets as security do specialized aircraft-asset-based lenders extend credit. Of course, superficial definitions of these highly complex assets hardly scratch the surface of the technical expertise necessary to be an aircraft finance professional. Parts 5 and 6 below do scratch the surface on some of the technical aspects of these assets, but be mindful that it is just a scratch.

There may be other types of credit and/or asset support that can serve as asset/credit support such as Letters of Moral Intent (LOMIs), Residual Value Agreements, and so on. See Part 12 for a listing of some other support documents.

Aircraft

An Aircraft is an Airframe together with the full complement of Engines that are associated with such Airframe. Here are the FAA definitions: a device that is used or intended to be used for flight in the air; FAR §1.1. Any contrivance invented, used, or designed to navigate, or fly in, the air; 49 USC §40102(6).

PERFECTION

The Chicago Convention points to the 'state of registration' of an Aircraft for its nationality in making the determination for jurisdiction that is relevant for purposes of perfection. Most jurisdictions have unique systems for perfecting in Aircraft. Happily, the system of perfecting interests in Aircraft is becoming standardized by the ever-growing stable of countries that have adopted the Cape Town Convention, which requires registration of interests at the International Registry. In the U.S., registrations must also be made at the FAA (a so-called 'entry point'). In non-Cape Town Contracting States, local counsel should be consulted on Perfection protocols for Aircraft. See Part 7, 'Blue Sky', for English law issues relating to Security Interests.

PRACTICE NOTE

Aircraft are financed *in toto*; other than Engines and In-flight Entertainment (IFE), individual parts while installed on Aircraft, such as seats or Winglets, cannot be separately financed without the express agreement of a lessor or mortgagee having an interest in that Aircraft (which will be very hard to come by). This is because interests in Aircraft, whether ownership, leasehold or mortgagee, attach to all Parts installed on or affixed to the Aircraft at the time of the creation of such interests, unless otherwise agreed.[1]

It is important to note that the International Registry is not a title registry.

[1] However, as discussed in 'Parts', certain Parts may be separately encumbered, and such encumbrance would not be lost upon attachment of such Parts to an Aircraft.

Aircraft Assets

Collectively, these are Aircraft, Airframes, Engines, Parts and Spare Parts. For each of these, the related Aircraft Records are, for the purpose of this definition as used in this Handbook, subsumed in the related asset.

Aircraft Records

These are the data, manuals and records relating to Aircraft, Airframes and Engines, including the history of all repair and maintenance tasks performed on Parts of the asset. The importance of: (i) an airline/operator maintaining full, complete and accurate Aircraft Records; and (ii) the Aircraft Financier having access to those records cannot be overemphasized. An Aircraft Asset with incomplete Aircraft Records is not eligible for an Airworthiness Certificate which, of course, very much affects the asset's value. Aircraft Records for all Life Limited Parts (LLPs) should have Back-to-Birth traceability. Accordingly, in repossession and other return contexts, it is imperative not only to repossess the financed Aircraft Asset, but to obtain all of the associated Aircraft Records in each case in complete and verifiable form. Reconstruction of missing Aircraft Records is extremely expensive.

PERFECTION

Perfection of Aircraft Records most often follows with the Perfection of the associated Aircraft, Airframe or Engine.

PRACTICE NOTE

Aircraft operators must, obviously, maintain proper records, but also keep them in English and in a format (paper or electronic) acceptable to one of the major Aviation Authorities that would permit operation of the relevant Aircraft Asset in that authority's jurisdiction.

Airframe

This is an Aircraft except for any Engines. Here is the FAA definition: the fuselage, booms, nacelles, cowlings, fairings, airfoil surfaces (including rotors but excluding propellers and rotating airfoils of engines) and landing gear of an aircraft and their accessories and controls; FAR §1.1.

PERFECTION

See 'Aircraft'.

Bank Accounts

These are accounts at banks, typically, into which Lease rentals or Security Deposits and/or Maintenance Reserves are deposited and thereafter held, subject to agreed disbursement arrangements.

PERFECTION

Under U.S. law, Bank Accounts are Perfected through possession and control. If the Finance Parties do not hold the Bank Accounts personally, they must Perfect their interest therein by means of an Account Control Agreement.

Engine

This is the component of the propulsion system for an Aircraft that generates mechanical power. Aircraft engines are almost always either lightweight piston engines or gas turbines. For FAA purposes, an Engine ('aircraft engine') is an engine used, or intended to be used, to propel an aircraft, including a part, appurtenance, and accessory of the engine, except a propeller; 49 USC §40102(7). An Engine without a Quick Engine Change (QEC) kit is known as a 'bare engine'. An Engine without its casing is known as a 'propulser' and can be financed in the same way as an Engine.

PERFECTION

See 'Aircraft'. However, unlike Aircraft, Engines are not 'registered' in any jurisdiction.

PRACTICE NOTE

Engines are assemblages of a gazillion[1] parts and pieces. That is all well and good when they are fully assembled and, better yet, installed on-wing. But when they are with a Maintenance, Repair and Overhaul provider (MRO) or at another repair station (including that of the operator), they may be disassembled with all those parts and pieces strewn about undergoing repair or in different boxes and/or locations. In such a case, repossession (and reassembly) may be quite the chore and entail great expense.[2] The MRO may also have a lien over such Engine (or parts) constituting a Mechanic's Lien, which may have priority over the interests of the lessor and/or mortgagee. In addition, absent application of the Cape Town Convention in any applicable jurisdiction (or other local law recognizing Engines as distinct from any Airframe to which they are attached), there may be title annexation issues with Engines where their individual identity is lost upon attachment to an Airframe. To address this concern, Aircraft lessors are usually required in lease documentation to waive any rights they may obtain in any Engines installed on their Airframes that are not 'their' Engines. Finally, Engines may be subject to individualized engine maintenance contracts (see Part 6, 'Total Care Package').

Engines are frequently financed on a pooled or portfolio basis.

[1] Maybe not a real word, but my kids use it, and the idea being conveyed should be quite clearly understood.

[2] Under Section 1110(c) of the U.S. Bankruptcy Code, an Engine subject to return in a bankruptcy need not be reassembled if it has been disassembled for, say, maintenance.

Gate

The Gate is the premises used for the purpose of holdroom seating and boarding space and related aircraft parking positions to enplane and deplane passengers at any airport or terminal that an airline conducts scheduled operations, arising under any lease, usufruct, use agreement, facility agreement or similar agreement governing the right to use that portion of the premises demised or covered by such lease, usufruct, use agreement, facility agreement or similar agreement.

PERFECTION

In the U.S., Gates are usually held by airlines/operators through a leasehold interest obtained from the airport terminal authority. These leasehold rights would be collaterally assigned to the financier and such Security Interest would be Perfected by: (i) a UCC-1; and (ii) if the leasehold was recorded in the real estate records, a filing in the real estate records office. The consent of the landlord would also be procured.

Insurance

This includes the hull and liability insurances (and other types of insurance, if applicable, such as political risk insurance) maintained with respect to Aircraft, Airframes and Engines (see Part 10).

PERFECTION

In the U.S., UCC-1. As well, in the case of hull insurances, the practice is to be named as loss payee (and additional insured).

Lease (and related contracts)

This is the agreement evidencing an Operating Lease or Finance Lease, including all related credit support documentation such as Guarantees, Standby Letters of Credit (SBLOCs) and LOMIs.

PERFECTION

Most jurisdictions have unique systems for perfecting in Aircraft Leases. Happily, the system of perfecting interests in Aircraft Leases is becoming standardized by the ever-growing stable of countries that have adopted Cape Town, which requires registration of interests at the International Registry. In the U.S., registrations must also be made at the FAA (a so-called 'entry point').[1] In non-Cape Town Contracting States, local counsel should be conferred with on perfection protocols for Aircraft Leases. In the U.S., most other credit support documentation is Perfected by the filing of a UCC-1.

Lessee Consents, discussed below, provide further protection to financiers of Leases by putting the related lessees on notice of the collateral assignment.

[1] In the U.S., Leases may be perfected by possessing the chattel paper original thereof. While, as a practice matter, secured parties seek to take possession of the chattel paper originals of a Lease, perfection is largely dictated by specific lease-perfection protocols of the FAA and Cape Town. See Part 7, 'Chattel Paper Original'.

Maintenance Reserves

These are periodic reserves on account of Aircraft and Engines utilization paid by a lessee to a lessor in an operating lease to cover anticipated maintenance costs. Maintenance Reserves are typically calculated on the basis of Aircraft/Engine usage. For a more in-depth discussion of Maintenance Reserves, see the related definition in Part 5.

PERFECTION

Same as 'Bank Accounts'.

PRACTICE NOTE

In a Back-leveraged Lease, the lessor may seek to retain possession of the Maintenance Reserves, as lessors thrive on this cash. Allowing the lessor to hold the cash may be risky for the lenders since their Lien will not be perfected without possession. For higher credit lessors, lenders may be willing to allow this, subject to requirements to turn-over of the Maintenance Reserves if there occurs a prescribed Trigger Event; provided that there may be associated Claw Back risks associated with such a turn-over.

Ownership Interests

Depending on the relevant legal entity, these are the indicia of ownership therefore. For example:

- Corporation – shares;
- Owner or Statutory Trust – beneficial interests;
- LLC – membership interests; and
- Partnership – partnership interests.

PERFECTION

In the U.S.: (i) if certificated, then by possession or a UCC-1 filing; and (ii) if not certificated, then by filing a UCC-1.

PRACTICE NOTE

In many Aircraft Financings, the Finance Parties seek to procure a pledge over the Ownership Interests in the Special Purpose Vehicle (SPV) that may own the Aircraft (as well as any intermediate lessors as in a Lease/Head Lease Structure), in addition to a mortgage on the Aircraft itself. This additional collateral may be, at enforcement time, an easier item to foreclose upon (as compared to the Aircraft itself). In addition, some financings may be done without a mortgage on the Aircraft due to prohibitively high mortgage registration fees, stamp duty or similar taxes, notary fees or other costs, or expediency issues, in which case the Finance Parties must rely on the pledge over the Ownership Interests in the Aircraft-owning SPV. However, a pledge of Ownership Interests taken in lieu of a mortgage may not qualify as an equivalent credit risk mitigant under Basel II and Basel III.

Parts

These are appliances, components, parts, instruments, appurtenances, avionics, accessories, furnishings and other equipment of whatever nature (other than complete Engines), which are incorporated or installed in or attached to an Airframe or an Engine, or removed from it.

PERFECTION

Parts, when attached to Aircraft Assets, are perfected when the related Aircraft Asset is perfected. However, certain discrete Parts that do not become accessories when attached, such as Auxiliary Power Unit (APU) and avionics, may be subject to pre-existing liens, and such liens would retain their prior position notwithstanding such attachment.

PRACTICE NOTE

Generally speaking, title to, or a Security Interest in, a Part removed from an Airframe or an Engine remains with the owner/mortgagee until replaced.

Purchase Documentation

This is documentation by which a purchaser may be acquiring an Aircraft Asset or entities that own, or have an interest in, the same. Purchase Documentation will likely include Bills of Sale and a Purchase Agreement, each of which may provide valuable Warranties and agreements from the seller.

PERFECTION

In the U.S., by filing a UCC-1, but see Perfection in 'Warranties'.

Route

This is the right, license, permit, or other authorization whereby an airline is entitled or permitted to fly between two or more points, either within one country or between two countries, including without limitation, applicable designations pursuant to any transport agreement between the U.S. and a foreign government, frequencies, exemption and certificate authorities and 'behind/beyond rights'.[1]

PERFECTION

In the U.S., by filing a UCC-1, but subject to Department of Transportation (DOT) regulation.

Security Deposit

This is a cash deposit given to a lessor (or lender) by a lessee (or borrower) as security for the obligations of the lessee (or borrower) under the related lease (or loan). Security Deposits may be replaced by SBLOCs in some transactions. Security Deposits are not the property of the pledgee thereof (unless foreclosed upon), and must be refunded to the provider thereof upon the completion of a transaction and complete performance thereunder.

PERFECTION

Same as 'Bank Accounts'. SBLOCs are Perfected by possession.

PRACTICE NOTE

See Practice note in 'Maintenance Reserves'.

Simulator

A Simulator is a device that simulates aircraft flight and various aspects of the flight environment. Simulators are used for, among other reasons, the training of pilots and other flight crew.[2] Simulator types range from simple part-task trainers that cover one or more aircraft systems to full flight simulators with comprehensive aerodynamic and systems modelling. This range reflects a spectrum of increasing complexity and sophistication as to physical cockpit characteristics and quality of software models, as well as various implementations of sound, motion, and visual sensory cues. The following training device types are sometimes financed.

- Flight Training Device (FTD) – used for either generic or aircraft-specific flight training. Comprehensive flight, systems, and environmental models are required. High level FTDs require visual systems but not the characteristics of a Full Flight Simulator.
- Full Flight Simulator (FFS) – used for aircraft-specific flight training under rules of the appropriate national civil aviation regulatory authority. Under these rules, relevant aircraft systems must be fully simulated. All FFSs require outside-world visual systems and a motion platform.

An FFS duplicates relevant aspects of the aircraft and its environment, including motion. This is accomplished by placing a replica cockpit and audio-visual system

Exhibit 2.1 Aircraft simulator (FFS)
Source: iStock: #126265872

13

on a motion platform. A motion platform using up to six electric and electric-pneumatic jacks is the modern standard. The full six degrees-of-freedom are required for the highest level flight simulator qualification as regulated by Aviation Authorities such as the FAA and EASA. Full Flight Simulator Levels[3] are as follows.

- FAA FFS Level A – a motion system is required with at least three degrees of freedom.
- FAA FFS Level B – requires three axis motion and a higher-fidelity aerodynamic model than does Level A.
- FAA FFS Level C – requires a motion platform with all six degrees of freedom. Also lower transport delay (latency) over levels A and B. The visual system must have an outside world horizontal field of view of at least 75 degrees for each pilot.
- FAA FFS Level D – the highest level of FFS qualification currently available. Requirements are for Level C with additions. The motion platform must have all six degrees of freedom, and the visual system must have an outside-world horizontal field of view of at least 150 degrees, with a collimated (distant focus) display. Realistic sounds in the cockpit are required, as well as a number of special motion and visual effects.

The most advanced simulators may weigh several tons and are necessarily affixed to the ground on which they sit.

PERFECTION

In the U.S., by filing a UCC-1 and a fixture (real estate) filing.

PRACTICE NOTE

Simulators may be located on the debtor's property or on leased premises. If the latter, a landlord consent is typically obtained that, among other things, would allow the financier to have access to, and remove, the Simulator in the event of a default and not allow the landlord to dispose of the Simulator for a specified period following default.

The software incorporated in a Simulator makes up a great deal of the asset's value. In order to make use of a Simulator, then, the operator must have rights in the software (and access to software updates). This software is not typically 'owned' by the owner of the Simulator, nor does the right to use it simply follow the owner. Rather, the operator typically licenses the software (often from a multiplicity of vendors) and such license must be examined to determine whether it is exclusive to the operator or can be transferred as part of a financing. Imagine the complexity in dealing with these licenses in a Leveraged

Lease of a Simulator: the operator must transfer them to the lessor, who must collaterally assign them to the lender, and then the owner and lender must re-transfer them back to the operator so that the operator can use the software – and the licensor may need to agree to all of this if it is an exclusive license.

Slot

A Slot refers to a take-off and landing slot of an airline at an airport. In the U.S., slots are regulated by the DOT, where a slot is defined as the authority for a single Aircraft arrival or departure. Two slots are required to effect a transit or turnaround flight at a given airport. In 1985, the DOT issued the 'buy-sell rule', where slots were grandfathered to the airlines that held them at the time, creating in theory, a free market system under which airlines could transfer slots among each other in order to allow new airlines to gain access to high-density airports and smaller communities could see service added from their origin to these airports. In reality, however, airlines were unwilling to give up their slots, as airlines consider them an extremely valuable commodity. To free up slots, the FAA adopted a 'use it or lose it' rule, whereby airlines were required to use 80 percent of their slots over a two-month period. Airlines were able to work this rule in their favor by leasing out their slots to other airlines in order to maintain an 80 percent utilization level as required.

Because of capacity controls, slots at certain high density airports are highly coveted (and valuable). It is no coincidence that these airports are among the most important business and leisure markets in the U.S., thereby creating barriers to entry for non-incumbent airlines or barriers to expanded service for those with the fewest amount of slots.

Slots are typically divided into three categories: 'High-Peak', 'Low-Peak' and 'Non-Peak'. Non-Peak slots are slots in an hour in which a carrier might obtain authority directly from the FAA (these would have an appraised value of U.S.$0 since they are available for the asking). Low-Peak slots are slots in an hour which is near or above the cap and would need to be obtained from another carrier having existing slots and one that is more likely used for domestic or North American destinations. High-Peak slots are slots in an hour which is near or above the cap and would need to be obtained from another carrier having existing slots and is likely to be of interest to domestic and international operators. The valuation of slots is dependent on a number of factors: (i) comparable transactions, including slot lease and swap transactions; (ii) policy and planning discussions with the FAA and relevant municipal airport authorities; and (iii) historical trends, including time-of-day adjustments and passenger movement metrics.

PERFECTION

In the U.S., UCC-1, but subject to DOT regulation.

Spare Engine

A Spare Engine is an engine that is not designated as being part of a particular Aircraft and is held by an operator as a spare. Spare Engines are typically held by operators on standby in case an on-wing Engine must be taken out of service for scheduled or unscheduled maintenance so that there is a Spare Engine available to replace the Engine that must be serviced.

PERFECTION

See 'Engine'.

Spare Parts

These are parts that are maintained as an inventory for the replacement of Parts no longer usable on an Aircraft Asset. For FAA purposes, 'an accessory, appurtenance or part of an aircraft (except an aircraft engine or propeller), aircraft engine (except a propeller), propeller or appliance that is to be installed at a later time in an aircraft, aircraft engine, propeller or appliance'.[4] Airlines and MROs maintain large inventories of Spare Parts in order to have the parts on-hand if an incorporated Part breaks down, needs refurbishment, times-out and so on. Spare Parts are integral to the functioning of an airline. While certainly not the same type of collateral as an Aircraft or an Engine insofar as there may be thousands of units (as compared with a handful of Aircraft Assets), parties that finance these assets take much solace from the fact that they are indispensable to the running of an airline.[5]

There are three basic types of Spare Parts (see Part 5 for definitions).

1   Rotable Parts – Repairable Parts.
2   Expendable Parts – Consumable Parts.
3   Life Limited Parts – Consumable Parts.

Spare Parts financed for a U.S. certificated airline are entitled to the benefits of Section 1110 of the Bankruptcy Code.

PERFECTION

In the U.S., perfection in Spare Parts owned by a certificated air carrier is accomplished by making a filing with the FAA that must specify the warehouse location at which the Spare Parts are held. Spare Parts may be located on the debtor's property or on leased premises. If the latter, a landlord consent is typically obtained that, among other things, would allow the financier to remove the Spare Parts in the event of a default and not allow the landlord to dispose of the Spare Parts for a specified period following default. In

---

addition, if a warehouse facility has inventory which is not intended to be secured by a Spare Parts facility, then the debtor needs to have in place appropriate signage indicating which inventory is subject to the financing facility.

PRACTICE NOTE

An Aircraft Financier of Spare Parts will want to ensure that there is a sophisticated tracking and inventory system for this collateral to ensure proper inventory control, and that it has access to such system. In the event of an impending bankruptcy (especially if the airline may not stay in business), it is imperative that the financier closely monitor the inventory and shut-down access, if necessary.

Warranties

These are assurances given by a Seller, Original Equipment Manufacturer (OEM) or MRO as to certain qualities of an Aircraft Asset. A seller of an Aircraft Asset may warrant that it has good title to an Aircraft Asset free and clear of liens (see Part 7, 'Full Warranty Bill of Sale'). An OEM may warrant the overall merchantability of an Aircraft, Airframe or Engine, that it is free from defects and so on. An OEM may also warrant that an Aircraft, Airframe or Engine will meet certain performance criteria, such as fuel-burn, maintenance costs, Aircraft range and so on. An MRO will warrant as to the repairs it has made to an Aircraft Asset. Warranties from OEMs and MROs typically expire after some period of time; and Engine-related Warranties usually have a shorter lifespan than Airframe Warranties.

Warranties may also include ongoing relationships that an airline may have with an MRO to maintain aircraft, particularly Engines. See Total Care Package.

PERFECTION

In the U.S., UCC-1.

PRACTICE NOTE

Insofar as many Warranties are not purportedly assignable, the consent of the warrantor may need to be obtained in order to collaterally assign the Warranties to a financier. OEMs are reluctant to assign any but the most basic of the Warranties.

Notes

1 Behind/beyond rights relate to routes prior to entry into a country and after leaving a country.
2 They are also used to entertain investors, bankers and contest winners.
3 See 14 CFR 60 (2012).
4 49 USC §40102(43) (2011).
5 That works, of course, only if the airline continues operations, or continues to operate the type of Aircraft Asset for which the Spare Part(s) are suited.

PART 3

DEAL TYPES, STRUCTURES AND ENHANCEMENTS

Introduction: basic transaction types

Aircraft Finance transactions come in many shapes and sizes but, fundamentally, they are either a lease financing or a mortgage financing (or some combination thereof). These two fundamental prototypes can be summarized as follows.

Lease Financing

The airline/operator of an Aircraft Asset (lessee) does not own the asset but borrows it from a third party who does own it (lessor) for a period of time (lease term) during which the lessee will have the right to possession and use of such asset in exchange for rent, and at the end of the lease term the airline/operator must return that asset to the lessor. While leases may be Operating Leases or Finance Leases (which are disguised Mortgage Financings), for the purposes of this book we are treating a 'Lease Financing' as a financing with the characteristics of an Operating Lease, where the lessor is treated as the true owner of the Aircraft Asset; that is, see 'Finance/Capital Lease', for an extensive discussion of the Finance Lease versus Operating Lease criteria determinants and see Exhibit 3.1 for a schematic depiction of a Lease Financing.[1]

Mortgage Financing

The owner of an Aircraft Asset (whether a lessor or airline/operator) borrows money from lenders to finance or refinance the purchase price (or value) of that asset and the lenders receive a Security Interest in that asset as collateral to secure repayment of the related loan. A Mortgage Financing includes a Finance Lease where the lessor is treated, not as an owner, but rather as a mortgagee. See Exhibit 3.2 for a schematic depiction of a Mortgage Financing.

The variations on these two prototypes keep Aircraft Financiers rather busy. Financings may involve multiple jurisdictions to take advantage of different tax regimes; they may involve public or privately placed securities to achieve the best pricing; they may be placed in the capital markets or the commercial bank/loan market; they may involve a single Aircraft Asset, or many of them; and they may be subject to airline or lessor risk or have the support of an Export Credit Agency (ECA).

Ultimately, the structure choice of Lease Financing or Mortgage Financing is determined by the operator in its 'lease versus own' analysis. The following criteria will have a hand in the operator's decision.

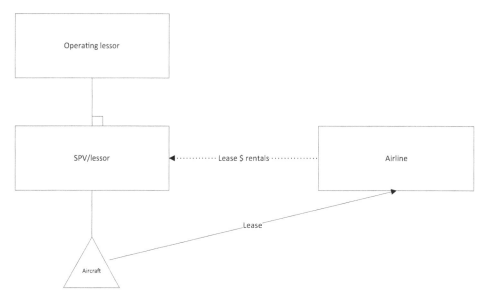

Exhibit 3.1 Lease Financing

Source: Author's own

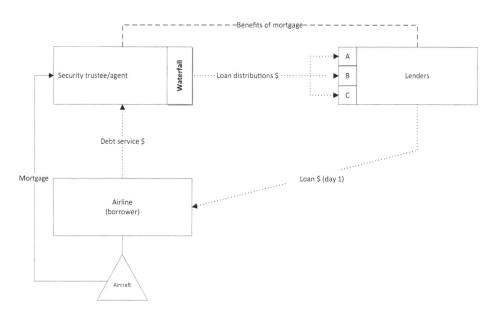

Exhibit 3.2 Mortgage Financing

Source: Author's own

- Operational flexibility.
- Cost:
  - lease rates versus debt rates; and
  - ability to take advantage of tax benefits.
- Residual risk:
  - long-term view of asset value appreciation;
  - risk appetite;
  - view on emergence of new aircraft models and technology;
  - view on current and future aircraft needs ('lift') and gaps (taking into consideration its own order book);[2]
  - future maintenance costs;
  - perceived future availability of asset model from other sources; and
  - Original Equipment Manufacturers (OEM) order book for future deliveries generally.
- Corporate policy:
  - keep fleet 'young';[3]
  - interest in de-leveraging (in a Lease Financing, the lessor would be the one to bear the burden of debt financing); and
  - accounting policy to minimize on-balance sheet debt.[4]
- Technological advancements:
  - cheaper maintenance for newer models;
  - better fuel burn for newer models; and
  - ability to retrofit these advancements on older models (for example, Winglets).
- Access to capital:
  - debt markets;
  - equity markets; and
  - lessor markets.
- Tax:
  - ability to take advantage of tax benefits available to an owner (for example, depreciation).

In this Part of the Handbook, the various structures used to finance Aircraft will be reviewed as well as different features that may be appended onto these structures.

Placement

Prior to delving into the various transaction structures, a word should be said about the *placement* of Aircraft Finance transactions in the financial markets. There are two considerations here: (i) the identity of the purchasers of the financing; and (ii) the nature of the placement itself.

The Purchaser

While the previous discussion largely revolved around the needs of the airline/operator, the other side to these matters is making sure the needs of the purchaser of the Aircraft Finance security are similarly satisfied. These purchasers can be broken down into two groupings:

1   the third party equity, being the lessor/owner in a transaction; and
2   the debt.

EQUITY

In lease financings, there will be an equity component. The purchaser of the metal, the person who would become the lessor/owner of the Aircraft Asset, is driven by its interest in the operating business of leasing and obtaining: (i) residual returns; (ii) current cash flow (rent); and/or (iii) tax benefits. Its economic return requirements based on these three criteria must be satisfied, taking into account the following (non-comprehensive) set of risks associated with any Aircraft Finance transaction ('Risk Factors').

•   Lessee performance (and other credit support providers) (credit risk).
•   Aircraft asset value stability (technological developments, obsolescence, life cycle, secondary markets, and so on) (residual value).
•   Jurisdictional issues (ability to enforce/repossess – country risk) (legal).
•   Airline industry performance (local and international; terrorism, SARS, and so on).
•   Economy performance (local and international).
•   Airline regulatory developments (airworthiness directives (ADs), noise rules and so on).
•   Fuel costs.
•   Change-of-law/accounting rules.

Of course, the longer the lease term, the greater the chance that any of these Risk Factors might surface into a problem, thus the tenor of a transaction will have a tremendous bearing on risk assessment (and, accordingly, return requirements). Investors in Aircraft Asset equity include:

•   traditional operating leasing companies (AerCap, GECAS, ACG, and so on);
•   specialized operating leasing companies (Vx, Apollo, Compass Capital, and so on);
•   tax-play investors (for example, MetLife);
•   hedge and private equity funds (Strategic Value Partners, Fortress, Wayzata, and so on);
•   finance companies (for example, Siemens Credit);
•   banks (for example, SMBC); and
•   pension plans.

The longer term trends show Aircraft leasing to be on the rise. Approximately 40 percent of new Aircraft deliveries are funded by Operating Lease Finance. The investor pool is constantly expanding as investors from all of the sectors described above are looking for the 'next big opportunity', with Aircraft ownership and leasing an increasingly visible opportunity. The historically strong performance of commercial aircraft investments should continue to attract sufficient equity (and any necessary leverage).[5]

DEBT

In Back-leveraged Lease financings and in mortgage financings there is a requirement for debt. Providers of debt have more simple criteria than equity for assessing their return which is primarily covered by: (i) up-front fees; and (ii) interest earnings. The economic return requirements of debt providers will necessarily factor in the Risk Factors spelt out above and, similarly, the tenor of a transaction will have a bearing on the risk assessments and return requirements. Investors in Aircraft Asset debt components include:

- commercial banks;
- non-bank lenders: insurance companies, finance companies, pension plans and funds (for example, Fidelity);
- public debt/capital markets;
- ECAs (directly or through guaranteed-debt); and
- OEMs.

The interest of these equity and debt participants in Aircraft Finance rise and fall with market developments and cycles. Boeing Capital, in their annual 'Current Aircraft Finance Market Outlook, 2013–2017' provides a useful snapshot of the current trend lines for the various Aircraft Finance funding sources; see www.boeingcapital.com/cafmo/.

Distribution; securities laws

While placement of the equity component is typically a straightforward matter of private contract law, the distribution of the debt component will require an examination of the securities laws. In the U.S., the securities laws (primarily the Securities Act of 1933, as amended (the 'Securities Act')) covers the offer and sale of any, *inter alia*, debt security, and, absent an exemption, requires the preparation and filing with the SEC of a registration statement in respect thereof (which is a rather costly undertaking that requires both public disclosure and ongoing compliance obligations) for any offer and sale of such security. Here is a list of the primary debt placement options and their relationship to the Securities Act.

- Bank financing – bank loans are not a security covered by the Securities Act, so may be placed without a registration statement.
- Rule 144A Financing – Rule 144A is a rule of the SEC that allows an exemption to registration for resales of debt securities to institutional investors previously acquired in a private placement. Such institutional investors must be Qualified Institutional Buyers (defined below). Holders and prospective purchasers of Rule 144A-placed securities must have the right to obtain from the issuer, upon request, certain bank information about

the issuer and the securities. For this reason, in a Rule 144A transaction, there is often drafted an offering memo describing the transaction.

- Private placement – under Section 4(a)(2) of the Securities Act, the obligation to register the offer and sale of securities does not apply to transactions by an issuer not involving a public offering. The SEC adopted Regulation D under the Securities Act to provide guidance as to what would constitute a private placement. Regulation D has varying criteria for exemption depending on the size of the offering. Rule 506 of Regulation D under the Securities Act is the particular private placement exemption that most Aircraft-secured debt securities would utilize because there is no limit on the dollar amount that may be raised under Rule 506. To have a valid private placement under Rule 506: (i) investors must be Accredited Investors (defined below); and (ii) neither the issuer nor any person acting on behalf of the fund may offer or sell interests in the fund by any form of general solicitation or general advertising.

- Government-guaranteed securities – Section 3(a)(2) of the Securities Act exempts securities guaranteed by the U.S. Insofar as U.S. Ex-Im-guaranteed transactions (see 'ECA Financings') have the full faith and credit of the U.S. standing as guarantor,[6] the related securities sold to investors are not subject to the registration requirements of the Securities Act.

- Registered offering – while, as noted above, there are significant expenses and ongoing compliance obligations for registered offerings, investors provide to issuers pricing benefits if a security is registered, since the registration significantly enhances the liquidity of the security (as offerings and sales of debt securities in secondary market trades are similarly subject to securities laws restrictions). Issuers of debt securities who are already 'reporting companies' under the Securities and Exchange Act of 1934, as amended (for example, companies that are publicly traded and which, accordingly, regularly make filings with the SEC) have already in place the mechanisms for their initial disclosure and ongoing compliance requirements, so many of these issuers will take the extra steps to have their debt securities registered. For this reason, many of the Enhanced Equipment Trust Certificates (EETCs) issued by U.S. publicly-traded airlines are registered.

The placement of Aircraft Asset-backed securities in the financial markets is not without aspects of financial engineering. Investment bankers and others tasked with this placement slice and dice the securities to minimize yields by tranching the debt so as to appeal to different investors' risk, yield and tenor appetite. It would not be unusual to see, for example, an EETC security structured as follows:

- Senior A Tranches:
  - A-1 – amortizing over 12 years;
  - A-2 – bullet maturity in year 7; and
- Junior B Tranche.

Each Tranche, then, is targeted to the needs of particular investors, thus expanding the market and optimizing yields for the issuer.

Balance of supply and demand

In the light of the foregoing review, we see that the need for Aircraft finance is, practically speaking, driven by the 'lift' needs of the airline/operators. Their need for Aircraft Assets drives the need to find the means for financing these expensive assets. The need for financing – the *demand* side of the equation – must, then, be satisfied by a *supply* of financing sources. Whether the supply and demand are in balance – whether there is enough product to supply the needs of Aircraft Financiers, and whether there is enough available financing to meet the needs of the airlines (and lessors) – is not always assured. On the one hand, there are times when Aircraft Financiers are bemoaning the fact that there are not enough deals out there; airlines and lessors may be paying (gasp!) cash[7] or turning to the capital markets for funding. On the other hand, there may be an overall lack of liquidity in the markets due to market developments, much as there was in 2008, 2009 and 2010 following the Lehman Brothers crash and resulting financial tailspin. In the aftermath of that market crash, a number of major international banks disbanded their aircraft finance teams and left the sector while other banks went on a lending 'hiatus' or had vastly reduced lending budgets. This phenomenon was the result of a number of factors:

- banks were looking to preserve capital, so as to minimize the addition of further liabilities at a time when their balance sheets are already rather stressed;
- banks were being cajoled, or even forced, by their new taskmasters and owners (that is, national and state governments) to redirect their available liquidity to local businesses and industries; and
- banks were having difficulty accessing capital with which to make loans in the light of restrictive credit exposure limitations imposed by their funding counterparties.[8]

The withdrawal of bank liquidity in the aircraft finance sector in the aftermath of the bank liquidity crisis led to much discussion as to whether there would be a 'Funding Gap'; that is, will there be enough available funding sources to finance new deliveries (as well as to refinance those aircraft the financings of which mature)?[9] The bottom line to the 'funding gap' debate is usually the question as to what degree the ECAs and the manufacturers[10] will step up to fill the gap so as to avoid, for new deliveries, the prospect of whitetails in the desert. In fact, in the aftermath of the 2009 funding crisis, the ECAs and the remaining market participants stepped up and no Funding Gap materialized. ECAs fill funding gaps in one of two ways. First, they can support with bank guarantees an increasing number of aircraft exports so as to tap the pool of ECA banks (which seem to have faced a less severe liquidity cutback than banks that are asset-based lenders). Second, some of the ECAs can issue loans on a direct basis if ECA banks are not willing to step up to the plate at competitive pricing (or at all). Importantly, the ECA financings can only support exports. So, due to that fact and agreements that exist among the ECAs, ECA financing is not available for purchasers/users of commercial aircraft located in the U.S., France, Germany, Spain or the UK (see Part 7, 'Aircraft Sector Understanding'). This lack of availability of ECA financing partially explains why airlines in these jurisdictions are more apt to turn to the capital markets.

Prospects

In the light of the foregoing, then, what are the industry prospects for these various debt funding sources?

ECA Financings

As discussed in 'Export Credit Agency (ECA) Financing', the ECA-guaranteed loan product remains an important source of Aircraft Financing, covering some 23 percent of new Aircraft Financings. The increasing usage of the capital markets to fund these (primarily U.S. Ex-Im) financings for both airlines and lessors. However, with the Aircraft Sector Understanding (ASU) rules requiring ECA financings to more closely approximate private market pricing, there will likely be a tail-off of usage for this financing source.

Bank Financings

With a number of years now having passed since the bank liquidity crisis of 2008–2009, the commercial bank market has become increasingly robust in appetite. While a number of European banks (especially German *landesbanks*) have permanently exited the market, the remaining European (primarily French and German) banks and the U.S. banks are being joined by new (or returning) market entrants from commercial banks in Japan, Australia, the Middle East and (in the case of financing Chinese airlines) China. Commercial banks currently fund approximately 28 percent of new Aircraft deliveries.[11]

Capital Markets

The prognosis for continued availability of Aircraft Financing in the capital markets is similarly positive. The EETC market continues to be strong, and that market's liquidity and improving pricing remains an important feature. In addition to airlines (who are accessing the capital markets with EETCs), lessors should continue to evolve their use of the capital markets through Collateralized Lease Obligations (CLO) type structures. The expansion of lessor penetration into the capital markets will need to coincide with an evolution of the credit rating agencies' understanding of the aircraft leasing business and the development of rational and consistent rating criteria for aircraft lessors. Rating agencies also have a role in introducing non-U.S. airlines to the efficiencies of capital markets financing and shaping the market for international EETCs relying on the Cape Town Convention.[12] While the U.S. capital markets are expected to continue to be the primary market for the origination and syndication of aircraft public debt, regional capital markets may start to play a growing role. The capital markets fund approximately 14 percent of new Aircraft deliveries.[13]

Non-bank Financing

Non-bank debt investors in Aircraft Finance are most prevalent in the purchasing of capital markets products. However, they are increasingly looking at this sector on a private basis for improved yields and palatable risk levels. This investor pool is constantly expanding, with Aircraft Finance investing an increasingly visible opportunity. The historically strong performance of commercial aircraft investments should continue to attract new investors to this sector, especially in the secondary markets (post-delivery) and with used[14] Aircraft (which will provide stronger yields).

Original Equipment Manufacturers

OEM financing support has not played a significant role in recent years, although many purchasers do require OEMs to provide 'back-stop' financing for new order deliveries. This means that the OEM will agree to finance its Aircraft on delivery on specified conditions if the purchaser is not able to source financing on its own or is not able to finance the Aircraft at desired commercial terms.[15]

Secondary market trading

The conversation on placement of Aircraft Asset-backed securities would not be complete without saying a word about secondary market trading. Trading these securities is subject to the securities laws rules, so, absent a registered security, one of the exemptions touched on above would need to be utilized. The trading of these securities has historically been rather light following initial placement, since the investors are usually making their initial acquisitions to fill a need in their portfolios. However, if there are (usually bad) market developments, there is often a rush to the exits, since many investors have internal policies on the holding of distressed securities. The obvious example of this is when a U.S. airline issuer enters bankruptcy. At the early stages of a bankruptcy, when there is tremendous uncertainty about the ultimate exit or exit strategy for the airline, these debt securities may sell at a deep discount. Hedge funds and other investors with a high risk tolerance (and a (hopefully) savvy market view) often snap up these securities at these times, looking to maximize value once the airline's exit strategy settles down. As well, these investors may buy the securities: (i) to negotiate with the debtor Section 1110(b) agreements (see Part 8, 'Section 1110') so as to restructure a transaction; (ii) to cash in on the Deficiency Claim aspects on these securities as the market stabilizes;[16] and (iii) to acquire Aircraft Assets on the cheap.

The buyer of these distressed securities may, accordingly, end up with three items of value:[17]

- a restructured lease with the debtor (with associated rent cash flow);
- deficiency claims; and
- the residual value of an Aircraft Asset at its lease termination.

Each of these items is subject to monetization in different ways: the Leases can be backleveraged or sold; the Deficiency Claims can be traded; and the Aircraft Assets can be sold on a current or forward basis. Mind you, the ability to be in a position to so monetize these assets may require Foreclosure (discussed in detail under Part 8, 'Foreclosure') and the taking of other enforcement action.

U.S. dollar supremacy

Aircraft Finance transactions are almost universally done in U.S. dollars. The primary reason for this is that Aircraft Assets are priced and traded exclusively in U.S. dollars. With transactions priced in this fashion: (i) airlines which operate in non-U.S. dollar environments (and whose revenue is earned in other than U.S. dollars) need to take into consideration currency risks on U.S. dollar-denominated liabilities such as leases and mortgage financings;[18] and (ii) financiers who would normally fund themselves in non-U.S. dollars must likely either hedge their exposure to U.S. dollar-denominated assets or obtain funding from U.S. dollar sources.[19]

Repossession risks

Finally, a word about repossession risks. Aircraft Finance is necessarily a discussion about asset-based financing where the fact that there is a valuable asset standing behind the obligation of a debtor is critical. The ability to repossess the financed Aircraft Asset, then, is of utmost importance in a distress situation. Section 1110 of the U.S. Bankruptcy, and comparable provisions under Cape Town, accordingly, are highly important features for Aircraft Financiers' assessment of repossession risk. The availability of these legal rights (or lack thereof) in any particular jurisdiction will, therefore, greatly color the view of Aircraft Financiers on doing business in that jurisdiction and acquiring any associated securities. The robustness of the U.S. EETC market is largely owing to the availability to the financiers of Section 1110 rights in respect of the financed Aircraft. See Part 8 for a discussion of repossession and other risks associated with defaults.

AFIC

Aircraft Finance Insurance Consortium (AFIC) is a credit insurance product developed by Marsh Aviation Aerospace Practice, an affiliate of Marsh & McLennan Companies, for financiers providing financing to operating lessors and airlines acquiring Boeing-manufactured aircraft. AFIC is intended to replicate, in large part, the credit support provided by U.S. Eximbank in its traditional support arrangements for Boeing-manufactured aircraft. This product was launched, for among other reasons, to provide an alternative to U.S. Eximbank financing support due to the freezing of U.S. Eximbank's export support programs because of political disputes over matters of U.S. Eximbank's role in the economy and corporate welfare concerns.

The AFIC product is a non-payment insurance policy; that is, it will pay the insured financier if the obligor under the related aircraft financing fails to make payments when due. The AFIC insurance policy is underwritten by international insurance companies that are affiliates of Allianz, Axis and Sompo International. These underwriting arrangements are on a several basis. Thus, an insured taking this policy is assuming credit risk of each of the underwriting insurance companies

– the individual underwriters do not backstop the underwriting obligations of the other underwriters. Insofar as the credit of the insurers is inferior to that of the U.S. government, combined with the risk inherent in the 'several liability', the insureds may have a greater interest in the aircraft collateral than they do in a U.S. Eximbank transaction.

At the time of this writing, the AFIC product is fairly new to the market, so it is difficult to ascertain how successful it will be at this stage. We would anticipate that Airbus will seek to replicate this product as an alternative to European ECA-supported financing transactions for Airbus aircraft.

Asset-backed Securities (ABS)

Asset-backed Securities (ABS) are issued in the private and capital markets secured by assets and lease (or debt) cash flows. The predominant forms of ABS in the realm of Aircraft Financing are transactions structured as CLOs and Collateralized Debt Obligations (CDOs). EETCs are not typically characterized as ABS since they are highly reliant on the single issuer credit (in addition to the Aircraft Asset collateral security).

Aircraft, Crew, Maintenance and Insurance (ACMI)

Aircraft, Crew, Maintenance and Insurance (ACMI) is a form of Wet Lease. The operator/lessor provides ACMI for the lessee. This arrangement is most prevalent in the cargo market.

Back-leveraged Lease

This is an Operating Lease which, together with its related Aircraft Asset, is collaterally assigned/mortgaged by a lessor in favor of an Aircraft Financier as security for a loan. The Debt Service requirements under such loan are serviced by the rentals under such Operating Lease, see Exhibit 3.3. Insofar as such rentals may not be sufficient to repay such loan, the related Aircraft Asset may need to be sold to pay off any resulting Balloon (and most Back-leveraged Lease facilities are Non-recourse). A Finance Lease is not well suited for a back-leveraging insofar as the 'lessor' (borrower) has no equity interest in the asset; accordingly, among other things, there can be no Balloon exposure (unless there is a matching balloon-type payment under the Finance Lease), and there is no 'equity cushion'.

From an airline's perspective, back-leveraging of an Aircraft should not be cause for too much concern as, other than a redirection of rental payments and amendments to the insurance certificates, the relationship between the operating lessor and the airline remains unaltered. However, that could change if the lessor defaults, if the airline defaults, or if the lessor does not refinance the debt balloon. In each case, the lender may be taking management of the lease and Aircraft away from the operating lessor. As a result, the airline may lose the operating lessor as a business partner with whom it has a larger relationship and for which waivers, workouts or management of Aircraft returns might yield a different result than with

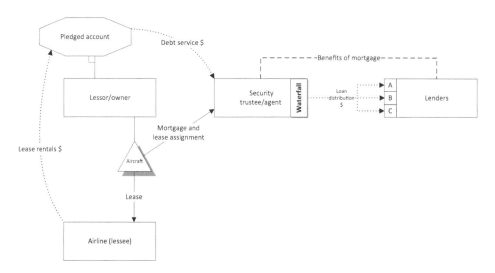

Exhibit 3.3 Back-leveraged Lease Facility

Source: Author's own

a new lessor (the lender) which has a different agenda. Of course, the ultimate identity of a lessor is never fixed as lessors always retain the right to trade their Aircraft positions. In addition, tapping bank financiers or the capital markets for the back-leveraging of Aircraft may absorb credit capacity that the same airlines wish to use for other bilateral or capital markets transactions. Finally, airlines should be mindful that certain 'bankruptcy remote' protections might conflict with the airline's interest to have the operating lessor guarantee the obligations of the lessor.

PRACTICE NOTE

Financiers taking Balloon risk in Back-leveraged Lease deals must be prepared to take over the asset and the lessor should be advising the financier in a timely manner if it is walking away from the asset so that the financier can prepare for potential repossession and remarketing at Lease maturity.

Bankruptcy Remote

This is a transaction using a Special Purpose Vehicle (SPV) that is structured in a manner to protect the financing from the bankruptcy of the originator, sponsor or servicer of the financed assets (see 'Securitization') or the entity that owns the entity that owns the financed assets. Transactions are characterized as Bankruptcy Remote rather than bankruptcy proof since there is no way to ensure 100 percent bankruptcy-proof protections (especially in the light of the fact that bankruptcy courts are courts of equity).

PRACTICE NOTE

In order to minimize bankruptcy risk, a number of features are typically employed: SPVs, Orphan Trust Structures and inclusion of provisions in organizational documents that require independent managers/members/ directors to agree to any bankruptcy filing and other significant corporate event or restructuring/reorganization. Many transactions structured as Bankruptcy Remote require from the lawyers of the originator a legal opinion that the Bankruptcy Remote entity will not be subject to Consolidation with the originator in the event of the originator's bankruptcy (see Part 8, 'Non-consolidation'). These opinions are reasoned (meaning that they are not absolute in conclusion), dozens of pages long and very expensive to procure.

Collateralized Debt Obligation (CDO)

Collateralized Debt Obligations (CDOs) are a type of structured asset-backed security, substantially similar to a CLO described below, with the primary difference being that the CDO securitizes *debt* obligations (that is, principal and interest debt service) and the CLO securitizes *lease* obligations (that is, rent).

An Aircraft Finance-related CDO, then, is the Securitization of multiple Aircraft-secured loans/bonds into a single facility. The principal and interest on the debt underlying the security is paid back to the investors regularly from the cash flow of the assigned equipment notes. Exhibit 3.4 displays a traditional CDO structure.

In a CDO, the *originator* assembles the debt securities that it wishes to securitize. Once a suitably large portfolio of assets is assembled or 'pooled', they are transferred to an SPV (the *issuer*), a tax-exempt company or trust formed for the specific purpose

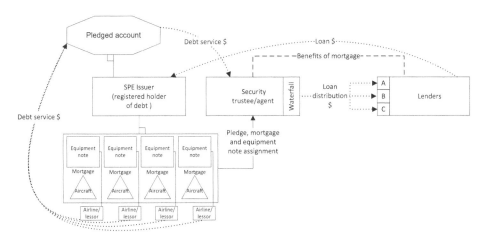

Exhibit 3.4 CDO Structure

Source: Author's own

31

of funding the assets. Once the assets are transferred to the issuer, there is normally no recourse to the originator. The issuer is designed to be Bankruptcy Remote.

Issuance

To be able to buy the debt securities from the originator, the issuer issues tradable securities to fund the purchase. Investors purchase the securities, either through a private offering (targeting institutional investors) or on the open market. Alternatively, this structure can be funded by bank lenders in the loan market. The performance of the securities is then directly linked to the performance of the assets. Credit rating agencies are often required to rate the securities which are issued as a matter of corporate policy (or regulatory necessity) for an investor, as well as to provide an external perspective on the liabilities being created and help the investor make a more informed decision.

The securities can be issued with either a Fixed Rate or a Floating Rate coupon, which will largely be driven by: (i) investor appetite for one or the other; and (ii) the nature of the cash flows (interest debt service) thrown off by the related loans.

Credit enhancement and tranching

Unlike conventional corporate bonds which are unsecured, securities generated in a CDO are 'credit enhanced', meaning their credit quality is increased above that of the unsecured debt of the obligor(s) of the underlying bundled debt obligations or the underlying asset pool. This enhancement is achieved by the availability of the Aircraft collateral, the related Loan to Value ratios (LTVs) and the other features described below. Such enhancements increase the likelihood that the investors will receive cash flows to which they are entitled or, at a minimum, recoupment of their principal investment on final maturity, and thus causes the securities to have a higher credit rating than such underlying obligor(s). Some Aircraft Finance CDOs use external credit enhancement provided by third parties, such as a Liquidity Facility.

Many Aircraft CDOs benefit from asset/obligor/geographic diversification, which are reflected in Concentration Limits (see Part 13, 'Concentration Limits') which seek to protect this diversity.

Individual securities issued by the CDO SPV are often split into Tranches, or categorized into varying degrees of subordination. Each Tranche has a different level of credit protection or risk exposure than another: there is generally a senior ('A') class of securities and one or more junior subordinated ('B', 'C' and so on) classes that function as protective layers for the 'A' class. The senior classes have first claim on the cash (including the proceeds from the liquidation of Aircraft Asset collateral) that the SPV receives, and the more junior classes only start receiving repayment after the more senior classes have been repaid. This cascade is effected through the Waterfall. In the event that the underlying loan pool generates insufficient debt service to make payments on the securities, the loss is absorbed first by the Subordinated Tranches, and the upper-level Tranches remain unaffected until the losses exceed the entire amount of the Subordinated Tranches.[20]

The most junior class (often called the *equity class*) is the most exposed to payment risk. In some cases, this is a special type of instrument which is retained by the originator as a potential profit flow. In some cases, the equity class receives no coupon (either fixed or floating), but only the residual cash flow (if any) after all the other Tranches have been paid. In addition to Subordination, credit may be enhanced through a reserve account, in which funds remaining after expenses such as principal and interest payments, charge-offs and other fees have been paid-off are accumulated, and can be used when debt service on the CDO exceeds the SPV's income.

A CDO may employ a *Servicer* to collect payments, to monitor the assets that are the subject of such Securitization and to provide remarketing and re-deployment services if the Aircraft are subject to return (due to scheduled or early (default) termination situations). A Servicer may be less likely in a CDO as compared with a CLO since the underlying debt securities typically used are full pay-out instruments, thereby minimizing redeployment risks.

PRACTICE NOTE

Securitizations may employ Turbo, Debt Services Coverage Ratio (DSCR) and LTV tests.

In contrast to a CLO where the investors have an opportunity to obtain value from the Aircraft Assets serving as collateral up to the entire value of those assets, the CDO investor is capped at the value of the underlying debt serving as collateral vis-à-vis the issuer thereof; with the airline or operator of whose debt is included in the CDO entitled to keep all of the equity in the Aircraft collateral.

Another difference with the CLO is that while the CLO would most likely have a diversity of underlying credits/lessees, CDOs may either have such diversity or may involve just a single credit (and, therefore, look more similar to EETC). In fact, CDOs are often masqueraded as EETCs to tap investor acceptability of that product, especially if Section 1110 or Cape Town remedies are available in connection with the underlying debt obligations.

Collateralized Lease Obligation (CLO)

A Collateralized Lease Obligation (CLO) is a type of structured asset-backed security, substantially similar to a CDO described above, with the primary difference being that the CDO securitizes *debt* obligations (that is, principal and interest debt service) and the CLO securitizes *lease* obligations (that is, rent).

An Aircraft Finance-related CLO, then, is the Securitization of multiple Aircraft leases packaged into a single facility. The principal and interest on the debt under-lying the security is paid back to the investors regularly from the cash flow of the assigned lease rentals. Exhibit 3.5 displays a traditional CLO structure.

In a CLO, the *originator*, typically a leasing company, assembles the Aircraft Assets that it wishes to securitize. Once a suitably large portfolio of assets is assembled or 'pooled', they are transferred to an SPV (the *issuer*), a tax-exempt company or trust formed for the specific purpose of funding the assets. Once the assets are transferred to the issuer, there is normally no recourse to the originator. The issuer is designed to be Bankruptcy Remote. Accounting standards govern when such a transfer is a sale, a financing, a partial sale, or a part-sale and part-financing. In a sale, the originator is allowed to remove the transferred assets from its balance sheet; in a financing, the assets are considered to remain the property of the originator. Because of these structural issues, the originator typically needs the help of an investment bank (the *arranger*) in setting up the structure of the transaction and placing the related securities.

Issuance

To be able to buy the Aircraft Assets from the originator, the issuer issues tradable securities to fund the purchase. Investors purchase the securities, either through a private offering (targeting institutional investors) or on the open market. Alternatively, this structure can be funded by bank lenders in the loan market as part of a loan (not securitization) facility.

The performance of the securities is then directly linked to the performance of the assets. Credit rating agencies are often required to rate the securities which are issued as a matter of corporate policy (or regulatory necessity) for an investor, as well as to provide an external perspective on the liabilities being created and help the investor make a more informed decision.

The securities can be issued with either a Fixed Rate or a Floating Rate coupon, which will largely be driven by: (i) investor appetite for one or the other; and (ii) the nature of the cash flows (interest debt service) thrown off by the related loans.

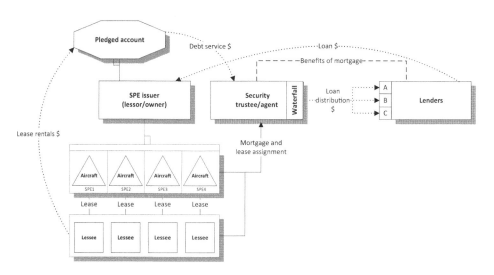

Exhibit 3.5 CLO structure

Source: Author's own

Credit enhancement and tranching

Unlike conventional corporate bonds which are unsecured, securities generated in a CLO are 'credit enhanced', meaning their credit quality is increased above that of the unsecured debt of the obligor(s) of the underlying bundled lease obligations or the underlying asset pool. This enhancement is achieved by the availability of the Aircraft collateral, the related LTVs and the other features described below. Such enhancements increase the likelihood that the investors will receive cash flows to which they are entitled or, at a minimum, recoupment of their principal investment on final maturity, and thus causes the securities to have a higher credit rating than such underlying obligor(s). Some Aircraft Finance CLOs use external credit enhancement provided by third parties, such as a Liquidity Facility.

Many Aircraft CLOs benefit from asset/obligor/geographic diversification, which are reflected in Concentration Limits (see Part 13, 'Concentration Limits'), which seek to protect this diversity.

Individual securities issued by the CLO SPV are often split into Tranches, or categorized into varying degrees of subordination. Each Tranche has a different level of credit protection or risk exposure than another: there is generally a senior ('A') class of securities and one or more junior subordinated ('B', 'C' and so on) classes that function as protective layers for the 'A' class. The senior classes have first claim on the cash lease rentals (including the proceeds from the liquidation of Aircraft Asset collateral) that the SPV receives, and the more junior classes only start receiving repayment after the more senior classes have been repaid. This cascade is effected through the Waterfall. In the event that the underlying lease pool generates insufficient debt service to make payments on the securities, the loss is absorbed first by the Subordinated Tranches, and the upper-level Tranches remain unaffected until the losses exceed the entire amount of the Subordinated Tranches.

The most junior class (often called the *equity class*) is the most exposed to payment risk. In some cases, this is a special type of instrument which is retained by the originator as a potential profit flow. In some cases the equity class receives no coupon (either fixed or floating), but only the residual cash flow (if any) and the residual value of the Aircraft Assets after all the other Tranches have been paid.

In addition to Subordination, credit may be enhanced through a reserve account, in which funds remaining after expenses such as principal and interest payments, charge-offs and other fees have been paid-off are accumulated, and can be used when debt service on the CLO exceeds the SPV's lease rental income.

In a CLO, a *Servicer* is often engaged to collect payments, to monitor the assets that are the subject of such Securitization and to provide remarketing and re-deployment services if the Aircraft are subject to a lease return (due to scheduled or early (default) termination situations). The servicer is often the originator, because the servicer needs very similar expertise to the originator (and the originator may be interested in the servicing fee income it charges for its undertaking of this role). The provision of these services is effected through a Servicing Agreement.

PRACTICE NOTE

In a CLO, investors have an opportunity to obtain value from the Aircraft Assets serving as collateral up to the entire value of those assets, in contrast to a CDO where the CDO investor is capped at the value of the underlying debt serving as collateral vis à vis the issuer thereof. The residual value of the Aircraft Assets subject to the CLO may be a substantial value component for the CLO 'equity' investor.

While CLOs would most likely have a diversity of underlying credits/lessees, they could be structured with a single lessee obligor. Either way, if a CLO's lease obligors are subject to Section 1110 or Cape Town remedies, then they may be packaged and marketed as EETC-type products, thereby taking advantage of the deep market for EETC securities.

CLOs may employ Turbo, DSCR and LTV tests.

Credit Default Swap (CDS)

A Credit Default Swap (CDS) is a credit derivative transaction that is a financial swap agreement pursuant to which a seller of the CDS will compensate the buyer in the event of a loan default or other credit event. The buyer of the CDS makes a series of payments (the CDS 'fee' or 'spread') to the seller and, in exchange, receives a payoff if the loan defaults. A CDS is linked to a 'reference entity' or 'reference obligor', usually (in our context) an airline or operating lessor. The reference entity is not a party to the contract. The buyer makes regular premium payments to the seller, the premium amounts constituting the 'spread' charged by the seller to insure against a credit event. If the reference entity defaults, the protection seller pays the buyer the par value of the bond in exchange for physical delivery of the bond, although settlement may also be by cash.

A default is often referred to as a 'credit event' and includes such events as failure to pay, restructuring and bankruptcy, or even a drop in the reference entity's credit rating.

Credit default swaps are often used to manage the risk of default that arises from holding debt. A bank, for example, may hedge its risk that a borrower may default on a loan by entering into a CDS contract as the buyer of protection. If the loan goes into default, the proceeds from the CDS contract cancel out the losses on the underlying debt.

A bank can also use a CDS to free up regulatory capital. By offloading a particular credit risk, a bank is not required to hold as much capital in reserve against the risk of default. This frees resources the bank can use to make other loans to the same key customer or to other borrowers.

There are other ways to eliminate or reduce the risk of default. A lender could sell (that is, assign) the loan outright or bring in other banks as participants. However, these options may not meet a particular lender's needs.

Club Deal

This is a financing where the debt participants are not led by a single arranger or agent but are on relatively equal footing when negotiating a financing with a borrower. Club Deals are, by their nature, somewhat more difficult to close since there are many 'captains' to co-ordinate.

Cross-border Tax Lease

This is a Finance Lease where the lessee and the lessor are domiciled in different jurisdictions and where, due to different accounting and/or tax treatment in the respective jurisdictions, both parties are able to obtain tax/accounting benefits of ownership in their respective jurisdictions. Such deal structures typically are successful where the lessor's jurisdiction treats the *form* of the transaction as the guiding principle and the lessee's jurisdiction treats the *substance* of the transaction as definitive. Cross-border Tax Leases have largely disappeared because many jurisdictions have shifted away from a 'form' to a 'substance' analysis.

Depository

This is a financial institution (typically a bank with a high credit rating) that holds the proceeds of loans in a financing transaction when such loans are pre-funded, such as in a pre-funded EETC. Loans are pre-funded so that an issuer can lock in current interest rates and take advantage of market liquidity in anticipation of scheduled future deliveries for specified Aircraft. The prefunded amounts are deposited with the Depository, are evidenced by escrow receipts and are made available to the EETC issuer only upon the delivery and financing of the earmarked Aircraft from the OEM. Such funds on deposit are not intended to be assets of the issuer.

Dry Lease

A Dry Lease is an Operating Lease that provides lease financing only for the equipment itself, and does not extend to personnel, maintenance, fuel and provisioning necessary to operate the asset. In contrast, there are the Wet Lease and ACMI arrangements.

Export Credit Agency (ECA) Financing

This is a financing supported by an Export Credit Agency (ECA). ECAs provide financing, or governmental-based guarantees to lenders, to support the export of aircraft, engines and other manufactured goods from the home country. ECAs supply billions of dollars of support annually for Aircraft and Engine exports alone. The primary ECAs involved in Aircraft Asset support are:

- Brazil – Banco Nacional de Desenvolvimento Economico e Social (BNDES) (supporting Embraer aircraft);

- Canada – Export Development Canada (EDC) (supporting Bombardier aircraft);
- France – Bpifrance Assurance Export (formerly known as COFACE) (supporting Airbus and ATR aircraft, among others);
- Germany – Euler Hermes Aktiengesellschaft (Euler Hermes) (supporting Airbus aircraft, among others);
- UK – Her Britannic Majesty's Secretary of State acting through the Export Credits Guarantee Department (ECGD), operating as UK Export Finance (supporting Airbus aircraft and Rolls Royce engines, among others); and
- U.S. – U.S. Ex-Im (supporting Boeing aircraft and CFM, IAE, GE and Pratt & Whitney engines, among others).[21]

ECA Financings by the ECAs described above are currently regulated by the ASU arrangements. However, it is worth noting other substantial ECAs, as follows, are not subject to the ASU rules. The roles being played by these ECAs are expanding and are increasingly factoring into export financings in the Aircraft Asset space.

- Nexi (Japan).
- China Ex-Im.
- Servizi Assicurativi del Commercio Estero (SACE S.p.a.) (Italy).

ECA Financings come primarily in two structures.

1 Guarantees issued by the ECA to lenders, whose loans allow an owner or operator to acquire the exported asset.
2 The ECA itself provides the financing to allow an owner or operator to acquire the exported asset.[22]

Most ECA Financings using the guarantee structure are constructed on the basis of the arrangements noted in Exhibit 3.6.

ECA guaranteed financing structures can support both privately placed loans funded (primarily) by banks or securities placed with investors in a public-type offering. In a typical public-style issuance the guaranteed loans would either be: (i) prefunded; or (ii) funded on a preliminary private basis. In the case of a *prefunding*, all of the funds necessary to fund the Aircraft would be drawn down at an initial closing, placed with a Depository and would be represented by escrow receipts until applied to the Aircraft's purchase price on delivery. In the case of a *private* funding, a bank funds each Aircraft as and when the Aircraft Assets are delivered, and once there is a sufficient sized pool of loans to support a placement in the public markets, the loans are converted to these public instruments and so placed. Capital market access for ECA guaranteed financings has been primarily available to financings supported by U.S. Ex-Im; U.S. Ex-Im was instrumental in the facilitation of transaction structures allowing for the issuance of securities in the capital markets that were comprised of U.S. Ex-Im guaranteed loans. This U.S. Ex-Im guaranteed bond structure is expected to evolve further, expanding the market breadth and improving its efficiency.[23]

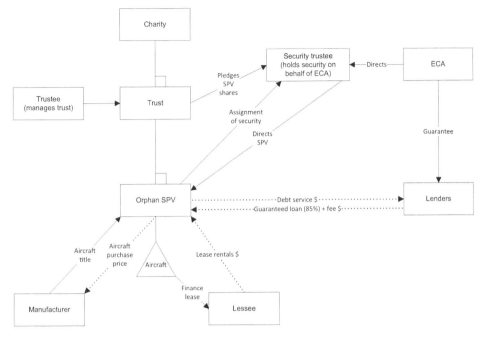

Exhibit 3.6 ECA Guaranteed Financing Facility

Source: Author's own

ECA guaranteed financing (absent some global shock akin to 9/11 or the Lehman collapse) is likely to reduce over time in the light of the ASU's requirements to impose market-level fees and rates on these financings, thereby making these types of financings less advantageous to Aircraft Asset purchasers.

ECA Co-commercial Financing

ECA Co-commercial Financing is the provision of commercial (non-guaranteed) funding by financiers secured by Aircraft on a transaction when such Aircraft also secure debt in an ECA guaranteed financing. In the light of the ECA's rather strict LTV advance rates and rapid paydown requirements, opportunities are created for extending commercial/non-guaranteed debt secured by the otherwise ECA-guaranteed debt. Based on these economics, certain financiers are willing to lend to airlines or lessors side-by-side with lenders that are receiving ECA support. Such loans may be effected at the time of the initial borrowing and/or may accrete over time as the ECA supported loans are repaid. The latter are called Stretched Overall Amortization Repayment (SOAR) loans, or 'mismatch loans'; however, the latest ASU has largely eliminated this particular financing structure. All such commercial lenders' loans are subordinate to the ECA's guaranteed loans in respect of the financed Aircraft and such commercial lenders have no power to direct the exercise of remedies unless and until the ECA-supported loans are paid in full.

If an airline or lessor elects to use a commercial loan to finance the difference between the purchase price of an Aircraft and the related guaranteed loan, the commercial lenders will receive proceeds from the exercise of remedies on an Aircraft-by-Aircraft basis, only after the related guaranteed loan and all other amounts due and owing to the ECA and the guaranteed lenders with respect to such Aircraft have been paid in full. However, the commercial lenders receive proceeds prior to the application of amounts toward other ECA-supported financings for that airline or lessor.

As long as either: (i) the related ECA guarantee in favor of the guaranteed lenders is in effect; or (ii) there are any amounts owing to the ECA under the related transaction; the ECA will have the sole right to direct the taking of any action under the transaction documents, including, without limitation, exercising any remedies after the occurrence of an event of default. If neither (i) nor (ii) applies, the commercial lenders would be the instructing party.

Even though these commercial lenders have limited rights with respect to the collateral, the commercial lenders often have the right to pursue claims directly against the airline or lessor for any amounts owed to them that are not covered by an ECA guarantee, provided, so long as the relevant ECA guarantee is in effect or the ECA has any exposure or is owed any amounts arising out of the transaction documents, that:

- no recovery may be had directly from, and no such suit shall assert any rights or claims against, the collateral or borrower;
- such action does not interfere with, or otherwise adversely affect, any restructuring, enforcement or other collection efforts by, or on behalf of, the ECA (other than by requiring payment of moneys then due to such person but only to the extent enforcement is on assets not constituting part of the collateral or part of the collateral securing any other obligation running to or for the benefit of or otherwise relating to any transaction involving the ECA);
- such commercial lender may not initiate any bankruptcy, suspension of payment or other insolvency proceedings against the borrower or the lessor/ airline in connection with such recovery; and
- the ECA has not sent a notice that such action materially interferes with the enforcement or remedial actions which are being taken or could be taken by or the direction of the ECA, or otherwise materially adversely affects the ECA.

Any amount recovered by any lender is required to be applied through the standard waterfall.

The above detailed description of the relative rights between an ECA and a commercial lender is based on a typical U.S. Ex-Im transaction. The other ECAs may adopt similar or more stringent or relaxed intercreditor terms.

ECA Co-financing

ECA Co-financing arrangements are arrangements made by different ECAs to supply credit support for Aircraft with components manufactured in multiple jurisdictions. The manufacture of large commercial jets is a global business. By way of example, for today's Boeing Aircraft, more and more of the components are manufactured outside the U.S. before they are shipped to, and installed on, an airframe in Boeing's States-side plants. This can and does significantly affect the strict eligibility requirements for U.S. Ex-Im financing. If a particular Aircraft has, for example, only 50 percent U.S. content, then U.S. Ex-Im's support would be limited to this amount. If the balance of the content would be eligible for support from another ECA, then it may be possible to arrange separate support from such other agency. To address this particular situation and streamline the financing process, U.S. Ex-Im has entered into a number of so-called 'co-financing' arrangements with other ECAs. This creates real value for the airline or lessor acquiring the aircraft by allowing 'one-stop shopping' for its export credit financing. Under a co-financing arrangement, there is normally a lead ECA which co-ordinates all negotiations, documentation, disbursements, administration and exercise of remedies. With respect to the airline/lessor borrower, its sole interaction is with the lead ECA who is typically given broad authority under the co-financing arrangement to service a transaction. As part of the co-financing arrangement, the other ECAs involved agree to insure the lead ECA for such other ECA's proportionate share of a transaction.

U.S. Ex-Im has active co-financing agreements for large commercial Aircraft with the UK's HM Export Credit Guarantee Department (ECGD), Japan's Nippon Export and Investment Insurance (NEXI) and Korea's Export-Import Bank, insofar as many Boeing aircraft components are manufactured in these countries. Similarly, the European export credit agencies (ECGD, COFACE in France and Euler Hermes in Germany) use co-financing arrangements frequently in supporting Airbus financings. In fact, in many European ECA-supported financings of Airbus Aircraft, a single of these ECAs may act as a 'fronting' ECA in a transaction, with internal arrangements as among all the participating ECAs behind the curtain to address risk-sharing protocols.

EETC Rating

An EETC Rating is a rating that assesses the likelihood that a borrower makes 'full and timely payment of interest and ultimate payment of principal'. This analysis treats interest and principal differently. As for interest, the requirement is for interest to be paid *currently*. This means that there cannot be any interruption of the interest payment even if the borrower is in default or bankruptcy. The primary means to ensure current payment of interest is to employ a Liquidity Facility (see 'Liquidity Facility') that will ensure payment of interest (only) if the borrower fails to pay interest. As for principal, *ultimate* payment of principal means that the principal needs to be paid in full by the relevant Final Legal Maturity Date (see 'Final Legal Maturity Date') which is a date, typically 18 months (in U.S. airline transactions), after the underlying debt is scheduled to mature. This extra period is the period that the rating agencies perceive is the maximum length of time it should take to repossess and liquidate a financed aircraft asset and apply the proceeds to outstanding balances of principal. This rating methodology is utilized by the rating agencies rating EETCs and certain other aircraft-secured debt securities.

Enhanced Equipment Trust Certificate (EETC)

An Enhanced Equipment Trust Certificate (EETC) is a publicly (but sometimes privately) issued, rated security that:

- relies on the credit of a single corporate issuer;
- is secured by Aircraft Assets as collateral;
- utilizes a Liquidity Facility to provide up to 18 (or, in some cases, 24) months of missed interest payments;
- utilizes structural enhancements to provide improved LTV ratios for the more senior levels, or Tranches, of debt securities; these structural enhancements include:
  - tranching
  - cross-default
  - cross-collateralization
  - cross-subordination
- relies on the certainty of remedies afforded by Section 1110 of the U.S. Bankruptcy Code or, in the case of non-U.S. airlines, comparable rights under Cape Town (or other local law);
- is rated based on an EETC Rating; and
- maintains a constant level of over-collateralization.

Exhibit 3.7 is a schematic that boils down the primary characteristics of an EETC. Exhibit 3.7 highlights the following primary features of an EETC:

- as a credit matter, there is a single full-recourse airline issuer on whose credit the financing rests; and
- as a collateral security matter, the Aircraft Assets serve as collateral and, as a matter of supreme importance to the structure, have the benefit of Section 1110 of the U.S. Bankruptcy Code (or the equivalent under Cape Town or local law), which provides certainty as to the ability to repossess the collateral (or be given assurances of continued performance).

Over $90 billion of EETC securities have been issued since the product was developed in 1994. While predominantly a financing vehicle used by U.S. airlines, non-U.S. airlines as diverse as Air Canada, DORIC/Emirates, LATAM, British Airways, Virgin Australia, Turkish Airlines (THY) and Norwegian Air Shuttle have utilized this product as well. The attraction of EETCs from the airline issuer's perspective are manyfold:

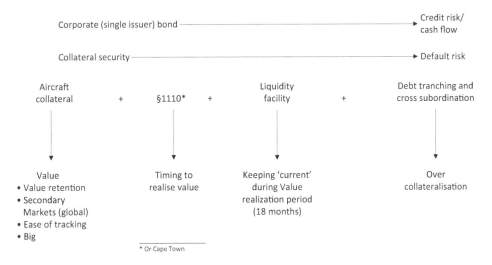

Exhibit 3.7 EETC Schematic

Source: Author's own

- Size: Given the capacity of the capital markets into which EETCs are issued, airlines may finance in a single financing upwards of $1.5 billion of aircraft debt in one fell swoop, which can cover dozens of aircraft. This saves airline treasury staff the trouble of needing to hit-up large numbers of financial institutions to finance a year's-worth of deliveries.
- Fixed rate: EETCs are typically issued at a fixed interest rate which is often desired by airlines.
- Ease of execution: While by no means simple to document (and rather expensive to do so given disclosure requirements typically subsumed in a detailed offering memorandum issued to investors), EETCs are subject to largely standardized documentation across the board (that is, from airline to airline) and, as to any single airline issuer, its transaction documentation will be entirely uniform across its EETC-financed fleet. This standardization applies as well to intercreditor terms, which can involve laborious negotiations in private transactions.
- Financing diversification: The access to the capital markets may be a very attractive option for airlines concerned with relying exclusively, say, on the bank market (with that market's difficulties such as risk of Market Disruption, Increased Costs and other costs).
- Pricing: The structural enhancements provided by an EETC allow the senior-most Tranches to be rated investment grade, with the attendant pricing advantages associated with such ratings.

Exhibit 3.8 provides a structural overview of a typical EETC transaction (with this indicative transaction including a financing done on a prefunded basis (utilizing a Depository); prefunded transactions are not universal to every financing).[24]

The EETC is fundamentally a secured corporate credit, not a securitization. The investor takes default risk of a single airline (and not a diversified borrower/lessee base). There are two primary financing structures for EETCs that create the underlying obligation of the airline to make its scheduled payments. The first is a Mortgage Financing. In this type of financing, the airline will issue promissory notes (commonly known as equipment notes) in Tranches, with the 'A' note Tranche being ranked ahead of the 'B' note Tranche, the 'B' note Tranche being ranked ahead of the 'C' note Tranche and so on. The equipment notes are issued on a per-aircraft basis, and are secured by that aircraft in the airline's fleet.

The second type of underlying financing is a U.S. Leveraged Lease Financing.[25] In this type of financing, an equity investor, acting through an owner trustee, will

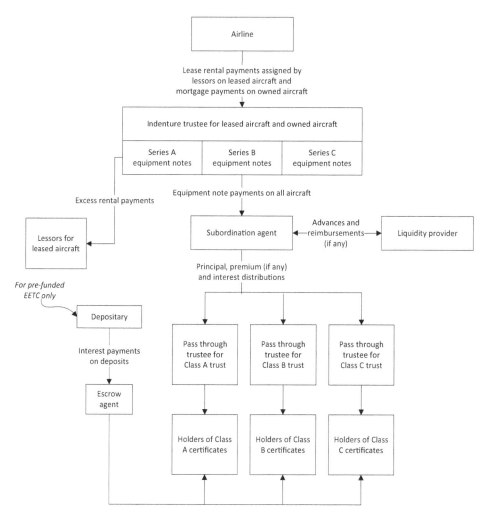

Exhibit 3.8 EETC Overview

Source: Author's own

purchase an Aircraft, borrowing up to 80 percent of the Aircraft Fair Market Value (FMV), with the balance provided by the equity investor. The borrowed funds will be obtained by the issuance by the owner trustee of equipment notes in Tranches as described above. The owner trustee will then lease the aircraft to the airline issuer. As collateral for the equipment notes, the owner trustee will grant the equipment note holders a mortgage on the aircraft and assign the lease to the holders. The equipment notes are issued by the owner trustee on a non-recourse basis and will be entirely serviced by the lease cash flows.

An EETC financing may involve 30 or more Aircraft. The equipment notes issued with respect to each aircraft will be aggregated and held by a separate pass through trust for each class of notes pursuant to a Pass Through Certificate (PTC) structure.[26] Thus, all of the A equipment notes for all of the Aircraft in a particular EETC are held by a class A pass through trust, all of the B equipment notes for all of the aircraft are held by a class B pass through trust, and so on. Each pass through trust, then, issues PTCs to investors. Each investor can then purchase PTCs that are issued at either the class A, B or C level. EETCs have historically been almost exclusively the domain of U.S. airline issuers, but with the advent of Cape Town (which satisfies the need for assured repossession), non-U.S. airlines have been making significant inroads with this product; see Practice Note below.[27]

Since 2005, EETC transactions have benefited from the Cross-default and Cross-collateralization across all of the Aircraft in the financed pool.[28] These two features allow issuers to combine diverse pools of Aircraft Asset collateral such as superior younger collateral pooled with older, less liquid aircraft types, since the weaker assets will be supported by the stronger ones. Putting it differently, the airline issuer will not be able to compel the return of Aircraft Assets under Section 1110(c) of the Bankruptcy Code if it wants to keep the stronger Aircraft Assets under Section 1110(a) of the Bankruptcy Code (see Part 8, 'Section 1110').

While the Mortgage Financing or Lease Financing facing an EETC issuance may be fairly standard, the transaction gets somewhat complicated among the investors, since they are often Tranched. The Tranched investors of different classes will have varying inter-creditor rights. An EETC will typically feature:

- Controlling Party rights transitions;
- Waterfalls;
- Minimum sales prices; and
- Buyout rights.

In addition, multi-Tranched EETCs, also feature, through the Waterfalls, Asset and Credit Subordination. An interesting feature of this subordination is that there are situations where the holders of the more junior Tranches of PTCs may be in a position vis-à-vis the airline-issuer worse than the holder of that airline's unsecured debt. Whereas the holder of a bankrupt airline's unsecured debt may receive pennies on the dollar (or equity in the reorganized airline), the holders of the more junior Tranches of PTCs have absolutely no direct claim at all. All Deficiency Claims flow through the Waterfall, even those attributable to the most junior classes of equipment notes. For example, if the controlling party were to sell all of the equipment notes in an EETC for, say, 60 percent of the outstanding principal balance of the equipment notes, the proceeds of such sale would, as per the Waterfall, be applied to the more senior Tranches leaving the holders of class C certificates with no further claim since the equipment notes, which represent the 'direct' claim against the airline (and would have the benefit of the Deficiency Claim), have been sold. Another interesting feature of EETC intercreditor terms which is rather unique to EETCs is that the Waterfall provides that the holders of the junior classes of the PTCs receive 'adjusted' interest[29] on their PTCs ahead of distributions to the holders of the senior classes of PTCs. This feature is intended to incentivize the holders of the senior PTCs to maximize recoveries post-default. Exhibit 9.3 summarizes the primary intercreditor terms in a traditional (2017) EETC.

All EETCs have the benefit of a Liquidity Facility, which assures EETC investors with continued payments of interest (only) for up to 18 months (or more in the case of certain non-U.S. jurisdictions) if the issuer ceased making debt service payments. This continued payment of interest for such period – being the period the rating agencies perceive as the maximum period required to repossess, market and liquidate the Aircraft Assets subject to an EETC – enables the rating agencies (together with the other structural features of the EETC) to provide enhanced ratings for the EETC, see 'EETC Rating' and 'Liquidity Facility'.

The enhanced ratings provided by the rating agencies tasked with rating a particular EETC issuance, of course, provide the EETC with improved marketability and liquidity, allow the issuer to obtain advantageous pricing and allow institutional investors to participate in such a type of transaction.

PRACTICE NOTE

While EETCs have traditionally been a financing vehicle enjoyed by U.S. airlines by reason of the availability of Section 1110 of the U.S. Bankruptcy Code in the U.S., with the advent of Cape Town's Alternative A, which is the Section 1110 equivalent for Contracting States, one would expect the EETC product to (theoretically) work for airlines in Contracting States that have adopted Alternative A. In fact, both Air Canada and Emirates have entered the EETC market on the basis of their jurisdiction's adoption of Cape Town (and Alternative A).[1] As well, Doric Nimrod Air Finance Alpha Limited, an aircraft leasing company, has issued EETCs on at least two occasions (with Emirates and British Airways as lessees). One should expect increasing attention paid to this market by non-U.S. carriers as ECA financings become more expensive as ASU requirements are reflected in loan pricing. See 'International EETCs'.

[1] British Airways has also utilized the EETC markets, notwithstanding that the UK is not a Contracting State, on the basis that rating agencies became comfortable with the UK's general creditor-friendly insolvency regime and structural enhancements offered by the EETC structure.

EETC documentation is somewhat voluminous, with two broad categories addressed. The first category is the aircraft financing documents. These would include:

- Note Purchase Agreement
- Participation Agreement
- Trust Indenture and Security Agreement (one per Aircraft)
- Equipment Notes

The second category covers the documents associated with the PTCs or trust-level documents. These would include:

- Underwriting Agreement
- Pass Through Trust Agreements
- Trust Supplements
- Pass Through Certificates (PTCs)
- Revolving Credit Agreements (Liquidity Facilities)
- Intercreditor Agreement
- Deposit Agreement(s) (for pre-funded facilities)
- Escrow Agreement(s) (for pre-funded facilities)

Equipment Trust Certificates (ETCs)

Equipment Trust Certificates (ETCs) are promissory notes issued by an airline or U.S. Leveraged Lease equity Owner Trust that evidence the loans made by lenders that are secured by Aircraft Assets and, in the case of a U.S. leveraged lease, the related lease. This term was primarily used as the name given to the debt securities issued as part of a public offering of the promissory notes to investors.

Finance/Capital Lease

This is a lease transaction that is treated as a Mortgage Financing for any particular purpose. The criteria for ascertaining whether a particular lease transaction is a Finance Lease will, in the first instance, depend on whether the determination is for tax, Uniform Commercial Code (UCC) or, in the case of U.S.-registered Aircraft, FAA filing purposes. As for accounting purposes, in light of newly adopted accounting rules under both GAAP and IFRS standards (ASC 842 and IFRS 16, respectively), the distinction between operating leases and finance leases has lost much significance for disclosure purposes insofar as the lessee, under any lease – howsoever characterized – is required to recognize (under GAAP): (a) assets and liabilities for all leases with a term of more than 12 months (unless the underlying asset is of low value); and (b) depreciation of leased assets separately from interest on lease liabilities in the income statement; provided that lessors will need to continue to classify their leases as either an operating lease or a finance lease, and to account for those two types of leases differently (see IAS 17).

Each of the tax, UCC and FAA disciplines has its own set of determination criteria.

Tax

Under U.S. tax rules, Revenue Procedure 2001–28 (Rev. Proc. 2001–28) sets out a guideline of criteria (the 'Guidelines') for classifying a lease as a true lease for U.S. federal income tax purposes.

The U.S. Internal Revenue Service (IRS) developed the Guidelines specifically for 'leveraged lease' transactions, which involve three parties: a lessor/owner, a lessee/user and a lender to the lessor. However, the Guidelines are also used to aid in structuring single investor leases, which involve two parties – a lessor and lessee. The Guidelines are not controlling as a matter of law, but provide a set of criteria by which the IRS decides the character of a transaction for purposes of providing advance income tax rulings. The application of the Guidelines to single investor leases, while useful, has largely served as a voluntary construct on which to conservatively structure a lease involving a lessor and lessee. The theory is that if the structure works for the more complex leveraged leases, the structure should work for single investor transactions.

A lessor under a U.S. Leveraged Lease should qualify as the tax owner of property if the following Guidelines criteria are satisfied.

- The lessor must maintain a minimum unconditional 'at risk' equity investment in the property being leased (at least 20 percent of the cost of the property) during the entire lease term. Within this general concept, a lessor must show that it expects the property at the end of the lease term to have a fair market value equal to at least 20 percent of its original cost. The lessor must also demonstrate that it expects that the equipment will have a useful life at the end of the lease of not less than 20 percent of its original useful life (or at least one year). These requirements are sometimes called the '20/20 tests'.
- The lessee may not have a contractual right to buy the property from the lessor at less than fair market value when the right is exercised.
- Lessee lease renewal options must be at fair rental value at the time of renewal.
- With exceptions, the lessee may not invest in the leased property.
- The lessee may not lend any money to the lessor to buy the property or guarantee the loan portion of a leveraged lease that the lessor uses to buy the leased property.
- The lessor must show that it expects to receive a profit apart from the tax benefits.

Importantly, the Guidelines are just that, guidelines. Leveraged (and single investor) leases may deviate from the Guidelines and still qualify as true tax leases for federal income tax purposes.

UCC/Commercial law

Under the Uniform Commercial Code (UCC) in the U.S., § 1–201(37) of the UCC provides guidance: whether a transaction creates a lease or security interest is determined by the facts of each case; however, a transaction creates a security interest if the consideration the lessee is to pay the lessor for the right to possession and use of the goods is an obligation for the term of the lease not subject to termination by the lessee, and: (i) the original term of the lease is equal to or greater than the remaining economic life of the goods; (ii) the lessee is bound to renew the lease for the remaining economic life of the goods or is bound to become the owner of the goods; (iii) the lessee has an option to renew the lease for the remaining economic life of the goods for no additional consideration or nominal additional consideration upon compliance with the lease agreement; or (iv) the lessee has an option to become the owner of the goods for no additional consideration or nominal additional consideration upon compliance with the lease agreement.

See 'Appendix C-2A' for the complete text of UCC § 1–201(37) which includes further elaboration on the above.

Importantly, the common 'commercial' understanding of the term 'Finance Lease' – that is, a lease intended for security, which is substantively equivalent to a mortgage financing – does not comport with the definition of that term under Article 2A of the UCC. Under Section 2.A-103(1)(g), a 'Finance Lease' is a lease with respect to which:

- the lessor does not select, manufacture, or supply the goods;
- the lessor acquires the goods or the right to possession and use of the goods in connection with the lease; and
- one of the following occurs:
  - (A) the lessee receives a copy of the contract by which the lessor acquired the goods or the right to possession and use of the goods before signing the lease contract;
  - (B) the lessee's approval of the contract by which the lessor acquired the goods or the right to possession and use of the goods is a condition to effectiveness of the lease contract;
  - (C) the lessee, before signing the lease contract, receives an accurate and complete statement designating the promises and warranties, and any disclaimers of warranties, limitations or modifications of remedies, or liquidated damages, including those of a third party, such as the manufacturer of the goods, provided to the lessor by the person supplying the goods in connection with or as part of the contract by which the lessor acquired the goods or the right to possession and use of the goods; or
  - (D) if the lease is not a consumer lease, the lessor, before the lessee signs the lease contract, informs the lessee in writing (a) of the identity of the person supplying the goods to the lessor, unless the lessee has selected that person and directed the lessor to acquire the goods or the right to possession and use of the goods from that person, (b) that

the lessee is entitled under this Article to the promises and warranties, including those of any third party, provided to the lessor by the person supplying the goods in connection with or as part of the contract by which the lessor acquired the goods or the right to possession and use of the goods, and (c) that the lessee may communicate with the person supplying the goods to the lessor and receive an accurate and complete statement of those promises and warranties, including any disclaimers and limitations of them or of remedies.

Insofar as this particular definition does not comport to our usage of Finance Lease for the purposes of this Handbook, it should be ignored for our purposes.

Bankruptcy

Under U.S. bankruptcy situations, the matter of true versus finance leases would most likely be determined by reference to the UCC tests.

FAA

The U.S. FAA pays attention to identify properly the entity in whose name an Aircraft should be registered. In 1990, pursuant to the so-called 'Leiter Letter',[30] the FAA's Chief Counsel issued a legal opinion setting out the relevant criteria for FAA purposes. According to that opinion, the FAA will recognize the lessee as the owner for Aircraft registration purposes under a lease with an option to purchase in three specific scenarios.

1   The purchase option price is 10 percent or less of the value of the Aircraft determined at the time the lease is executed.
2   The purchase option price is above the 10 percent bright line, but contains a requirement that if the option is not exercised, the lessee nevertheless is obligated to pay a residual value or termination sum equal to or exceeding the purchase option price.
3   The purchase option price is higher than 10 percent and there is no mandatory full payout if the option is not exercised, but the option price is less than the lessee's reasonably predictable cost of performing under the lease if the option is not exercised.[31]

In addition to one of the above, both of the following factors must also be present:

•   the lessee must have the obligations of maintenance, insurance, taxes, operations and risk of loss with respect to the Aircraft; and
•   the lease must not permit the lessee the unilateral right to terminate the lease without an economic penalty.[32]

As the Leiter Letter does not have the binding nature of a statute or regulation, it is prudent to consult with the FAA and obtain an opinion from the Aeronautical

Center Counsel in advance of a closing in order to determine the propriety of registering an Aircraft in the name of a lessee.

PRACTICE NOTE

While the lessor in a Finance Lease holds 'title' to the asset, holding such title is akin to having a mortgage interest in the asset with the lessee being treated as the true owner of the asset insofar as the benefits and burdens of ownership ultimately rest with the lessee. Transactions are structured as Finance Leases, rather than mortgages, for many reasons, including to obtain benefits of a Cross-border Tax Lease or for the financier to achieve 'lessor' status under local law in overseas jurisdictions (which almost universally entitles the financier to a more beneficial status in the exercise of remedies than it would have had it been classified as a mortgage).

The mischaracterization of a lease as a true lease can have major ramifications.

In the context of *tax* matters, a mischaracterization for a lessor would result in recoupment by the IRS of tax benefits previously taken plus fines and penalties.

In the context of *commercial matters*, failure to obtain true lease characterization would eliminate any residual recovery and potentially leave the lessor with a substantial economic downside loss. In addition, if a lease is construed to be a loan, for usury purposes, the lessee may argue that the lessor violated applicable usury law and should suffer the penalties of overcharging the lessee for the imputed 'interest' paid by the lessee on the financing.

In the context of a *lessee bankruptcy* proceeding, a lessee may challenge a lessor's true-lease characterization and rights under a purported lease. Consequences of losing a true lease challenge include the following.

- *Loss of Section 365(d)(10) payments (if Section 1110 not applicable).* A lessor has meaningful rights to payment on leases under Section 365(d)(10) of the federal Bankruptcy Code. Section 365(d)(10) requires debtors-in-possession to 'timely perform all of the obligations of the debtor . . . first arising from or after 60 days after the order for relief in a . . . Chapter 11 . . . under an unexpired lease of personal property . . . until such lease is assumed or rejected'.
- *No required cure of defaults under Section 365(b).* Under Section 365(b) of the Bankruptcy Code, a debtor-lessee must cure all monetary and non-monetary defaults before the debtor-lessee can assume the lease. A lessor does not enjoy this right if the court characterizes the transaction as financing instead of a true lease.
- *Potential loss of lien and/or priority.* If for any reason a lessor fails to make a timely and correct filing of financing statements under the UCC or other priority-creating statute to perfect a security (versus ownership) interest, the lessor may not achieve expected priority with

---

51

respect to the leased property. In that instance, another secured party may take priority and leave the lessor with little or no recovery in a bankruptcy. Even if a valid, secured claim exists due to a proper grant of a security interest, the security interest may still be subject to:

- o the 'cram down' provisions under Section 1129 of the Bankruptcy Code (producing lower payments to the lessor than contractual rent);
- o a reduction of the value of the collateral (leased property) under Section 506 of the Bankruptcy Code; or
- o the 'strong-arm' powers of the trustee in bankruptcy to avoid a lien of the lessor against the leased property Section 544(a) of the Bankruptcy Code.

- *Loss of favorable lessor remedies.* When a lessor properly structures a 'lease' as a true lease, the lessor obtains remedies as an owner, instead of remedies as a secured party under Article 9 of the UCC. For example, Part 6 of Article 9 of the UCC requires a secured party (even if called a 'lessor') to give reasonable notice of a disposition of foreclosure of collateral, and to sell the collateral in a commercially reasonable manner, paying the excess proceeds to the debtor (even if called the 'lessee'). The true lessor is not bound by these rules. Lessors can cancel a lease, recover the equipment, collect discounted rents and even require the payment of liquidated damages (a genuine pre-estimate of damages, including lost residual value, that cannot be easily or accurately calculated). The remedies available under a true lease provide several advantages to a lessor over those of a secured party with respect to the same property.[1]

---

[1] Mayer, DG, 'True leases under attack: lessors face persistent challenges to true lease transactions', *Journal of Equipment Lease Financing 23(3)*, Fall 2005/Part B.

A Finance Lease financing is, as noted, the economic equivalent of a Mortgage Financing, insofar as the 'lessor' has no upside (or downside) risk in the financed Aircraft. The utilization of Finance Lease structures is ubiquitous in the financing of Aircraft world-wide in the situation where a mortgage financing would otherwise suffice; that is, the financiers are providing debt financing with no residual interest in the Aircraft. This financing structure is so highly favored for, among others, the following reasons:

- The laws, and the courts, of the local jurisdiction of the airline may provide greater rights in an airline default situation to lessors over mortgagees (notwithstanding the substantive equivalent of the two structures).
- Procurement of a mortgage in the local jurisdiction may be prohibitively expensive (by reason of, for example, stamp taxes on recorded mortgages).

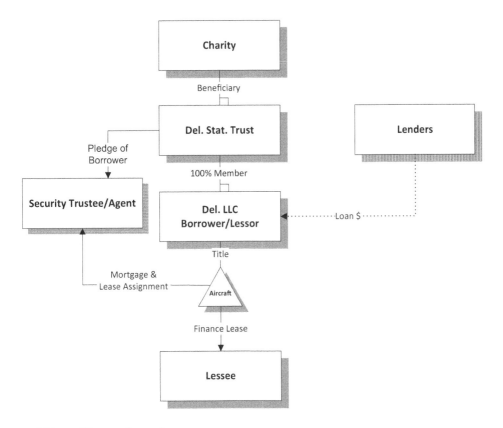

Exhibit 3.9 Finance Lease Structure

In fact, nearly all ECA financings utilize the Finance Lease structure. Exhibit 3.9 is a schematic illustration of a standard Finance Lease structure utilized to finance Aircraft.

Final Legal Maturity Date

In an EETC or other aircraft-secured facility that is rated using the EETC Rating, the Final Legal Maturity Date is the date that is 18 months (or, in certain cases, 24 or 36 months) after the date that is the economic maturity date of the underlying security. This 18-month period (or other period, as explained below) is the period determined by the rating agency providing the EETC Rating on a given security to be the longest length of time it should take to repossess and liquidate aircraft collateral starting on the last possible day the security can be in default.

While 18 months is the norm in deals for U.S. airlines that utilize the EETC Rating methodology, rating agencies may assess longer periods in other jurisdictions based on their assessment of the time it might take to repossess and liquidate aircraft collateral.

French Lease

This is a Cross-border Tax Lease where the lessor is domiciled in France. These types of leases are increasingly difficult to arrange in the light of EU investigations into whether these amount to an illegal subsidy. French commercial banks have historically been the investors in these transactions, with their subsidiaries acting as lessors.

Interchange

This is an arrangement whereby an Aircraft changes operators and Registration when its jurisdiction of operation changes.[33] Interchange arrangements are a matter of increasing interest with the conglomeration of airline operators across jurisdictions, especially in Central and South America. There is the tension between Cabotage rules, on the one hand and the desire by operators to maximize Aircraft utilization, on the other.

PRACTICE NOTE

The change of Registration of an Aircraft as it bounces from jurisdiction to jurisdiction has critical effects on Aircraft Financiers for all the reasons noted in Part 7, 'Re-registration'. Documentation to deal with this can be rather arduous and complex.

International EETCs

This is the issuance of an EETC by a non-U.S. airline/issuer. As discussed in the Practice Note in 'EETCs', EETCs are no longer the sole domain of U.S. airline issuers. EETCs issued by non-U.S. airlines are, absent any particular local or issuer requirements, structured in the identical fashion as those issued by U.S. airlines. The Rating Agencies' willingness to rate non-U.S. airline issuers will turn almost exclusively on the built-in requirement of the EETC to have legal and practical certainty to the return of the Aircraft Asset collateral in a default situation. Section 1110 is the lynchpin to this in the U.S. The adoption of Cape Town in non-U.S. jurisdictions may provide the necessary support for non-U.S. carriers. You will note that the preceding sentence uses the word 'may'. Whether a jurisdiction which has adopted Cape Town will satisfy the Rating Agencies will depend on the following criteria.

- Has the jurisdiction adopted the 'right' Cape Town Declarations?
- How do the Rating Agencies evaluate the rule of law in that jurisdiction?

A guide to what the 'right' declarations are is in Appendix II to the ASU. That appendix identifies the Cape Town declarations necessary for a country to receive a discount from the minimum premium rate in respect of export financing support from the applicable ECAs. Key declarations like Alternative A, choice of law, IDERA, self-help and timely remedies are must-haves under the ASU rules and are usually must-haves for Rating Agencies as part of their analysis. The Organization of Economic Co-operation and Development (OECD) assesses whether countries that have adopted Cape Town have adopted the proper 'qualifying declarations' under Appendix II to the ASU. The OECD website, www.oecd.org/tad/export credits/ctc.htm, provides a status list of countries qualified for the ASU discount. That list details those countries eligible for the ASU discount, those countries currently under review or subject to future review, those countries which did not adopt the qualifying declarations and those countries that have adopted the qualifying declarations but on which there are implementation issues.

The adoption of Cape Town with all of the right declarations may not be enough, however. An analysis will be made by the Rating Agencies as to the degree that the attendant repossession laws will be enforced under local law and by law enforcement authorities. This analysis follows the Country Risk analysis discussed in Part 7, 'Country Risk'.

Finally, it should be noted that Cape Town, and the adoption of all of the right, qualifying, declarations is not dispositive for the Rating Agencies. As they have done in a British Airways (England) and a Virgin Australia (Australia) EETC issuance, the applicable Rating Agencies have concluded that the legal system in the relevant jurisdiction, even absent Cape Town or Section 1110 like laws, is creditor-friendly enough to allow a positive rating assessment because creditors in that jurisdiction are assured of repossessing their Aircraft Assets in a timely manner.

PRACTICE NOTE

If Rating Agencies conclude that a particular jurisdiction, even with the proper legal framework, would not allow a timely (for example, 60 to 90 days) recovery of an asset, rather than excluding that jurisdiction altogether, they may add on more time to the requisite Liquidity Facility to cover the longer recovery period. Thus, 24 month Liquidity Facilities have been required in the DORIC 2013 1 EETC which was a United Arab Emirates based transaction (with Emirates Airlines as the underlying credit).

Islamic Finance

Islamic Finance in the context of Aircraft Finance is the financing of Aircraft Assets in compliance with Sharia'a requirements. Sharia'a is the religious law forming part of the Islamic tradition. This law is derived from the holy book of the Muslims – the Koran – and the *Hadith* (the body of statements or actions ascribed to Mohammed, the Prophet) and *Sunnah* (the verbally transmitted record of the teachings of Mohammed, the Prophet). The particular item in the Sharia'a impacting matters of aircraft finance is the prohibition against the charging or receipt of interest. Accordingly, for a transaction to be Sharia'a compliant, it cannot employ the payment of interest as a return factor for investors. The matter of whether a particular transaction is compliant with Sharia'a precepts is often passed upon by a panel of experts in Islamic law.

While there are multiple forms of Sharia'a compliant forms of financing for commercial aircraft, three types stand out as worthy of mention:

Ijarah – Ijarah is, in concept, akin to the leasing of an aircraft.

Murabaha – Murabaha is a financing method where a financial institution will take title to an asset and then sell that asset to its customer – which may be over time. The repayment to the investor may include a component for a reasonable profit. Such profit element serves as compensation to the investor for having taken title to the asset – and the associated risk of ownership.

Sukuk – See 'Sukuk' herein.

Japanese Leveraged Lease (JLL)

A Japanese Leveraged Lease (JLL) is a Cross-border Tax Lease where the lessor is MOST COMMONLY domiciled in Japan. This structure was a frequently used financing mechanism in the 1980s and 1990s, but became largely untenable when the Japanese tax authorities shifted to a more substance-based approach in their analysis of Aircraft 'ownership'. JLLs are no longer a financing option.

Japanese Operating Lease (JOL)

A Japanese Operating Lease (JOL) is an operating lease fully or partly funded by a Japanese investor or equity sourced from Japan.

Japanese Operating Lease Call Option (JOLCO)

A Japanese Operating Lease with Call Option (JOLCO) is a long term operating lease with a call option in the form of an early buyout option for the benefit of the lessee. Such call option provides to the lessee the right to purchase the subject Aircraft at a fixed purchase price at a fixed date prior to the expiry date of the related lease.

The JOLCO is a tax/financing structure that can provide airlines with 10 to 12 years of low-cost aircraft funding. An investor with Japanese tax liability puts up a minority portion of equity funding (the balance is debt) in exchange for the tax benefits associated with the ownership and debt financing of the Aircraft. The tax investor must take some actual asset risk to receive this tax benefit.

When the Japanese tax authorities required that an investor take some portion of asset risk for most transactions, the JOLCO was developed. A key requirement for any JOLCO is the '90 percent test'. This means the overall lease rentals payable during the life of the transaction cannot exceed 90 percent of the lessor's acquisition cost of the financed aircraft, together with debt costs and other costs such as fees. The lessee's exercise of the fixed price call option must not be economically compulsory. To demonstrate that the lessee's purchase option is not compulsory, that purchase option price is supported by third party appraisals. The call option purchase price in a JOLCO would be an amount sufficient to repay the JOLCO debt and provide a return to the equity investor.

The debt component of the JOLCO is typically bank debt. The JOLCO debt is repaid from the rents under a JOLCO lease. The JOLCO debt may amortize to zero prior to the maturity of the related JOLCO lease. If it has a balloon payment due at the final maturity of the JOLCO lease, the balloon would be repaid from the sale of the financed Aircraft.

From an investors' perspective, there is the risk that the lessee under a JOLCO will not exercise its call option for the leased Aircraft (leaving the Japanese equity and, if applicable, financiers with a balloon on their financing, with the concomitant asset risk). This risk can be minimized by maximizing the lessor's investment recovery to the extent that the transaction is not classified as a finance lease for the Japanese tax purposes combined with more stringent redelivery conditions (e.g. payment of maintenance accruals by the lessee when the aircraft is returned at the lease expiry). In practice, the risk of a non-exercise of the call option has not materialized.

Debt of a JOLCO lease is in principle booked by the lessor domiciled in Japan to avoid Japanese withholding taxes.[34] The Japanese equity investor can earn fixed income during the lease term and make capital gains by disposal of the Aircraft at the lease expiry. These investors can only take depreciation of the Aircraft to the extent of their invested equity amount if the transaction is partly funded by non-recourse debt. The typical equity investor in a JOLCO is a small and medium-size Japanese company that is privately held.

There are some disadvantages associated with the JOLCO. It is a long-term commitment – typically the transactions are of about 10 years' duration. JOLCO structures can be relatively inflexible – once they are up and running it can be difficult to revisit the terms of a JOLCO if anything changes. An airline is not dealing with an operating lessor in the traditional sense; it is really dealing with Japanese investors. Over the life of a transaction, an airline may need to go to those investors and seek their consent from time to time to do various things. That process needs to be handled quite carefully in order to make sure the investors fully understand the issues and the airline gets an answer within the time frame it needs.

Kommanditgesellschaft (KG)

Kommanditgesellschaft (KG) is an Aircraft Asset financing by individuals[35] participating as limited partners (directly or by employing a trustee) in single purpose companies which are organized in the legal form of a German limited partnership (that is, a *Kommanditgesellschaft*). The KG has one general partner and one or several limited partners. In the majority of cases, the general partner of a German KG is a limited company (that is, *Gesellschaft mit beschränkter Haftung* (GmbH)). The limited partners may be participating in the KG directly or via a trustee. The KG may have an advisory board representing the investors' interests by monitoring the management of the general partner. The purchase of the Aircraft

Asset[36] acquired by the KG is partly financed by the equity provided by the investors and partly leveraged by bank loans. The KG receives income from leasing the Aircraft Asset to a lessee. The KG is liquidated after the financed Aircraft Assets have been sold.

Lease Financing

See introductory provisions to Part 3.

Lease/Head Lease Structure

This is the lease financing of an Aircraft whereby the owner of the Aircraft (head lessor) leases an Aircraft Asset to an intermediate lessee/lessor and that intermediate lessee/lessor subleases such Aircraft Asset to the operator/lessee.

PRACTICE NOTE

These structures are used typically to allow the operator/lessee to make payments to the intermediate lessee/lessor without having to pay withholding taxes on the net payments, taking advantage of favorable tax treaties between the jurisdiction of the operator/lessee and the intermediate lessee.

Lease-in Lease-out (LILO)

A Lease-in Lease-out (LILO) structure is akin to a Lease/Head Lease Structure.

Liquidity Facility

A Liquidity Facility is a facility appended to a larger financing that provides for payment of debt service (typically interest only) by a bank or other creditworthy entity if the underlying obligor defaults in the making of such payment. A Liquidity Facility is usually a revolving credit facility, but could be structured as a Standby Letter of Credit (SBLOC). Liquidity Facilities are utilized in EETCs to provide (typically) 18 months of interest coverage. The provider of the Liquidity Facility would typically have a super-priority position in the Aircraft Assets securing the larger financing. As such, principal and other amounts owing to the beneficiaries of the Liquidity Facility are subordinated to recovery by the Liquidity Facility provider of amounts it has paid. Effectively, then, the beneficiaries are trading timely payment of interest for future offsets to recoveries in the amount of such interest plus a financing cost payable to the Liquidity Facility Provider. A Liquidity Facility, therefore, is of value only if the investor (or the Rating Agency) places a value on current, timely receipt of principal, since the investor will ultimately have to repay the Liquidity Facility provider from collateral liquidation proceeds ahead of repayment of principal.

The coverage period of 18 months is used in EETCs as this is the perceived (by rating agencies) maximum period it should take to liquidate the Aircraft Assets subject to an EETC. However, this period may be lengthened if the EETC issuer is located in a jurisdiction in which the repossession time frame for financed Aircraft Assets may exceed that in the U.S. Liquidity Facilities may be drawn if the liquidity provider has its Credit Ratings downgraded to a specified threshold level or it does not renew the facility by its renewal date.

Mortgage Financing

See introductory provisions to Part 3.

Non-recourse

This is an arrangement in a Back-leveraged Lease where the financier agrees that recourse for its debt is expressly limited to the related Lease, the mortgage on the leased Aircraft and other pledged collateral. In other words, the lessor is not otherwise personally liable for the debt (except, perhaps, the Rats and Mice), and none of its other assets are at risk.

PRACTICE NOTE

Some Back-leveraged Lease financings do provide for recourse to the operating lessor, but those are atypical. Allowing for some level of recourse may provide an operating lessor with better pricing and other terms, as well as access to a broader lender pool. Also, in the case of an especially strong operating lessor as to which there is full recourse, that lessor may be able to retain more (exclusive) control over any Back-leveraged Lease that serves as security for the transaction.

Operating Lease

Relating to any particular discipline (for example, tax, accounting, commercial law or bankruptcy), this is a Lease Financing which does not meet the criteria of a Finance Lease under the rules of such discipline. See a detailed discussion of these criteria in 'Finance/Capital Lease'. This would capture the structure where the owner of the equipment (the operating lessor), rather than the lessee, retains most of the benefits and risks of asset ownership. Operating Leases do not usually provide the lessee with an option to terminate the lease prior to its scheduled termination date.

Orphan Trust Structure

This is a financing structure where the ultimate owner of the asset is a charity. Orphan Trust Structures are employed to provide Bankruptcy Remote protection and as a method for certain banks to avoid aggregating lessor/borrower exposure for internal or regulatory lending-limit purposes. Exhibit 3.10 is a schematic showing the parties' relationships in an Orphan Trust Structure.

PRACTICE NOTE

While the idea behind Orphan Trusts is to place ultimate ownership of Aircraft Assets with a charity for the reasons outlined above, truth-be-told, the charity never (intentionally) benefits from such ownership (other than by earning an accommodation fee). By means of residual interest certificates in a CLO (which entitle the holders of such certificates to residual value interest in the Aircraft Assets) or a bargain purchase option granted to an operating lessor of the financed Aircraft Assets at transaction conclusion, all upside and residual value is engineered away from the charity.

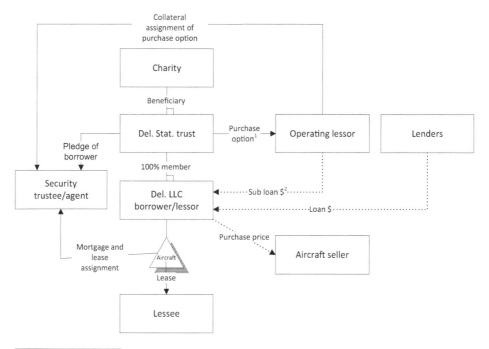

---

[1] Bargain Purchase Option to purchase membership interest in Borrower/Lessor exercisable when loan repaid.

[2] Equivalent to equity investment.

Exhibit 3.10 Orphan Trust Structure

Source: Author's own

Owner Trust

This is a grantor trust established by a lessor (as beneficiary) to own an Aircraft. In the U.S., Aircraft may be registered in the name of an owner trust.[37] This is usually accomplished by having the trust company/department of a bank establish a trust to hold legal title to the Aircraft on behalf of one or more beneficiaries.[38] An Owner Trust has historically been the ownership vehicle of choice in U.S. Leverage Leases, and for many other U.S. transactions. The benefits of using an Owner Trust are comparable to other SPVs, with the added benefit that they will enable non-U.S. lessors to own U.S.-registered Aircraft. In addition, under U.S. bankruptcy laws, grantor trusts are not a type of legal entity that is susceptible to a bankruptcy filing. The trustee of such a trust (often the trust department of a U.S. bank) must be either a U.S. citizen or a resident alien. If any beneficiary is not a U.S. citizen, the trust may still register the Aircraft in the U.S. provided the trustee also files an affidavit stating that the trustee is not aware of any reason, situation, or relationship the result of which would be that those persons who are not U.S. citizens together would have more than 25 percent of the aggregate 'power' to influence or limit the exercise of the trustee's authority.

If the beneficiaries of a trust consist of non-U.S. citizens that do have such power, the trust instrument itself must provide that those persons together may have no more than 25 percent of the aggregate power to direct or remove a trustee. This is ordinarily accomplished by adding limitation language to the trust agreement to the effect that the beneficiary will have no rights or powers to direct, influence or control the trustee's absolute and complete discretion concerning the ownership and operation of the Aircraft. To protect the non-U.S. citizen's interests, this limitation language is normally subject to the requirement that the trustee exercise due regard for the interests of the beneficiary and agree not to sell, mortgage or otherwise encumber the Aircraft without the prior consent of the beneficiary.

PRACTICE NOTE

On 18 June 2013, the FAA published in the Federal Register 'Notice of Policy Clarification for the Registration of Aircraft to Citizen Trustees in Situations Involving Non-U.S. Citizen Trustors and Beneficiaries' as an official rule of the FAA. This Notice made clear that all owners of U.S.-registered Aircraft, whether or not Owner Trusts and regardless of onshore or offshore beneficiaries, have obligations 'to communicate [to the FAA] critical safety information'. To meet these obligations, an owner must maintain current information about the identity and whereabouts of the actual operators of an Aircraft and location and nature of the operation on an ongoing basis, thereby allowing that owner to provide the operator with safety critical information in a timely manner, and to obtain information responsive to FAA inquiries, including investigations of alleged violations of FAA regulations.

The FAA expects that an owner trustee of Aircraft on the U.S. registry, in carrying out the above-described obligations, normally should be able to respond to a request by the FAA for the following information about the Aircraft and its operation within two business days.

- The identity of the person normally operating, or managing the operations of, the Aircraft.
- Where that person currently resides or has its principal place of business.
- The location of maintenance and other Aircraft records.
- Where the Aircraft is normally based and operated.

The FAA further expects that an owner trustee of Aircraft on the U.S. registry normally should be able to respond within five business days to a request by the FAA for more detailed information about the Aircraft and its operations, including:

- information about the operator, crew, and Aircraft operations on specific dates;
- Maintenance and other Aircraft records; and
- the current airworthiness status of the Aircraft.

In the event of an emergency, the FAA may request a trustee to provide information more quickly than the timelines noted above.

The FAA will in most cases, go directly to the air carrier or similar operator through FAA personnel (for example, principal operations or maintenance inspectors) to obtain information about the Aircraft and its operation. The FAA will, however, always reserve the right to seek information from the registered owner of an Aircraft on the U.S. registry.

Participation

This is a mechanism for a bank (or other debt financier) to off-load the obligations and benefits of holding a loan (or credit commitment) by having a third-party lender agree to take on the risks and rewards of such loan (or a portion thereof).[39] As compared with an assignment, where the lender holding the loan assigns its interest to a new lender and is out of the picture once the assignment becomes effective, the bank selling the Participation remains in the transaction as the lender of record with the borrower. The purchaser of the Participation, the participant, in these transactions, therefore, has no privity with the borrower, and is entirely reliant on the selling bank to perform on the contract evidencing such Participation. Accordingly, the participant is not only taking borrower risk, but also the risk of performance by the bank from which it acquires the Participation. Also, it is the subject of negotiation as to what instructions the facing bank must comply with coming from the participant, and other intercreditor issues as between the two parties may come into play. There are two types of Participations: funded and unfunded. In a funded

Participation, the participant pays the facing bank the full amount of the loan (or portion thereof) in which it has purchased the participation, and such participant receives distributions of principal and interest, at the contracted or agreed rate,[40] made by the borrower. In an unfunded Participation, the participant pays the facing bank the amount of the loan (or portion thereof) in which it has purchased the Participation if and when a default occurs,[41] and in the ordinary course such participant receives a participation fee out of interest payments received by the facing bank. Many Participations are sold such that the borrower is not aware of it (as the selling bank may not want the borrower to know); these are called 'blind participations'.

While most loan agreements permit the granting of a Participation interest in the related loans, restrictions are often placed on the ability of the participant to have consent rights on other than the basic economics of the transaction.

PRACTICE NOTE

The term 'participation' is what I consider a loaded term. It may be used in this context as a simple, straightforward term, but actually it needs a lot of fleshing out to get to intended allocation of rights and obligations.

Pre-delivery Payment (PDP)

A Pre-delivery Payment (PDP) is a facility for the financing of the incremental (progress) payments required to be made by Aircraft purchasers to OEMs prior to the delivery of Aircraft. Aircraft production, for obvious reasons, has long lead times. Procuring parts and construction of a particular Aircraft can take 6 to 18 months. During this construction period, the OEM is making large expenditures for labor and materials. To offset these ongoing pre-delivery costs, the OEM typically requires in the related Purchase Agreement for its customer to make periodic progress payments. These progress payments may be for as much as 20 to 30 percent of the Aircraft's purchase price.[42] Insofar as these progress payments represent a large outlay of cash by the airline or lessor Aircraft purchaser, financing is often sought to maintain liquidity. An impediment to a financing is that there is no hard asset (yet) that can be used as collateral. This is where the PDP comes into play. A PDP is a full recourse financing by the purchaser secured by a collateral assignment of the rights of the purchaser in the Purchase Agreement as relates to the delivery positions of the Aircraft whose progress payments are being financed. The PDP financier, therefore, can step in to the delivery of an Aircraft if the borrower defaults. The crux of this benefit is the purchase price at which the financier can take delivery. Insofar as Aircraft purchasers often receive significant discounts off OEM list prices, this right of the financier to obtain delivery of an Aircraft at below market prices and thereafter 'flip' the Aircraft at a higher price would enable the financier to recover its loan. In addition, the financier does not typically fund 100 percent of the progress payments, so there is so-called equity of the borrower in the Aircraft which improves the LTV of the Aircraft's delivery position.

Importantly, the OEM's co-operation in the PDP is critical since the Aircraft Purchase Agreements are not assignable without consent. The OEM, however, while normally co-operative, typically does not provide to the financier the full discounted purchase price under the Purchase Agreement, but rather a higher number. It is this higher OEM offered purchase price (taking into account the borrower's 'equity') that the financier must assess relative to market values when analyzing this financing. As well, the financier will need to be mindful of, and build in restrictions concerning:

- Aircraft price escalation;
- engine pricing and credits, if applicable;
- change orders to the Aircraft Purchase Agreement;
- Buyer Furnished Equipment (BFE) costs of the Aircraft; and
- OEM-required buy-back options (and the price the OEM will pay).

A further consideration for these types of financings is the Claw Back Risk inherent in this structure. See Part 8, 'Claw Back'.[43]

PRACTICE NOTE

PDPs are more credit-oriented than an Aircraft Financing in the light of the lack of a hard asset serving as collateral.

As noted above, the OEM plays a major role in PDP financings. Here is a summary of items the PDP financier and the OEM would normally agree to.

- PDP financiers may buy PDP-financed Aircraft for an agreed purchase price from the OEM and thereafter exercise the borrower's right to purchase PDP-financed Aircraft under the Purchase Agreement.
- PDP financiers may step into the borrower's shoes (to the exclusion of the borrower) under the Purchase Agreement within a specified time frame following a borrower default and not to sell the PDP-financed Aircraft if a related 'Event of Default' is continuing under the Facility.
- OEM may buy out PDP financiers for the debt balance of the PDP facility (subject to agreed caps on breakage, expenses and interest) within a specified time frame following a borrower default.
- OEM waives set-off rights.
- PDP financiers can assign rights under Purchase Agreement, subject to customary conditions; the assignee will be entitled to certain warranties, customer support and so on.
- OEM will apply PDPs against PDP-financed Aircraft purchase price.
- OEM will not refund PDPs to borrower (only to PDP financiers).

- OEM will not release the borrower from its obligations under the Purchase Agreement or accept cancellation of the Purchase Agreement unless OEM repays PDPs to PDP financiers.
- Subject to customary exceptions, OEM will not amend the Purchase Agreement without PDP financier consent.
- Upon request, the OEM will notify the PDP financiers of any outstanding breaches by the borrower which would entitle the OEM to terminate the Purchase Agreement and, if the OEM intends to exercise any right of termination it may have, the OEM will agree not to exercise any such right without giving the PDP financiers the opportunity to elect to cure the breach that gave rise to the right of termination or complete the purchase of the relevant PDP-financed Aircraft within an agreed period. In addition, the OEM will promptly notify the PDP financiers of any cancellation by the borrower of the Purchase Agreement or any part thereof.

Certain PDP financiers may require that there be in place at the time the PDP facility is entered into definitive and binding arrangements for the PDP-financed Aircraft to be financed on delivery to the borrower. In such a case, the financier can rest assured that the borrower will have access to funds for repayment of the related PDP loan on that Aircraft's delivery.

Finally, PDP financiers wishing to create a relationship with an airline located in a jurisdiction with relatively high repossession risk may turn to PDP financing because the collateral (that is, the Purchase Agreement) is *per se* located in the OEM's jurisdiction.

Pfandbrief

This is an internationally recognized type of covered bond as used in Germany by German banks to leverage their aircraft loan portfolios. *Pfandbrief*-covered bonds are highly secure securities that have wide appeal and a high level of liquidity. They are regulated by statutory law (Pfandbrief Act and supporting directives). The *Pfandbrief* product was only recently approved in Germany to be available for aircraft loans (having long been available for shipping, government-backed and real estate portfolios); the aircraft-secured loans in a *Pfandbrief* program are known as the covered pool. This *Pfandbrief* program can be used by qualifying financial institutions to finance aircraft loans on their books. *Pfandbrief* bonds offer high levels of security as a result of a combination of safety mechanisms. Under a *Pfandbrief* program, a fiduciary agent (*Treuhänder*) and at least one deputy are appointed by the Financial Supervisory Authority (*Bundesanstalt für Finanzdienstleistungsaufsicht*, commonly referred to as 'the BaFin'). The most important duty of the fiduciary agent is to monitor the prescribed cover of the *Pfandbrief* on a day-by-day basis. The *Pfandbrief* program contains detailed provisions on requirements for a 'day one' LTV of 60 percent with an accelerated amortization profile required,[44] a going forward 'normal'

LTV reduction requirement through the amortization of the loan and additional obligations to provide additional collateral in case of a severe LTV disturbance. In addition, the Pfandbrief Act imposes detailed specifications for qualifying collateral and a risk management system in order to further improve protection for *Pfandbrief* creditors. Furthermore, the transparency provisions of the Pfandbrief Act are intended to permit investors to assess the composition of the *Pfandbrief* cover pool on a quarterly basis. Finally, anything relevant for the *Pfandbrief* program of a *Pfandbrief* bank is supervised and monitored by the BaFin, which makes random checks on the cover of the *Pfandbrief*. The separation of cover pool assets from the general estate of a *Pfandbrief* bank (as issuer) in the event of a *Pfandbrief* bank's insolvency is achieved by the registration of cover assets in a cover register which is – under statutory law – sufficient to secure the segregation. This is a fundamental difference to other covered bond programs where the cover assets are transferred to SPVs. As a result of such segregation, the claims of the *Pfandbrief* creditors are not affected by the commencement of insolvency proceedings against the assets of the *Pfandbrief* bank. *Pfandbrief* debt may be issued in the U.S. capital markets as a 144A transaction.

Pooling

Pooling is the contribution by two or more aircraft operators of Spare Parts and/or Engines into a common pool for utilization by the parties to such pooling arrangement. Pooling is a way to reduce, through economies of scale, the carrying costs of Spare Parts and Spare Engines. Participants in a pooling arrangement are entitled to withdraw items from the agreed stock (pool) held by any participant.

PRACTICE NOTE

Financiers are rather reticent to allow Pooling of their Aircraft Asset collateral since tracking and traceability are greatly impaired once the asset leaves the possession of the operator that is the beneficiary of the financing. Pooling arrangements might have impacts on ownership and security-interest rights in collateral.

Pass Through Certificates (PTCs)

Pass Through Certificates (PTCs) are securities issued to investors by, usually, a trust established by an issuer, that represent a fractional undivided interest in that trust. The main assets of these types of trusts as utilized in Aircraft Finance transactions are promissory notes (also known as 'equipment notes') of an airline issued pursuant to Mortgage Financings (or, in the old days of U.S. Leveraged Leases, equipment notes of Owner Trusts issued pursuant to Lease Financings). By utilizing these trusts, equipment notes of a comparable economic nature from multiple Aircraft can be pooled into a single security. EETCs use the PTC structure. Holders of a PTC have only such rights and benefits as flow through to it from the equipment notes held in the related trust.

Securitization

This is the financial practice of pooling various types of contractual payment obligations to investors which, in the context of Aircraft Finance, would typically be Lease rental receivables or debt obligations in a CLO or CDO, respectively, as discussed above.

Spare Parts Financing

This is the financing of an airline's Spare Parts inventory (see Part 2, 'Spare Parts').

Special Purpose Vehicle (SPV)

A Special Purpose Vehicle (SPV)[45] is a legal entity (corporation, partnership, owner trust, limited liability company) established for the sole purpose of participating in a particular financing, typically as the owner/lessor/borrower of a particular asset or group of assets. In a typical structured finance transaction of Aircraft Assets that involves a Bankruptcy Remote entity, the legal entity is created whose purpose is limited to acquiring Aircraft Assets (and related Leases) to be financed and undertaking ancillary obligations. The SPV acquires the Aircraft Assets and enters into a financing arrangement, pledging the equipment and the leases as collateral. Some of the customary features of an SPV are: (i) organizational documents and transactional document covenants that limit the entity's business to a single, specific and narrow purpose (generally, acquiring, leasing, financing, refinancing and eventually liquidating the Aircraft Assets); (ii) organizational documents and financing documents that contain 'separateness' covenants that require the subsidiary to be managed and operated in a manner that is distinct from the assets and business of the related lessor or transaction originator (among other entities); (iii) Non-recourse (recourse is limited only to the financed assets); and (iv) the appointment of one or more independent managers or independent directors of the SPV who meet certain requirements that provide some comfort to financing parties that the independent director is not overly sympathetic to the interests of the related lessor or transaction originator. The independent director's favorable vote or consent is typically necessary for the SPV to approve a voluntary bankruptcy filing, to consent to an involuntary bankruptcy or to engage in a corporate restructuring.

PRACTICE NOTE

There are a variety of reasons SPVs are so frequently employed to own Aircraft Assets on behalf of an operating lessor.

- If the Aircraft Asset is subject to a financing, removing the asset from the operating lessor:
    - makes the Non-recourse nature more simple to document;
    - provides Non-consolidation protection for the lender;
    - insulates the lender from Lessor Liens that may spread to all the assets of an operating lessor;
    - provides the lender with another avenue for foreclosure by having a pledge over the SPV's Ownership Interests; and
    - allows inclusion of an Orphan Structure if required by a lender.
- Insulates the operating lessor from Lessor Liability.

Importantly, independent directors appointed by creditors – notwithstanding that their appointment is to protect creditors – must be subject to normal director fiduciary duties. Thus, if it is in the best interest of the SPV to file for bankruptcy, a director is required to authorize the same. This legal requirement should not impair the intended purpose to protect an SPV from bankruptcy as a matter of consolidation with a parent entity, but does present risk when the underlying financing is at risk due to, for example, an impaired lessee in a Back-leveraged Lease situation.

Statutory Trust

A Statutory Trust is a trust formed by statute (for example, Delaware Statutory Trusts (DSTs) formed pursuant to the Delaware Statutory Trust Act, 12 Del. C. § 3801 et. seq. (the 'DST Act')). Due to the inherent vagaries of common law and fact-specific case law, certain states felt that they could clarify the benefits offered by trusts and provide more certainty as to the availability of such benefits (and in certain circumstances actually enhance the benefits) by codifying the benefits under state law, creating Statutory Trusts. The laws governing Statutory Trusts vary slightly by state jurisdiction. Any such variations will typically concern the wording on the topics of: (i) the description of the subject-matter of trusts; (ii) the nature of the writing required; (iii) the party who must sign or subscribe; (iv) the use of an agent or attorney; and (v) whether there must be a signature or subscription. For example, Delaware defines a Statutory Trust as an unincorporated association which:

1   is created by a governing instrument;
2   [is a trust] under which property is or will be held, managed, administered, controlled, invested . . . and/or operated, or business . . . activities . . . are carried on or will be carried on, by a trustee or trustees . . . for the benefit of such person or persons as are or may become beneficial owners or as otherwise provided in the governing instrument; and
3   files a certificate of trust pursuant to § 3810 of [the DST Act].

An additional requirement is that of trustee residency. Therefore, the trustee must be a citizen residing in the state under which the Statutory Trust will be formed. If the trustee is a natural person, such individual must reside in such state; or if the trustee is a business entity, such entity must have its principal place of business in such state. Once formed, a Statutory Trust will have perpetual existence, except to the extent otherwise provided in the governing instrument (commonly referred to as a Trust Agreement). With respect to common law Grantor Trusts, on the other hand, the rule against perpetuities continues to apply, and common law Grantor Trusts must have a limited life.

Structurally, the beneficiary holds the beneficial interest in the trust and thereby indirectly owns all the assets therein. In Aircraft transactions, the trustee owns the aircraft on behalf of and for the benefit of the owner of the trust and must follow the direction of the owner. A key feature of Statutory Trusts is that they protect the assets held in trust so that no creditor of a beneficial owner of the trust has any right to obtain possession of, or exercise legal or equitable remedies with respect to, the assets of the trust. Unless otherwise provided in the governing instrument, the beneficial owners of a Statutory Trust have the same limitation of liability as would a stockholder of a corporation.

A validly formed Statutory Trust offers many other attractive features. It has broad power to provide indemnification to a trustee, a beneficiary and any other person, except as restricted by the governing instrument. Furthermore, the risk of bankruptcy may be minimized by vesting the decision of whether to voluntarily commence bankruptcy proceedings in an appropriate decision maker, such as an independent trustee or other manager.

Sukuk

Sukuk is an Islamic law-compliant financing. A *Sukuk* (which works in a broadly similar way to a conventional securitization) is evidenced by a certificate that provides an investor with ownership or part-ownership in the underlying asset, usufruct, or service. The *Sukuk* represents beneficial ownership of the underlying assets and therefore entitles its holders to receive a pro rata share of profits generated by the asset (not a fixed return tied to their face value). A *Sukuk* can also be issued in tradable form and listed on investment exchanges. A *Sukuk* transaction is intended to comply with Sharia'a prohibitions on the charging of interest and, therefore, represents an equity, not a debt, interest. Other forms of Islamic finance structures include *ijara*, *mudaraba* or *murabaha* financings.

Synthetic Lease

This is an off-balance sheet (or 'synthetic') lease financing structure treated as a lease for accounting purposes but as a loan for tax and commercial finance purposes. This structure is used by corporations who may be seeking off-balance sheet reporting of their asset-based financing, and who can nevertheless efficiently use the tax benefits of owning the financed asset.

To achieve off-balance sheet treatment, a 'lease' financing must satisfy the standards set out in Statement of Financial Accounting Standards No. 13 (SFAS 13). SFAS 13, as modified by various amendments, interpretations and technical bulletins issued by the Financial Accounting Standards Board (FASB), applies fairly objective tests between operating leases (the desired result) and capital leases. The goal of the lessee in a synthetic lease is operating lease treatment for financial accounting purposes, such that neither the asset nor the liability from the transaction is reflected on its balance sheet. To reach this goal, the lease must:

- not automatically transfer ownership of the leased property to the lessee by the end of the lease term;
- not contain a bargain purchase option (that is, a purchase price not less than the reasonable estimate of fair market value at the end of the lease term);
- have a term of less than 75 percent of the estimated economic life of the property; and
- be priced such that the present value (typically discounted at the debt rate) of the rentals and other minimum lease payments is less than 90 percent of the fair market value of the leased property.

These SFAS standards are subject to ongoing review and will likely change in the near term. In order to effect an off-balance sheet lease financing, the borrower/ lessee sells the assets to be subject to the financing to a lender/lessor for the amount of the intended financing (typically for an amount of 80 to 100 percent of its then current fair market value; the applicable percentage to be based on credit and asset considerations). The lender/lessor immediately leases the financed assets back to the borrower/lessee. The amortization (that is, the amount of 'rent' earmarked for the principal component of debt service), back-end balloon (that is, the unamortized amount of the loan) and tenor of the lease must satisfy the guidelines set out above. On maturity of the synthetic lease, the lessee will have the option either to purchase the leased assets for the amount of the unamortized portion of the loan (the Balloon) or to pay an amount (the 'deficiency guarantee amount') to the lessor (and receive nothing in return). The at-risk portion for the lessor/lender, the difference between the Balloon and the deficiency guarantee amount (the At-Risk Amount), is a risk the lender/lessor is willing to take because it would make economic sense for the lessee/borrower to walk-away from its asset by paying the deficiency guarantee amount only if the value of the financed assets are less than the At-Risk Amount. In fact, at the outset of an off balance sheet synthetic lease financing, the projected value of the financed assets is typically many multiples in excess of the At-Risk Amount.

An off balance sheet lease financing is, in substance, a mortgage financing. For tax and commercial (UCC) law purposes, the lessee is viewed as the true owner of the asset. This is because, except for title, indicias of ownership reside with the lessee:

- the lessor/lender's return on the transaction is equivalent to a loan financing; there is no upside available;
- the lessee/borrower retains the burdens (maintenance, insurance, risk of loss and so on), as well as the benefits (use, upside, and so on), of ownership; and

- the lessee/borrower may not unilaterally terminate the lease without recouping for lessor/lender its entire investment.

The conveyance of title to the financed asset and leaseback structure is, under commercial law terms, a 'lease intended for security' with the holding of title to the asset simply a security device akin to a mortgage.

By way of illustration, consider the following example. Company X desires to finance on an off-balance sheet basis a sizeable asset on its books, such as a corporate jet. An appraisal shows that the asset is worth U.S.$110 million has a remaining economic life of 9 and one third years and has a projected value of U.S.$75 million in 7 years (75 percent of 9 and one third years). A lessor/lender may be willing to advance U.S.$100 million on the asset (based on its assessment of both the asset and Company X). Based on the 90 percent test, Company X can agree to a Deficiency Guarantee Amount of U.S.$66 million and a Purchase Option Payment of U.S.$82 million (assuming a U.S.$700,000 a month rent payment, interest at 7 percent per annum and interest and principal allocated on a mortgage-style basis). The At-Risk Amount is U.S.$16 million (U.S.$82 million minus U.S.$66 million), which is well below the projected value at the end of the lease term.

PRACTICE NOTE

In 2002, the Financial Accounting Standards Board (FASB) promulgated a number of 'interpretations' that largely rendered the Synthetic Lease untenable.

U.S. Leveraged Lease

This is a long-term leveraged Operating Lease where the lessee brings together both the equity investor owning the lessor and the lender (in contrast with a Back-leveraged Lease where the lessor sources the lender) to finance an Aircraft. The lessor owns the equipment and will generally provide a portion of the purchase price (20 percent) while borrowing the remainder, usually on a non-recourse basis, from the lender. The lessor's economic return in these transactions includes the tax benefits accruing to an owner of Aircraft (depreciation and so on), and such benefits are factored in the rentals payable by the lessee thereby reducing the lessee's rental obligations. As compared with traditional Operating Leases, U.S. Leveraged Leases allow for an early termination for obsolescence (which includes the lack of a need by the lessee for the Aircraft). U.S. Leveraged Leases were the preferred financing structure for the major U.S. airlines (and a number of non-U.S. airlines) in the 1980s and 1990s, and, accordingly, billions of dollars of Aircraft were financed by this structure. In the aftermath of 9/11 and with so many of the U.S. airline majors filing for bankruptcy in the 2000s, the traditional lease-equity sources of capital (Philip Morris, Disney, Ford Motor Credit, NYNEX, AT&T and so on – large corporations with income to shelter) dried up when they suffered heavy losses (and

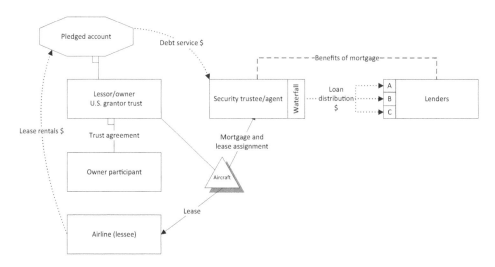

Exhibit 3.11 U.S. Leveraged Lease Structure

Source: Author's own

tax recapture) when their deals soured. Today, only the strongest credits can obtain lease equity and, accordingly, the U.S. Leveraged Lease is rarely employed.

Exhibit 3.11 shows the parties' relationships in a U.S. Leveraged Lease.

PRACTICE NOTE

Since U.S. Leveraged Leases are tripartite transactions (and not beholden to market standards for documentation terms, much in the way ECA transactions are), these are among the few types of transactions where the parties actually sit down in a conference room to negotiate documentation.

Warehouse

A Warehouse is a financing facility typically supplied by banks for an operating lessor: (i) to enable the lessor to accumulate a sufficient number of owned (and leased) Aircraft Assets to effect an eventual Securitization of the accumulated pool; or (ii) as a financing vehicle for Aircraft to 'warehouse' them until they are traded, parted-out and/or an alternative long-term financing solution is found. Warehouses may also be employed as a short-term facility to accommodate fast acquisitions by a lessor who would then turn to long-term financing at a later date.

Warehouse facilities may be structured with an identified pool of Aircraft Assets to be financed, or may be more open-ended with no identified pool. In the latter case, the bank will identify the eligible equipment, eligible jurisdiction and eligible lessees that may be part of the facility, and impose Concentration Limits, Mandatory

Document Terms and Mandatory Economic Terms as well. In either case, significant due diligence is performed on the financed leases, lessees and jurisdictions.

Wet Lease

A Wet Lease is an Operating Lease where the lessor provides the crew and, depending on the deal, fuel and maintenance services, to the lessee, in addition to the Aircraft itself. An ACMI is a form of Wet Lease.

Notes

1 The vast majority of leasing transactions employ an SPV structure, but that is not required. The lessor operating company can certainly be the direct owner of the leased Aircraft Asset.
2 Known as 'interim lift'.
3 Some airlines, for example, have arbitrary policies of, for example, 50 percent leased and 50 percent owned, and use the shorted term operating leases to rejuvenate their fleets.
4 But see discussion on 'Operating Lease', where accounting rule changes may result in Operating Lease financing to be 'on' balance sheet.
5 Boeing Capital Corporation, 'Current Aircraft Finance Market Outlook, 2013–2017', available at www.boeingcapital.com/cafno/ (The BCC Outlook).
6 See Nicholas Katzenbach's opinion as Attorney General, 30 September 1966.
7 The 'four letter word' in Aircraft Finance.
8 Scheinberg, R, 'The bank liquidity crisis and aircraft finance: a sector review', *Commercial Lending Review 24(5)*, 2009, pp. 17–23.
9 The answer to this question has major ramifications in the aircraft finance sector. If there is a gap, then the major aircraft manufacturers, if they are unwilling to supply the requisite financing, may be left placing 'whitetails' in the desert insofar as their customers will not be able to purchase the aircraft as they roll off the assembly line. In addition, if refinancings cannot be done because of the lack of financing, airlines or operating lessors who own balloon payments on loan maturities may face bankruptcy or, at a minimum, may be forced to turn over aircraft collateral in satisfaction of debt (if a non-recourse financing). A plethora of whitetails and repossessed aircraft would serve to place pressure on aircraft values over and above the pressures placed on those values as a result of the economic downturn.
10 In a bid to obtain financing from the aircraft (and engine) manufacturers, airlines may condition new aircraft orders on new (and immediate) financing. As the treasurer of one U.S. major told an OEM (repeatedly) as it was negotiating a financing package for a new order: 'We don't buy airplanes; we buy financing and it comes with airplanes.'
11 The BCC Outlook.
12 Non-U.S. airlines accessing the EETC market will likely rely on private (versus public registered) offering securities exemptions such as Rule 144A so as to avoid the necessity to become a reporting entity subject to the Securities and Exchange Act of 1934.
13 Non-U.S. airlines accessing the EETC market will likely rely on private (versus public registered) offering securities exemptions such as Rule 144A so as to avoid the necessity to become a reporting entity subject to the Securities and Exchange Act of 1934.
14 Thankfully, no one is calling them (yet) 'pre-owned'.
15 See endnote 9, for some further color on this topic.
16 At the beginning of the American Airlines bankruptcy (filed November 2011), the related Deficiency Claims were selling at 22 cents on the dollar. In June 2013, they were selling at 93 cents on the dollar.
17 Of course, if the debtor agrees to perform under Section 1110(a), the investor simply has a performing security. And, if the debtor returns the Aircraft Asset under Section 1110(c), there would be no lease attached to the security.

18 Similarly, they need to take into account currency issues that relate to their oil needs since oil is similarly traded only in U.S. dollars.

19 When the financial markets melted down in 2008–2009, many French banks, for example, lost U.S. dollar funding from U.S. money market funds. They had to scramble to find alternative U.S. dollar funding sources (or had to pull back on U.S. dollar lending).

20 As an intercreditor matter, the holders of the Subordinated Tranches would not typically have the right to accelerate the debt, but would have a Buy-Out Right. See Part 9, generally.

21 These ECAs are all parties to the ASU.

22 While U.S. Ex-Im actually offers both of these financing structures, many of the other ECAs only offer one or the other.

23 Boeing Capital Corporation, 'Current Aircraft Finance Market Outlook, 2013–2017', available at www.boeingcapital.com/cafno/.

24 There was a period of time when many EETCs also utilized a 'wrap' structure; the most senior Tranche was, in essence, 'guaranteed' by a monoline insurance company, such as MBIA or AMBAC. However, with the demise of the monoline insurance industry in the aftermath of the 2008 financial meltdown, this feature is not currently available.

25 As noted in U.S. Leveraged Leasing, this product is largely unavailable due to the widespread losses suffered by the equity investors who funded this market. We take the trouble to make note of it here because of the large number of older EETCs still outstanding that have embedded U.S. Leveraged Leasing structures.

26 A pass through trust is a grantor trust established under state law that is a 'disregarded entity' for federal income tax purposes. It operated just like it sounds: all money received by it is passed through the holders of the PTCs.

27 Scheinberg, R, 'A guide for the perplexed: exogenous elements to consider when investing in enhanced equipment trust certificates (EETCs)', *Journal of Structured Finance*, Winter 2005, pp. 46–54.

28 Cross-default/collateralization was lacking historically largely due to the fact that in EETCs having embedded U.S. Leveraged Leases, the equity investors would not permit cross-default/collateralization.

29 'Adjusted' interest is interest on junior class PTCs in respect of unimpaired pool balances.

30 Treatment of Leases with an Option to Purchase for Aircraft Registration, 5 Fed. Reg. 40,502 (3 October 1990). The legal opinion was addressed to Peter Leiter, General Counsel of Bank of America.

31 Treatment of Leases with an Option to Purchase for Aircraft Registration, 5 Fed. Reg. 40,502 (3 October 1990). The legal opinion was addressed to Peter Leiter, General Counsel of Bank of America.

32 Treatment of Leases with an Option to Purchase for Aircraft Registration, 5 Fed. Reg. 40,502 (3 October 1990). The legal opinion was addressed to Peter Leiter, General Counsel of Bank of America.

33 There are other definitions of 'Interchange Agreement' that are used that are more likely to be used in the context of business aircraft.

34 Some banks with no Japanese booking office that nevertheless want to participate in this product may engage a fronting bank with a Japanese booking office and purchase a silent participation in the JOL loan. There are tax risks in taking this approach.

35 Most often referred to as 'doctors and dentists'.

36 The KG market in the shipping and real estate sector is far larger than the aircraft sector.

37 14 CFR. §47.7(c). In the U.S., Aircraft are registered in the name of the owner, rather than the name of the operator.

38 The FAA has interpreted 14 CFR. §47.7(c)(2)(iii) as requiring registration of the applicable aircraft in the name of the trustee and not in the name of the trust itself.

39 This term can certainly have other meaning, and may be used to indicate that a party is simply taking a part of a facility (but for that, you do not need a definition).

40 The selling bank may retain a skim.

41 This trigger is subject to negotiation.

42 While this range covers a large portion of aircraft orders, there are variations above and below the range depending on the nature of the purchaser, its order book and so on.

43 Gee, CA, 'Aircraft pre-delivery payment financing transactions', *Journal of Structured Finance*, Fall 2009, pp. 11–21.

44 Mortgage-style amortization is not fast enough, and 'bullet' maturities are not permitted.

45 Sometimes referred to as a special purpose corporation (SPC).

PART 4

PRICING PROVISIONS

Introduction

As a topic, 'Pricing' is the practice area of making sure the financiers are getting their agreed financial return – neither more nor less. Pricing involves mathematical calculations and attention to detail to ensure that, for example: (i) Debt Service matches Lease rentals; and (ii) underlying fixed rate interest cash flows under a Lease or a mortgage match up to the fixed rate payor delivery obligations on any corresponding Interest Rate Swap. Pricing is benefited by numerous 'conventions', which address things such as: (i) adjustment to period end dates (for example, if a payment date falls on a Saturday and that payment is due on the following Monday, does that payment include interest for Sunday?); (ii) business days (for example, if a payment falls due on a Saturday, is it due the next Monday, or on Friday?); and (iii) the denominator in an interest accrual period (360 or 365 days?). It is important not only to understand these conventions, but also to know when they are to apply. Other pricing provisions Aircraft Financiers deal with include pricing options, costs associated with early (unscheduled) repayments, default interest and unique issues arising for particular types of lenders, like banks.

Since so many Aircraft Finance transactions are funded by banks, much of the Pricing material pertains to bank finance, and the primary pricing yardstick in bank finance is LIBOR (the London interbank offered rate). In contrast to financial institutions like insurance companies which have access to large pools of cash generated by insurance premiums that they seek to deploy, the banks in this market (theoretically) obtain funds to make, and maintain, loans to their borrower customers primarily by borrowing funds from other banks; they borrow from Peter to lend to Paul. They make their money, then, by charging their borrowers a margin over the banks' own borrowing costs, which margin reflects their borrowers' credit risk and the banks' return requirements. This is called 'match funding', which model of bank funding serves as the theoretical basis for bank lending and pricing of Eurodollar-priced loans. Having said that, the theory does not usually follow bank practice, since banks by and large fund themselves on a portfolio basis. Insofar as borrowers want an objective publicly available benchmark rate for determining the interest they are required to pay, and banks do not always care to reveal their own cost of funds, a proxy market standard for assessing the banks' cost of funds for U.S. dollar-based borrowings is the published LIBOR rate.

Exhibit 4.1 summarizes the usual allocation of payment conventions and other pricing terms as between banks, on the one hand, and institutional investors (such as insurance companies), on the other.

Exhibit 4.1 Payment conventions and other pricing terms between banks and institutional investors

Convention	Banks (Column A)	Institutional investors (Column B)
Day count	Actual/360 (unless Fixed Rate, in which case as per Column B)	30/360
Business day	Modified Following Business Day Convention	Following Business Day Convention
Breakage	LIBOR Breakage, Prepayment Fee (and Hedge Breakage, if Fixed Rate)	Make-Whole
Default rate	Interest Rate + X% (if Fixed Rate, X% + higher of (i) LIBOR plus Margin and (ii) Interest Rate)	Interest Rate + X%
Period end dates	Adjusted for Business Days	Not adjusted for Business Days
Increased costs	Yes	No
Market disruption/ cost of funds	Yes	No
Withholding tax protection	Yes	No

Source: Author's own

30/360

See 'Bond Basis'.

Actual/360

This is a method of calculation of interest, which is done on the basis of a year of 360 days and actual number of days elapsed. If the interest rate is 6 percent per annum on a U.S.$100 loan, and the loan is outstanding for 26 days, the calculation would be: U.S.$100 $\times$ .06 $\times$ 26/360.[1] LIBOR-based loans are priced using this convention.

Actual/365

This is a method of calculation of interest, which is done on the basis of a year of 365 days and actual number of days elapsed. If the interest rate is 6 percent per annum on a U.S.$100 loan, and the loan is outstanding for 26 days, the calculation would be: U.S.$100 $\times$ .06 $\times$ 26/365. Prime Rate-based loans are priced using this convention.

Adjustable Rent

This is rent payable under a Lease Financing that periodically adjusts based on specified market indicia.

Average Life

This is the average period before a debt is repaid through amortization. Average life is calculated by multiplying the date of each payment (expressed in number of years (or fractions thereof) from the calculation date) by the percentage of total principal that is to be paid on such date; the sum of the results is the average life.

Example 1: a loan is repayable according to the following profile:

Year 1	100	100/1,000	× 1 =	100/1,000
Year 2	200	200/1,000	× 2 =	400/1,000
Year 3	300	300/1,000	× 3 =	900/1,000
Year 4	400	400/1,000	× 4 =	1,600/1,000
	1,000			3,000/1,000 = 3 years Average Life

Example 2: a loan is repayable according to the following profile:

Year 1	0	0/1,000	× 1 =	0
Year 2	0	0/1,000	× 2 =	0
Year 3	0	0/1,000	× 3 =	0
Year 4	1,000	1,000/1,000	× 4 =	4,000/1,000 = 4 years Average Life

Balloon

This is the final payment of an amortizing term loan in a Mortgage Financing, which is usually substantially larger than the other payments.

Base Rate

This is an interest rate frequently used in U.S. (money market bank) credit agreements as an alternative to LIBOR pricing; this rate is usually defined as the highest of: (i) the Prime Rate; (ii) the Federal Funds Rate plus one half of 1 percent; and (iii) one-month LIBOR plus 1 percent. Margins on Base Rate loans are usually 100 bps less than the Margin for LIBOR-priced loans. In U.S. money market bank credit agreements, Base Rate is often used as a fallback to LIBOR-priced loans in case of a Market Disruption. While the Prime Rate will almost always be the highest of the three legs, on certain days of the year (quarter ends) bank cost of funds may spike due to surges in capital needs, making the other legs applicable. The Base Rate is intended to reflect the all-in lending costs for U.S. money center banks.

Basel II

Basel II is the framework for measuring the capital adequacy of banks in the form set out in the paper entitled 'International Convergence of Capital Measurement and Capital Standards, a Revised Framework' issued by the Basel Committee on Banking Supervision in June 2004. As at the date of writing, Basel II is in the process of being superseded by Basel III; Basel III is only being phased in over time, and the timing for such phasing-in varies by jurisdiction.

The 'advanced methodology' approach of Basel II was a primary driver for the rapid expansion of commercial bank aircraft financing over the last few decades. This approach has allowed banks to use their own internally generated estimates of 'probability of default' and 'exposure at default', thereby allowing them to take into account, for purposes of calculating capital requirements for transactions, the value and strength of Aircraft Assets serving as collateral for Aircraft Finance transactions. See 'Basel III' for a more detailed examination of Capital Adequacy rules.

Basel III

Basel III is a comprehensive set of reform measures superseding Basel II promulgated by the Bank for International Settlements, the Basel Committee on Banking Supervision, designed to improve the regulation, supervision and risk management of banks; provided that Basel II is still, in many cases, the applicable Capital Adequacy Standard insofar as Basel III is being phased in over time. The Basel Committee on Banking Supervision published its proposals under two documents.

1   Basel III: a global regulatory framework for more resilient banks and banking systems, December 2010, revised June 2011.
2   Basel III: International framework for liquidity risk measurement, standards and monitoring, December 2010.

Largely in response to the credit crisis, banks are required to maintain proper leverage and 'Core Tier One' capital ratios and meet certain liquidity and funding structure requirements. Basel III is part of the continuous effort made by the Basel Committee on Banking Supervision to enhance the banking regulatory framework. It builds on the Basel I and Basel II documents, and seeks to improve the banking sector's ability to deal with financial and economic stress, improve risk management and strengthen the banks' transparency. A focus of Basel III is to foster greater resilience at the individual bank level in order to reduce the risk of system wide shocks. This capital framework consists of three pillars: minimum capital requirements, a supervisory review process, and effective use of market discipline. With regard to minimum capital requirements, some banks are permitted to use internal credit ratings and portfolio models to provide a more accurate assessment of a bank's capital requirements in relation to its particular risk profile.

These measures aim to:

*   improve the banking sector's ability to absorb shocks arising from financial and economic stress, whatever the source;
*   improve risk management and governance; and
*   strengthen banks' transparency and disclosures.

The reforms target:

- bank-level, or microprudential, regulation, which will help raise the resilience of individual banking institutions to periods of stress; and
- macroprudential, system wide risks that can build up across the banking sector as well as the procyclical amplification of these risks over time.

Basel III is implemented within the EU (that is, its member states) pursuant to a regulation named Capital Requirement Regulation (CRR)[2] (which is directly applicable) and a directive named Capital Requirements Directive (CRD IV),[3] which requires separate transformation legislation in each EU member state.

PRACTICE NOTE

The adoption of Basel III will most likely raise the operational costs for banks, which they will necessarily need to pass on to their customers in the form of higher Margins and shorter maturities. First, as the leverage ratio under Basel III does not recognize the quality of assets since it is not credit-sensitive, it will require a more conservative approach to capital in relation to good quality assets which will adversely affect pricing. As a consequence, whereas, under Basel II, a fully secured aviation financing facility would attract a low capital requirement relative to its notional amount (in part due to the low risk-weighting generated by the high expected recovery rate), under Basel III, the increased capital requirement introduced through the non-risk sensitive leverage ratio is likely to raise the cost of financing.[1]

Second, the long useful life of Aircraft Assets is likely to make them more expensive for banks to fund because of the impact of the new liquidity standards. The Basel III requirements with respect to liquidity are likely to increase the cost of long term funding and as a consequence to increase the cost of aviation financing facilities.

It is widely assumed that most banks have already priced in their Basel III costs to their customers, even if it has not been fully implemented in a particular jurisdiction. Nonetheless, for purposes of ascertaining whether a bank may obtain indemnification for capital adequacy and liquidity costs, Basel III is typically treated as not constituting current law; the result being that banks may request such indemnification as a result of Basel III's introduction and implementation.

[1] The loss of the benefits associated with a secured financing under Basel III is not necessarily the case in every jurisdiction. In the EU, for example, the CRR does make allowance for benefits associated with 'other physical assets' (that is, any assets other than real estate).

Basel IV

Basel IV is an anticipated revision to the internationally recognized capital adequacy guidelines governing banks that is commonly referred to as Basel III, which revision, at the time of this writing, is uncertain as to content. However, one proposal under consideration under Basel IV that is relevant for Aircraft Finance is to eliminate the ability of banks to use their own proprietary modelling of transaction risk, requiring them instead to apply broad-category risk levels for the purpose of calculating capital adequacy charges. The ability of some banks to use these risk models currently allows them to generate a lower risk profile than the general category of risk, thereby lessening their capital charges for aircraft finance transactions and giving them leeway, then, to offer cheaper pricing to airline and aircraft lessor borrowers.

Basis Points

Basis Points (bps or bips) are gradations of interest rates, margins or fees characterized in terms of multiples of 1/10,000, with 100 basis points being 1 percent.

Bond Basis

This is an interest rate calculation methodology for, usually, fixed rate transactions whereby interest is calculated on the basis of a 360-day year consisting of 12 30 day months. A formula for calculating Bond Basis Interest would be as follows:

Interest (U.S.$) for any Particular Period =

$$\frac{[360 \times (Y_2 - Y_1) + 30 \times (M_2 - M_1) + (D_2 - D_1)] \times IR\% \times \$P}{360}$$

Where:

$Y_1 M_1 D_1$ = the Year, Month and Date of the start date for accrual
$Y_2 M_2$ and D_2 = the Year, Month and Date of the end date for accrual
$IR\%$ = Interest Rate
$\$P$ = Principal

The formula above is subject to the following adjustment rules (more than one may take effect; apply them in order, and if a date is changed in one rule the changed value is used in the following rules).

- If the investment provides that interest is payable on the last day of the month and D_1 is the last day of February and D_2 is the last day of February, then change D_2 to 30.
- If the investment interest is payable on the last day of the month and D_1 is the last day of February, then change D_1 to 30.
- If D_2 is 31 and D_1 is 30 or 31, then change D_2 to 30.
- If D_1 is 31, then change D_1 to 30.

As an example, to compute the amount of interest on a Bond Basis for a U.S.$100 loan at a 5 percent interest rate from 3 January 2013 to 16 November 2014 would be:

$$\frac{[360 \times (2014 - 2013) + 30 \times (11 - 1) + (16 - 3)]}{360 \times 5\% \times \$100 = \$9.35}$$

Business Day

This is a day on which banks in a particular jurisdiction are open for business and not authorized to be closed (such as a Saturday or Sunday).

PRACTICE NOTE

While a seemingly obvious meaning, the tricky part is identifying the appropriate jurisdiction(s) for use.

- London is relevant in LIBOR-based transactions because that is where the rate is set.
- The borrower's or lessee's jurisdiction is relevant if it originates payments from that jurisdiction.
- The paying agent's jurisdiction is relevant since it must be open for receipt.
- Other jurisdictions may be relevant if a Lease-in Lease-out (LILO) structure is employed and cash assets physically flow through the intermediary jurisdictions.

Capital Adequacy

These are rules relating to the adequacy of capital and liquidity that affect regulated financial institutions, such as banks and insurance companies, that are imposed to promote institutional and financial sector stability. Capital Adequacy is a component of Increased Costs. See 'Basel III'.

PRACTICE NOTE

Capital Adequacy rules do not assess costs on a loan-by-loan basis, but rather over a bank's entire portfolio. Accordingly, in assessing any costs to be passed on to a borrower, a bank seeking indemnification for Capital Adequacy must employ allocation methodologies across all of its loans. Also, Capital Adequacy costs for a particular bank may be dictated by that bank's financial condition. Finally, Capital Adequacy rules are in fluid development; rules implementing this regime have not been implemented in many jurisdictions. Accordingly, banks necessarily treat Basel III as a matter for which they seek current (and future) protection, even though its general guidelines are known.

Clear Market

This is an agreement by a borrower not to have any other credit facilities in the market during a period of debt syndication. This is a provision often requested by banks or investment banks acting on behalf of a borrower to syndicate a credit facility or place securities in the open market. The object of this provision is to avoid having to compete with the borrower or other borrower-engaged-syndicators at the same time, especially if there is a limited market for that borrower's securities.

Close-out Amount

This is a method for determining Hedge Breakage Loss/Gain as embodied in the 2002 Master Agreement produced by ISDA. In calculating its Close-out Amount, the non-defaulting party is required to calculate the amount of its losses and costs that are, or would be, incurred (similar to the 1992 'Loss' measure) in replacing or providing the economic equivalent of the payments under the terminated transactions that would have been required but for the early termination (similar to the Market Quotation measure). The definition of Close-out Amount provides a number of factors that the determining party may consider in assessing its losses and costs, including:

- quotations from third parties, such quotations being permitted to take into account the creditworthiness of the determining party;
- relevant market data provided by third parties;
- information of the types described in the preceding bullets from internal sources if that information is of the same type used in the regular course of business for the valuation of similar transactions; and
- application of different valuation methods depending on the type, complexity, size or number of the terminated transactions.

Commitment

This is the obligation of a financier to participate in a transaction pursuant to which the financier promises to supply money or other credit in the future. Such obligation may be subject to the satisfaction of identified conditions precedent (including a Material Adverse Change (MAC) or a Funding MAC).

Commitment Fee

This is the fee charged by a financier for keeping available to a borrower its Commitment. This fee is usually assessed on an Actual/360 day basis, for the period of the Commitment's availability (ending on the Commitment Termination Date).

Commitment Termination Date

This is the last day on which a Commitment may be utilized. In establishing a Commitment Termination Date, the parties must be cognizant of potential delays by Original Equipment Manufacturers (OEMs)[4] in the manufacturing process that may push out deliveries beyond anticipated/scheduled delivery dates.

Cost of Funds

This is a per annum interest rate advised by a bank as the cost for it to supply a loan (often for discrete Interest Periods) addressing Market Disruption periods or stub interest periods for which no Screen Rate is available. Banks insist on not providing any borrower (or anyone else, for that matter) with any ability to second-guess their costs of funds. That information comes from the bank's treasury desk, and its determination is necessarily conclusive; as one can imagine, a bank's assessment of its cost of funds is a complex proprietary evaluation of its funding sources. For this reason, banks require documentary formulations such as: 'The report by any bank of its cost of funds shall be conclusive and shall constitute a certification by such bank that the interest rate so provided is an accurate and fair calculation of its Treasury-assessed funding costs for such period.' Borrowers, on the other hand, are not often happy with being force-fed these rates; see 'Market Disruption', for borrower reactions.

Debt Service

This is the sum of interest on and principal of a loan that is payable at any time (or a specified period of time) on that loan.

Default Rate

The Default Rate is the rate of interest charged in respect of defaulted loans. The Default Rate (also called the 'Past Due Rate' or 'Late Rate') is usually quoted as an additional margin over the current interest rate. The Default Rate may be assessed on amounts that are past due or, in certain facilities, it may be assessed on the entire principal amount of the facility once there is an 'Event of Default'.[5] In Fixed Rate deals, Floating Rate Lenders may require that the Default Rate be X% plus the higher of: (i) the fixed rate; and (ii) LIBOR plus the Spread. This formulation: (i) prevents the borrower from arbitraging rates if floating rates are way above the contracted fixed rate; and (ii) protects the lender in that circumstance since the lender may be funding itself in the floating rate markets.

Federal Funds Rate

The Federal Funds Rate (FedFunds) is the rate per annum equal to the weighted average of the rates on overnight federal funds transactions with members of the Federal Reserve System arranged by federal funds brokers on such day, as published by the Federal Reserve Bank of New York on the business day next succeeding such day. The Federal Funds Rate is the interest rate at which depository institutions actively trade balances held at the Federal Reserve, called federal funds, with each other, usually overnight, on an uncollateralized basis. Institutions with surplus balances in their accounts lend those balances to institutions in need of larger balances. FedFunds is often utilized in the definition of Base Rate.

FINA01

This is a Reuters Brokers page Screen Rate that displays a short-term money market interest rate. Certain banks are willing to use this rate as a proxy to their cost of funds in the event of a Market Disruption.

Fixed Rate

A fixed (as opposed to floating) rate of interest. Fixed Rates are usually calculated on a 30/360 basis.

PRACTICE NOTE

Loans bearing interest at a Fixed Rate bear particular complications. First, is the matter of setting the rate. Insurance companies and other institutional investors may be pleased to quote a rate at a Spread Over (Average Life) Treasuries. Banks offering a Fixed Rate as a Swapped-to-Fixed Rate may have a more black box approach to the rate fix, insofar as that rate may simply be the rate their Treasury Desk tells them to charge. In the light of the uncertainties facing borrowers on a swapped-to-fixed Fixed Rate, they may require their lenders to enter into an Interest Rate Swap with a Hedge Counterparty selected through a swap auction.[1] Second, is the matter of pre-payments, where Make-Whole or Hedge Breakage may apply. Make-Whole and Hedge Breakage costs may be enormous. Accordingly, whereas in a low rate environment, borrowers would typically want to refinance high coupon debt, they may be disincented from doing so by sizeable Make-Whole or Hedge Breakage liabilities.

[1]   An auction of sorts conducted by the borrower where it asks multiple potential Hedge Counterparties (with whom the lender has lines to do business) to offer its best fixed rate quote at an appointed time based on the loan economics.

Fixed Rate Lender

This is a lender, typically an institutional lender such as an insurance company, which has access to long-term funds and lends on a Fixed Rate basis. Fixed Rate Lenders also include hedge funds and finance companies.

Floating Rate

This is an interest rate that is not fixed, but adjusts over time based on market indices such as LIBOR (which is 'fixed' for discrete Interest Periods) or Prime Rate. Floating Rates calculated by reference to LIBOR are calculated on an Actual/360 basis and Floating Rates calculated by reference to the Prime Rate are usually calculated on an Actual/365 basis.

Floating Rate Lender

A Floating Rate Lender is a lender (typically a bank) which (theoretically) obtains its funding for a loan in the short-term credit markets, and lends on a Floating Rate basis.

Following Business Day Convention

This is a payment convention that provides that if a payment date falls on a day that is not a Business Day, the payment is made on the first day following such non-Business Day that is a Business Day.

Forward Fix

This is the fixing of an interest rate on a forward basis, typically in advance of the delivery of an identified to-be-financed Aircraft. Rates are Forward Fixed: (i) to address concerns that rates are moving up; and (ii) to lock-down a debt rate relative to a fixed rental rate, the concern being that if interest rates increase, the rentals may be insufficient to service any debt.

PRACTICE NOTE

There are a number of issues to contend with when Forward Fixing an interest rate associated with a future Aircraft delivery.

- The lender providing the Forward Fix would need to Hedge its interest rate exposure in order to provide the Fixed Rate. If rates go down over the interim, the Hedge would be out of the money on a Marked to Market basis. If the Aircraft does not deliver during the Hedge Period or the lender's Commitment to fund terminates before the anticipated financing, the lender would be required to break its Hedge, and a loss may be incurred. Since there is no Aircraft collateral at this point, the lender must rely on the credit of the borrower or other credit party (on a full recourse basis, of course), or require the posting of cash collateral in respect of some or all of any out-of-the-money exposure.
- Aircraft deliveries are, at best, targetable to a particular month, while Hedges are usually crafted with precise dates. The Hedge, therefore, must necessarily allow for a range of dates on which it would commence (the 'Hedge Period') and payments would start to flow.[1] This feature makes a Forward Fix more expensive.

[1] Alternatively, a forward hedge can be fixed to commence on a date past the last likely delivery date of an Aircraft, with interest for the interim period calculated at a Floating Rate.

Fuel Hedge

A Fuel Hedge is a Hedge employed by airlines to provide protection against oil price increases. Aircraft fuel is a very large component of an airline's cost (in many cases, second only to labor costs). In the light of the rather high volatility of oil prices, many airlines use Fuel Hedges to lock in prices for some or a large component of their anticipated fuel needs. Some airlines may pledge Aircraft to their Fuel Hedge Swap Counterparties as collateral in lieu of having to provide cash cover in case oil prices fall.

Funding Material Adverse Change

A Funding Material Adverse Change (MAC) is a condition precedent to the disbursement of a loan that forgives a lender for not making a loan if there is a material adverse change in the financial markets in which such lender funds itself.

PRACTICE NOTE

The Funding MAC would have been relevant during the 2008 market meltdown when many banks were unable to procure U.S. dollars in the inter-bank markets. This type of provision, for obvious reasons, tends to make borrowers nervous that the financing on which they were counting will not be available.

Hedge

This is an Interest Rate Swap or other derivative transaction by which a borrower or a lender hedges exposure to changes in interest rates or commodity prices. In the context of Aircraft Finance transactions, this is typically a transaction undertaken to minimize the risks associated with the movement of floating interest rates (for a Floating Rate Lender) relative to fixed rental rate Leases or to allow a Floating Rate Lender to provide a Fixed Rate of interest to a borrower seeking the same (to lock-in interest at low rates, for example). Also, Fuel Hedges are employed by airlines to provide protection against oil price increases.

Hedge Breakage Loss/Gain

This is the cost or gain obtained from the termination of an Interest Rate Swap or other Hedge, typically associated with the unwinding of a swap/hedge utilized in a Swapped-to-Fixed fixed rate loan transaction. Such loss or gain may be determined by having the Swap Counterparty (or, for an internally Hedged deal, the lender itself) provide a quote of its cost/gain from unwinding the related Interest Rate Swap or requiring the unwinding to be calculated based on Close-out or Market Quotation methodologies.

86

Hedge Counterparty

The Hedge Counterparty is a counterparty on an Interest Rate Swap that is providing the floating rate leg of an Interest Rate Swap. In the context of Aircraft Finance transactions, the Hedge Counterparty may be acting in one of two roles.

- From the perspective of a Floating Rate Lender that is supplying a Swapped-to-Fixed Fixed Rate, the Hedge Counterparty is the party to which the fixed rate payments of interest taken in by the Floating Rate Lender are passed on to and that pays to such Lender an amount equivalent to interest on a floating rate basis. In this situation, the Hedge Counterparty is taking the credit risk of the Floating Rate Lender. The counterparty here may be Floating Rate Lender's own Treasury desk.
- From the perspective of a borrower that must satisfy a Hedging Requirement, the Hedge Counterparty is the party effectively converting fixed rentals received by the borrower into floating rentals. In this situation, the Hedge Counterparty is taking the credit risk of the borrower (and any collateral pledged for its benefit).

Hedging Requirement

This is a requirement by Floating Rate Lenders for a borrower to Hedge the incoming fixed (typically) rent stream so that increases in floating rates will be matched by income from the related Hedge. Hedging Requirements (for the purpose described in the second bullet above in Hedge Counterparty) may be necessary if Floating Rate Lenders are unwilling or unable to effect a Fixed Rate facility by offering a Swapped-to-Fixed/Hedged Fixed Rate.[6] In the event one or more Hedge Counterparties is effecting a Hedge with the borrower directly in connection with, say, a Back-leveraged Operating Lease Facility, the Hedge Counterparty would typically require its exposure to the borrower to be secured by the facility's Aircraft and other collateral, and the Hedge Counterparty would be interested in certain voting rights and placement in the Waterfall distributions (both ongoing and default).

Illegality

This is a clause in loan documentation that typically addresses two general areas of concern.

1   In transactions where the bank and the borrower reside in different countries, political developments may arise that cause cross-border lending transactions to become illegal (for example, developments with South Africa and Iran of the not too distant past). Such developments would, obviously, require a prepayment of the loan.
2   LIBOR-priced transactions in the U.S. are, by their very nature, funded on a cross-border basis. These transactions work only so long as the funding arrangements are themselves not disrupted by international events that could, for example, affect the availability of dollars in the London market.

For this reason, banks often seek to be able to reprice their LIBOR loans to an unaffected pricing basis if it shall become unlawful for them to obtain their funding in the LIBOR markets, or, alternatively, to cause their loans to be repaid.

PRACTICE NOTE

Illegality events are largely theoretical, especially in the second instance described above. Should the West be taken over by the Caliphate, with Sharia'a becoming the governing law, lending with interest would be illegal (and therefore, the Illegality clause, operative), but we suspect we would have other, more weighty, concerns at the time.

Increased Costs

Increased Costs are costs potentially incurred by banks due to changes in law or regulation relating to bank-oriented regulatory regimes, including Capital Adequacy, reserves and so on. In bank financings, borrowers are required to indemnify the banks for Increased Costs.

PRACTICE NOTE

While Increased Cost indemnification has been in place for dozens of years, we are unaware of any instance when a bank has actually made a claim for it.

Interest Period

This is a discrete period of time during which the interest rate relating to a Floating Rate, such as LIBOR, may be established. At the beginning of each such period, the Floating Rate is established or re-established for such period. Certain Floating Rates are indexed to interest rate benchmarks that may change every day, such as the Federal Funds Rate and the Prime Rate, so these rates do not use the Interest Period concept (other than as a means to establish payment dates and accrual periods). Interest Periods are usually of a one-, two-, three- or six-month duration.

Interest Rate Swap

This is an agreement between two parties to exchange cash flows, with one party (the floating rate payor) agreeing to pay the cash flows based on an indexed (floating/variable) interest rate (usually LIBOR) calculated by reference to a specified notional amount(s) with the other party (the fixed rate payor) agreeing to pay fixed/specified cash flows based on such notional amount. See Exhibit 4.2 for a schematic representation of an Interest Rate Swap.

Interest Rate Swaps may be used for Hedge purposes.

Interpolation

This is a method to establish a price or yield of a security by utilizing known values that are located in sequence (higher and lower) with the unknown value. Interpolation methodologies are often employed in the determination of off-screen LIBOR Interest Periods and to determine the Treasury Rate for Make-Whole amounts.

International Swaps and Derivatives Association (ISDA)

International Swaps and Derivatives Association (ISDA) is the organization that develops the forms used for Hedges.

KLEIMM

This is a Reuters Brokers page Screen Rate that displays a short-term money market interest rate. Certain banks are willing to use this rate as a proxy to their cost of funds in the event of a Market Disruption.

London Interbank Offered Rate (LIBOR)

London Interbank Offered Rates (LIBOR) are short-term fixed rates quoted for Interest Periods of, typically, one, two, three and six months; overnight and 12-month Interest Periods are also available. Insofar as these rates are set for discrete periods of time, they are good for the duration of those periods and are reset at the

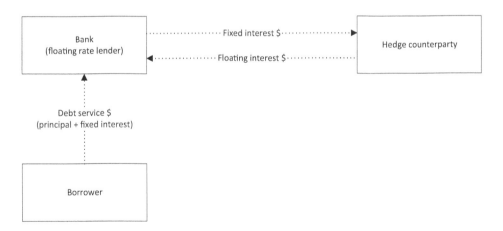

Exhibit 4.2 Interest Rate Swap

Source: Author's own

end of the respective periods to the then available rates reflecting market conditions. LIBOR rates used in the Aircraft Finance market are Screen Rates posted by, among others, Bloomberg (see Exhibit 4.3 for a sample screenshot). The determination of LIBOR is managed by the Intercontinental Exchange (ICE), which is the parent of the New York Stock Exchange and a number of other exchanges and markets around the world. ICE took over administration of LIBOR from the British Bankers Association (BBA) in early 2014, following a rate-fixing scandal involving LIBOR interest rates under the auspices of the BBA.

LIBOR rates are produced for five currencies (Swiss Franc, Euro, Pound Sterling, Japanese Yen and U.S. Dollar) with seven maturities quoted for each – ranging from overnight to 12 months, producing 35 rates each business day. These rates provide an indication of the average rate at which a LIBOR contributor bank can obtain unsecured funding in the London interbank market for a given period, in a given currency. Individual LIBOR rates are the end-product of a calculation based upon submissions from LIBOR contributor banks.

ICE Benchmark Administration (IBA), which is the body that oversees the rate setting processes, maintains a reference panel of between 11 and 17 contributor banks for each currency calculated. Every contributor bank is asked to base their ICE LIBOR submissions on the following question: 'At what rate could you borrow funds, were you to do so by asking for and then accepting interbank offers in a

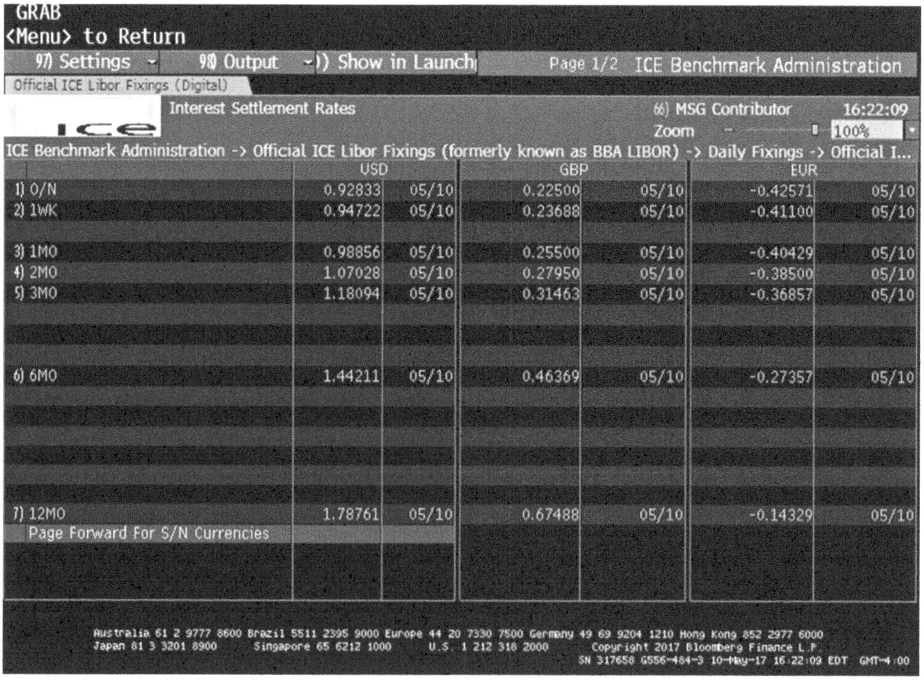

Exhibit 4.3 LIBOR screen snapshot

Source: Bloomberg

reasonable market size just prior to 11 am London time?' Therefore, submissions are based upon the lowest perceived rate at which a bank could go into the London interbank money market and obtain funding in reasonable market size, for a given maturity and currency. LIBOR rates are quoted as an annualized interest rate.

LIBOR rates are calculated using a trimmed arithmetic mean. Once each submission is received, they are ranked in descending order and then the highest and lowest 25 percent of submissions are excluded. This trimming of the top and bottom quartiles allows for the exclusion of outliers from the final calculation. The remaining contributions are then arithmetically averaged and the result is rounded to five decimal places to create a LIBOR rate. This is repeated for every currency and maturity, producing 35 rates every business day.[7]

LIBOR Breakage

These are the costs assessed to a borrower for repaying a loan, the interest rate of which is based on LIBOR, if that repayment is made on other than the last day of the then current Interest Period. Such costs will be assessed if LIBOR rates at the time of prepayment have gone down in relation to what they were at the beginning of the Interest Period. Interest rates on LIBOR priced loans are 'set' for Interest Periods of, typically, one, two, three or six months. This means that the bank making the LIBOR loan is itself (theoretically) funding the loan by taking a deposit ('Deposit') from another bank (the 'Deposit Bank') for the same interest period as its loan. The bank, therefore, is counting on the LIBOR component of the interest charged to its borrower to pay the Deposit Bank interest on the Deposit. If, however, the borrower prepays the loan in the middle of an Interest Period, the bank runs the risk that it will be unable to reinvest the loan proceeds at a rate at least equal to the rate borne by the Deposit made by the Deposit Bank (where the Deposit is not subject to prepayment). For example, if a LIBOR-priced loan was priced at a six-month LIBOR rate of 6 percent per annum (excluding the spread), and in the

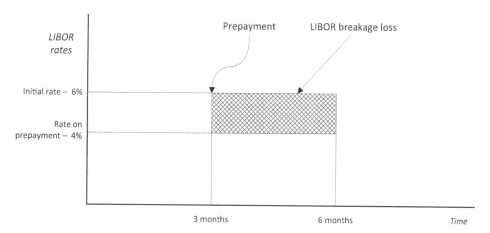

Exhibit 4.4 LIBOR breakage

Source: Author's own

third month of the interest period, the borrower prepays the loan, the bank will need to reinvest the prepaid proceeds of the loan in an investment (that is, a eurodollar deposit) yielding at least 6 percent per annum so that it will have sufficient interest income to pay the Deposit Bank interest on the 6 percent Deposit when it matures. If the three-month reinvestment rate is less than 6 percent per annum, say, 4 percent, then the bank will seek break funding compensation from the borrower upon prepayment to make up the difference, see Exhibit 4.4 for an illustration. Banks offering LIBOR (and other short-term fixed rate) pricing options, therefore, either require that all prepayments be made on the last day of an interest period, thereby avoiding any broken funding, or require any prepayment made on a day other than the last day of an interest period to have paid with it the appropriate break-funding indemnity.

PRACTICE NOTE

Banks assessing LIBOR Breakage charges will usually look at London Interbank Bid Rate (Libid) as its reinvestment rate; this rate is usually one-eighth of 1 percent less than the comparable LIBOR rate. Other than in some very unusual circumstances, there is no LIBOR Breakage 'gain' payable to a prepaying borrower if rates have gone up.

Liquidity Bank

A Liquidity Bank is a bank (most likely German) that, due to governmental regulation, needs to lock-in the availability of LIBOR at a specified price (margin) with a counterparty (the 'Liquidity Counterparty') over the life of a transaction. In other words, these banks must find a counterparty that is willing to lend to it at LIBOR (at whatever the underlying transaction's Interest Period periodicity) for the period commencing on the anticipated closing date through the anticipated maturity date. So, in a 10-year quarterly pay Aircraft-secured loan financing, a Liquidity Bank will contract with another institution to provide to it funding for each three-month period at the then-current LIBOR rate (plus the Liquidity Margin) over 10 years. A bank employing this model will have obviated the risk of not being able to find a counterparty during the life of the particular transaction from which to borrow U.S. dollar funds for the requisite Interest Periods. The Liquidity Bank, of course, is taking long-term credit risk on its chosen Liquidity Counterparty.

PRACTICE NOTE

Since a Liquidity Bank has locked-in its LIBOR accessibility, it should not be entitled to charge Market Disruption. A Liquidity Bank will likely charge an incremental margin over the credit spread in the amount of the Liquidity Margin and may require Liquidity Breakage protection.

Liquidity Breakage

These are the costs assessed to a borrower for making an unscheduled repayment of a loan to a lending bank that has funded itself with a Liquidity Bank. Such costs would be assessed if a Mark-to-Market Liquidity Margin for such Liquidity Bank in respect of the remaining notional loan term would be lower than the current Liquidity Margin. Liquidity Breakage is typically calculated at the present value of the rate differential times the amortizing loan balance over the remaining term of the related loan.

Liquidity Margin

This is the margin or cost over LIBOR (typically expressed in Basis Points) assessed to a Liquidity Bank by the Liquidity Counterparty for agreeing to lock-in LIBOR availability over a specified period. Such cost will be a function of, among other things, the tenor of the specified period and the credit standing of such Liquidity Bank. The Liquidity Margin is a supplemental margin over LIBOR passed on by a Liquidity Bank to its borrower in addition to the Margin (or may otherwise be subsumed in the Margin).

Make-Whole

These are the costs assessed to a borrower for repaying a Fixed Rate loan owing to Fixed Rate Lenders other than on the scheduled payment dates. Make-Whole is a pre-payment charge typically assessed by non-bank lenders, and represents such lender's reinvestment opportunity cost losses if interest rates have declined. A traditional Make-Whole is typically calculated as the excess, if any, of: (i) the present value, as of the date of the relevant prepayment, purchase, or acceleration, of the instalments of principal of and interest on such affected debt that, but for such pre-payments, purchase, or acceleration, would have been payable after such prepayment, purchase, or acceleration; over (ii) the principal amount of such debt then being prepaid, purchased, or accelerated. Such present value is determined by discounting the amounts of such instalments semi-annually or other applicable periodicity (on a 30/360 basis) from their respective payment dates to the date of such prepayment, purchase, or acceleration at a rate equal to the weighted Average Life Treasury Rate (or a swap rate) *plus* a spread. The size of the spread over Treasuries (or swap rate) is an important business consideration, and is typically solved at the term sheet stage, with the lower the 'spread' (or no spread, in some deals) the more advantageous is the resulting calculation for the lender.[8] Make-Whole is 'one-way', meaning that the lender would not have to pay over any 'gain' if interest rates go up. In the typical deal where Make-Whole is used, *it is not payable*: (i) in connection with a prepayment occurring following an Event of Loss; and (ii) in connection with an acceleration following an event of default.

Make-Whole formulations are typically used in the institutional lender market with Fixed Rate Lenders where these lenders do not hedge their fixed rate fundings; these institutional lenders are sitting on piles of cash that they need to invest at long-term fixed rate yields. The theory, then, for no Make-Whole to be payable in the two instances noted above is (i) in the case of an event of loss, the lender should not 'profit' on a catastrophe and (ii) in the case of an acceleration, U.S. courts may find that the Make-Whole is a 'penalty', and, therefore, unenforceable.

Mandatory Costs

These are costs associated with banks complying with certain regulatory funding requirements (as opposed to capital adequacy costs which will be within the margin). The Bank of England, in common with central banks in many jurisdictions, requires banks to place non-interest bearing deposits with it; the income from investing these deposits is used to help fund the Bank's supervisory role. The amount of the deposit is calculated as a percentage of a bank's 'eligible liabilities' (essentially sterling deposits made with that bank). Historically, UK-based banks have required their borrowers to indemnify those banks for these costs, but this requirement has recently been dropped by most institutions.

Margin/Spread

This is the spread (expressed in Basis Points) over an agreed market indicia, such as LIBOR or Treasury Rates, for determining the interest rate for a borrower. The Margin/Spread is, obviously, a critical economic component of any Aircraft Financing transaction.

Market Disruption

For bank-funded loan transactions (whether a Fixed Rate or Floating Rate), this is a disruption in the markets such that either: (i) quotes for LIBOR (whether on a screen basis or as quoted by reference banks) are not available; or (ii) the quoted rates for LIBOR are not reflective of the cost of funds of the bank (or banks) making such loans. If a Market Disruption is invoked, the lender is entitled to charge the borrower more interest (see below). Insofar as the Screen Rate-based LIBOR rate is an average of quoted rates, any particular bank's cost of funds for any interest period may naturally be higher or lower than the screen rate. Whether higher or lower will depend, largely, on such bank's credit quality; the better its perceived credit, the lower the interest rate other banks will charge it. The banks that are not of the highest credit quality will typically adjust (higher) the margins they charge borrowers to take account of their own higher borrowing costs relative to the LIBOR Screen Rate. Historically, higher interbank borrowing costs have hit whole classes of banks; in the early 1990s, Japanese banks were almost universally assessed an interbank market premium of 20 to 100 basis points over screen-rate LIBOR rates due to credit concerns endemic to the Japanese economy.

The liquidity crisis of 2008–2009 has radically challenged the presuppositions on which LIBOR-based lending is based. Not only may banks' borrowing costs be far higher than the quoted Screen Rates, but many banks during liquidity-crunch times are literally unable to borrow altogether for any of the monthly standard interest periods. During the most difficult days of the liquidity crisis in fall 2008, many banks were largely unable to access funds in the London interbank market other than on an overnight basis; that is, rather than borrowing for prescribed interest periods in the interbank market, banks were extending credit to each other in the interbank market on a day-to-day basis only. Banks were extremely wary of other banks' credit risks and not willing to grant other than overnight loans. Some banks were even finding that overnight funds were not available altogether or, even if available, they were unwilling to place themselves at risk of a failed overnight rollover or constantly shifting interest rates, which may move adversely relative to a Screen Rate for a contractually prescribed Interest Period.

Charging for any incremental cost of funds over the LIBOR Screen Rate, is one that has been seen in various guises in the bank financing market. In transactions documented in Europe (primarily documentation governed by English law), there exists an industry standard as adopted by the London-based Loan Market Association (LMA) that provides, in part, that a bank may charge its cost of funds if 'the cost to it of obtaining matching deposits in the relevant interbank market would be in excess of LIBOR'. In a number of U.S.-based transactions, primarily widely syndicated unsecured loan facilities, the failure of the LIBOR screen rate to adequately cover the cost of funds would kick over the interest rate basis to Base Rate loans. The LMA approach is deficient insofar as it presumes availability of matching deposits in the London interbank market; as noted above, many banks were unable to access deposits in the interbank market only on an overnight basis. The U.S.-based approach is deficient for most European-based lenders as neither U.S. Prime Rate nor Federal Funds Rates (components of the Base Rate) have any operational meanings for them insofar as they have no ability to access funds at those rates. Accordingly, the evolving standard has become that borrowers will indemnify banks for their incremental cost of funds over the LIBOR screen rate for any applicable interest period.

PRACTICE NOTE

It is important to note that the cost-of-funds issue is an issue for banks even if a particular transaction is quoted on a fixed-rate basis. This is because in a fixed-rate deal, the bank is swapping the fixed-rate interest-related cash flow it is receiving from its borrower for LIBOR under an Interest Rate Swap. Therefore, the banks offering Fixed Rate interest are seeking cost-of-funds protection that would entitle them to the differential of cost of funds over the LIBOR applicable to each Interest Period. Borrowers abhor Market Disruption provisions, especially because it places uncertainty on their borrowing costs notwithstanding a pre-agreed Margin and, if looking to Cost of Funds, unverifiable. In addition to attempting to eliminate this type of Yield Protection clause outright, the borrower may seek: (i) to cap the supplemental costs; (ii) require the bank to 'eat' the first X number of basis points; or (iii) require the bank to come up with a verifiable screen rate (such as FINA01 or KLEIMM).

Market Flex

In an underwritten bank financing, Market Flex is the right given to the arrangers of such financing to seek changes (especially to things like Margin) to the financing terms if such arrangers are unable to syndicate an agreed portion of the related facility on the approved terms.

Market Quotation

This is a method for determining Hedge Breakage Loss/Gain in respect of an Interest Rate Swap as embodied in the 1992 ISDA Master Agreement. This methodology required the parties to obtain quotations from at least three reference market makers of the cost for their entering into replacement transactions comparable to such Interest Rate Swap. If only three quotes are obtained, the high and the low quotes are tossed and the remaining quote is determinative. If four or more quotes are obtained, all of the quotes are averaged. Market Quotation has been largely superseded by the Close-out Amount methodology.

Modified Following Business Day Convention

This is a payment convention that is the same as Following Business Day Convention, except that if, by virtue of applying the Following Business Day Convention, payment would be made in a subsequent month, then payment will follow the Preceding Business Day Convention. This is the standard Payment Convention for LIBOR-based loans.

Mortgage/Annuity-style Amortization

This is a method for determining the amortization profile of a loan, which provides that each payment of debt service (interest plus principal) is the same. The utilization of this methodology results in a profile whereby, in the earlier years in respect of any debt service payment, the largest component is interest and, in the later years in respect of any debt service payment, the largest component is principal. In order to develop an amortization schedule on this basis, a single interest rate must be established for this purpose. Mortgage/Annuity-style Amortization is a useful methodology to match a fixed (level) Lease rental with a level debt service. Mortgage/Annuity-style Amortization may also have a Balloon as part of the final payment.

Payment Convention

This is the methodology for determining when payments fall due if the scheduled due date is not a Business Day. The Payment Convention will be any of: Preceding Business Day Convention, Following Business Day Convention or Modified Following Business Day Convention.

Period End Date (Adjusted versus Non-adjusted)

For establishing an accrual period for interest on a loan, this is the designation of whether the last day of the period, if it falls on a non-Business Day, is not adjusted or is adjusted to end on the Business Day determined in accordance with the Payment Convention. In Bond Basis transactions, the Period End Date is not adjusted. In most other transactions, the Period End Date is adjusted. For example, if the Period End Date (the day on which interest is scheduled to be paid) falls on a Saturday: (i) if the Period End Date is to be adjusted, the interest that is payable on the next Business Day, Monday, would include interest accrued over the weekend; and (ii) if the Period End Date is unadjusted, such weekend accrued interest would not be payable on that Monday (but rather at the end of the next payment period).

Preceding Business Day Convention

This is a payment convention that provides that if a payment date falls on a day that is not a Business Day, then payment is made on the first day *preceding* such non-Business Day that is a Business Day. Many Leases employ this convention.

Prepayment Fee

This is a fee assessed to a borrower (in addition to LIBOR Breakage, Liquidity Breakage, Hedge Breakage and/or Make-Whole) by a lender for voluntarily repaying a loan prior to its scheduled maturity date. Prepayment Fees typically apply over a period of the first three to five years and decline annually over the applicable period.

Prime Rate

The Prime Rate is the rate of interest announced from time to time by a bank as its 'prime rate'. The Prime Rate of a bank is not necessarily the lowest rate at which the quoting bank will lend to its best (or 'prime') customers.

Screen Rate

This is an interest rate the source of which is its posting on an electronic screen by Reuters, Bloomberg or other financial reporting service.

Spread over (Average Life) Treasuries

This is the method by which institutional and other fixed rate lenders often quote the pricing for transactions in which they intend to participate. These quotes consist of a Margin and the yield on a designated U.S. Treasury security that matches the average life of the related transaction. If there is no such match, the parties will often obtain the yield based on an Interpolation of yields from next higher and lower designated U.S. Treasury securities. Treasuries are often quoted on the basis of the Board of Governors of the Federal Reserve System's weekly statistical release designated as H.15(519). Fixed Rates may also be quoted on a Spread over a screen-derived swap rate based on the Swap Curve.

Stub Period

This is an incomplete Interest Period, often occurring at the beginning or end of a financing due to the misalignment of Period End Dates to the closing and/or Maturity Date. Interest due for a Stub Period may need alternative pricing mechanisms since the period may not be one that matches any Screen Rate. Accordingly, they may be priced at Cost of Funds or may be based on an Interpolated rate taking into account the next higher and lower available Screen Rate.

Swap Breakage

See 'Hedge Breakage'.

Swap Counterparty

See 'Hedge Counterparty'.

Swap Curve

This is the plotted curve showing the fixed (LIBOR-swapped) interest rates for designated time periods. In determining Swapped-to-Fixed interest rates, the parties often look to the Swap Curve for determining the base rate (that is, the interest rate less the agreed Margin) for a loan (by reference to such loan's Average Life).

Swapped-to-Fixed/Hedged

In the case of loans made by a bank (or other Floating Rate Lender), this is the Fixed Rate determined by the applicable lender based on an Interest Rate Swap entered into, or other Hedge utilized, by such lender.

Syndication

This is the act of procuring lenders by transaction arrangers or underwriters to supply the Commitments for a financing transaction.

T + [X]

In a Fixed Rate financing where the loan pricing is quoted on a Spread Over Treasuries ('T') basis, this is the shorthand method for describing the settlement period from pricing to funding. In this case, 'X' is the number of days in the settlement period, that is, the period between the date the benchmark U.S. Treasury securities are priced and the day when settlement on the financing or trade is to occur. This methodology may also be used to describe the settlement period for an LSTA (or other) secondary-market trade.

Tranche

This is a designation that identifies discrete portions of loans in a loan or Securitization facility in relation to seniority/subordination ranking (and associated economic terms) or relate to particular Aircraft or other specified assets. 'Class' or 'series' are also commonly used terms for this purpose.

Yank-a-Bank

This is a documentation provision that provides to a borrower the ability to replace a bank that has invoked Illegality or is charging it for certain Yield Protection (Rats and Mice) associated costs or is unwilling to give its consent to a revision sought by the borrower.

Yield Protection

This is the mantra of bank financiers as to their need to protect their yield (typically Margin) in a particular transaction due to changes of law or regulation, or in market events. This covers things like Capital Adequacy, Increased Costs, Withholding Taxes and Market Disruption matters. Borrowers are required to indemnify the bank financiers for these matters.

PRACTICE NOTE

Borrowers dislike having to provide Yield Protection, but the provision of this protection to the banks is the price of doing business with banks. With all of the fuss on Yield Protection, it should be noted that (aside from Market Disruption) it has seldom (ever?) been invoked.

Notes

1 It is a mystery how this convention, which short-changes borrowers for those five missing days, got started.
2 Regulation (EU) No 575/2013 [of the European Parliament and of the Council] of 26 June 2013 on prudential requirements for credit institutions and investment firms and amending Regulation (EU) No 648/2012.
3 Directive 2013/36/EU of the European Parliament and of the Council of 26 June 2013 on access to the activity of credit institutions and the prudential supervision of credit institutions and investment firms, amending Directive 2002/87/EC and repealing Directives 2006/48/EC and 2006/49/EC.
4 Labor strife, parts unavailability or other supply-chain problems, regulatory problems, new Airworthiness Directive (AD) adoptions and so on.
5 The latter formulation may face enforceability issues insofar as the assessment of a default rate on the entire principal amount may be construed as punitive. Of course, once the loan is accelerated,

then the entire amount of principal is past due and may be assessed the Past Due Rate without enforceability issues.

6 In the case of multiple Floating Rate Lenders, if they are not all of the same credit quality, for example, they may be unable to achieve a uniform fixed interest rate by doing individual hedges.

7 For more on the matter of LIBOR funding, see Scheinberg, R, 'LIBOR: a finance lawyer's assessment', *The Banking Law Journal*, February 2013, pp. 122–128.

8 These spreads are usually 0 to 100 basis points. There may be a built-in profit element to the spread since the equivalent spread for the related borrower may be well in excess of this range.

AIRCRAFT/ENGINE TERMS

Introduction

This Part 5 details terms used in describing certain technical aspects relating to Aircraft Assets.

Airworthiness Directive (AD)

An Airworthiness Directive (AD) is a mandatory order issued by the FAA or other Aviation Authority, usually applying to specific types of Aircraft Assets, when an unsafe condition exists and that condition is likely to exist or develop in other Aircraft Assets of the same design. An AD usually requires some maintenance action (possibly only an inspection) within some specified time in order to ensure continued safety and airworthiness, and no Aircraft may be operated in contravention of the requirements or limitations of an AD. Airworthiness Directives may be issued against any component on an Aircraft which has failed in its designed performance and for which an immediate adjustment or change must be made. The initial notification for this type of AD would be by telegram to all affected operators – possibly grounding all Aircraft depending on urgency. An AD could also be less crucial, requiring as little as a visual inspection of a particular area that has shown a defect in its designed structure, such as a crack, for which certain limits are permitted. The FAA may request that the operator perform a visual inspection of the area at specified intervals.

Airworthiness Directive (AD) Cost Sharing

This is the sharing of the cost of an AD as between a lessor and a lessee under a lease. A typical AD cost-sharing formula would require the cost of the AD to be allocated to the lessor and the lessee on the basis of an assumed useful life for the AD (typically 8 to 10 years), with the lessee being obligated to pay for the proportion that the remaining lease term bears to such assumed useful life and the lessor being obligated to pay the remainder.

Airworthiness Directive (AD) Termination

The termination of an AD is achieved by accomplishing a physical repair or modification to the Aircraft Asset, thereby eliminating (usually) the need for periodic inspection of the defect. Termination of an AD can be a costly procedure and is often avoided as long as possible by the airline in order to defer or avoid the cost. Most ADs will require termination by a specific date, which means that the current operator, the lessor or a future operator will eventually have to pay for termination of such AD. There may be multiple ways to terminate an AD, and certain termination methodologies may not be acceptable in all jurisdictions. For example, an AD may be terminated by a single (expensive) fix or terminated by adopting a long-term inspection program. The latter may not only be more expensive for a subsequent operator of an Aircraft, it may not be an acceptable termination approach in that operator's jurisdiction. Accordingly, return conditions as relate to AD Terminations need to be carefully crafted to achieve any preferred outcome.

Aircraft Parts

An Aircraft Part is an accessory, appurtenance or part of an Aircraft or Engine installed in or removed from an Aircraft, Airframe or Engine. See Part 2, 'Parts'.

Aircraft on Ground (AOG)

Aircraft on Ground (AOG) is the term typically used for an Aircraft in either long- or short-term Storage.

Avionics

This is the array of sophisticated and expensive electronic equipment in an Aircraft's cockpit. Basic airline avionics for jet Aircraft include VHF and HF communications

Exhibit 5.1 Aircraft avionics
Source: iStock: #136939778

radios, VHF navigation radios, auto pilot, radar altimeter, flight director, instrument landing system, transponder and marker beacon. Overwater Aircraft will also have long range navigation equipment, a global positioning system and an inertial navigation system.

Auxiliary Power Unit (APU)

The Auxiliary Power Unit (APU) is the on-board generator that powers an Aircraft when power is not drawn from the Engines. See Exhibit 5.2.

Buyer Furnished Equipment (BFE)

Buyer Furnished Equipment (BFE) is any equipment which under a Purchase Agreement is to be provided by the purchaser of an Aircraft (whether actually provided by the purchaser as buyer-furnished equipment or the Original Equipment Manufacturer (OEM) as seller-purchased equipment).

Certificate of Airworthiness (C of A)

A Certificate of Airworthiness (C of A) or Airworthiness Certificate is a certificate issued by the FAA or the Aviation Authority of an EASA Member State (or other Aviation Authority) for an individual Aircraft when that authority is satisfied that the Aircraft conforms to the Type Certificate and is in a condition for safe operation.

Exhibit 5.2 Auxiliary power unit
Source: iStock: #90628757

103

In the U.S., the Airworthiness Certificate is issued to the registered owner, and is transferred with the Aircraft. It remains in effect as long as the Aircraft is maintained (or altered) according to the appropriate FAA regulations and continues to be registered in the U.S. (See Part 7, 'FAR 21').

In EASA jurisdictions, the non-expiring Airworthiness Certificate is only valid for as long as the Airworthiness Review Certificate (ARC) is in force. The validity of the ARC is 12 months and may be issued by the operator of the Aircraft, if appropriately approved, or the incumbent national air authority (NAA).

PRACTICE NOTE

It is mandatory for the original Airworthiness Certificate and, if applicable, the ARC to be carried within the Aircraft at all times.

Consumable Part

This is a Part that is not a Repairable Part.

Cowling

The removable part of the Nacelle.

Derating

Derating relates to an Engine, this is the operation of such Engine's thrust at levels below maximum power. Operating an Engine on a Derated basis improves Engine performance economics by reducing maintenance costs. Certain conditions may warrant Derating, such as: (i) lighter than maximum take-off weight; (ii) operation at cool temperatures; (iii) high altitude destinations; and (iv) short flight segments. Derating an Engine typically requires only software changes, however, doing so may require an updated C of A and a change of the Engine's model designation.

Escalation Charges

These are charges assessed by OEMs to their customers on Aircraft and Engine orders to protect the manufacturers against inflation of labor costs, commodities and parts when customers order Aircraft and Engines for future delivery. With long lead times typically required in the Aircraft manufacturing process, these Escalation Charges may add significant costs to an Aircraft acquisition.

Extended-range Twin-engine Operations (ETOPS)

Extended-range Twin-engine Operations (ETOPS) applies specifically to twin-engine Aircraft such as the Boeing 777 operated at a distance more than 60 minutes flying time with one engine inoperative from a suitable airport. In the U.S. there are approvals for ETOPS at distances from suitable airports, varying from 75 minutes up to 180 minutes (or possibly more under the 'Beyond 180' designation) at the engine-out cruising speed. In general, Aircraft used in ETOPS must have an ETOPS-type design approval which may require the incorporation of a substantial number of equipment options and Service Bulletins, and the Aviation Authorities may approve specific operators, Aircraft and Routes based upon various qualifications, demonstrated reliability and competence. See 14 CFR part 25, and App K (Note 5) for a discussion of the U.S. requirements for ETOPS.

Expendable Parts

These are Parts for which no authorized repair procedure exists, and for which cost of repair would normally exceed that of replacement. Expendable Parts include nuts, bolts, rivets, sheet metal, wire, light bulbs, cable and hose.

Export Certificate of Airworthiness

An Airworthiness Certificate is a certificate issued by the Aviation Authority of a particular country so that an Aircraft can be exported from that country to another country, and in the meantime actually engage in flight operations for that purpose.

Fly by Wire

This is the transmission of a control input by electrical cabling rather than mechanical controls. In the flight deck, transducers translate the physical displacement of a control input on an electronic signal which is transmitted via various flight control computers to electrically controlled, hydraulically actuated power control units attached to the flight control surface. This type of control system is also used by the powerplant to control thrust.

Hush Kit

This is a piece of equipment that makes Engines more quiet, usually to meet Stage 3 Aircraft Noise Standards.

Level Rent

This is rent payable under a Lease Financing that is fixed for the duration of a lease term. These rents may nevertheless have scheduled increases and decreases during the lease term.

Landing Gear (LG)

Landing Gear (LG) constitute the Parts supporting the landing, taxiing and parking function of an Aircraft. There are usually three Landing Gear: (i) nose landing gear;

(ii) rear landing gear (right); and (iii) rear landing gear (left). LG usually have their own assessed Maintenance Reserves and Return Conditions.

Life Limited Parts (LLPs)

Life Limited Parts (LLPs) are Parts which, when listed on the Aircraft or Engine Type Certificate data sheet or the OEM's instructions for continued airworthiness, must be permanently removed from service and discarded before a specified time (for example, Hours, Cycles or calendar limits) is achieved. Among the most significant life-limited items are engine disks and shafts.

Minimum Equipment List (MEL)

The Minimum Equipment List (MEL) is a list of systems and certain critical parts that a particular model of Aircraft has installed, and the minimum number of these parts which must be operative for dispatch. The MEL also considers the effect of multiple system failures and provides restrictions regarding the number of related systems that can be inoperative at dispatch.

Maintenance Planning Document (MPD)

The Maintenance Planning Document (MPD) is the document in the OEM's baseline maintenance manual for a particular Airframe or Engine that contains all required maintenance checks and inspections necessary to maintain continued airworthiness of the Aircraft Asset. An airline's Maintenance Program will be based on the OEM's MPD and must be approved by the local Aviation Authority.

Maintenance, Repair and Overhaul (MRO)

A Maintenance, Repair and Overhaul (MRO) is a service provider (which may include an OEM) that can perform maintenance, repair and overhaul functions on an Aircraft or Engine. An MRO must be certified by an applicable Aviation Authority.

Maximum Take-off Weight (MTOW)

The Maximum Take-off Weight (MTOW) is a design limitation placed on an Aircraft as to such Aircraft's take-off ability.

Nacelle

The Nacelle is the part of the Aircraft that houses an Aircraft Engine. A subpart of the Nacelle is the Cowling, which is the removable portion of the Nacelle.

OEM Parts

Aircraft Parts manufactured by the relevant OEM (in contrast to PMA Parts).

Original Equipment Manufacturer (OEM)

The primary Original Equipment Manufacturers (OEMs) are: (i) in the case of Airframe manufacturers, Boeing, Airbus, Embraer and Bombardier; and (ii) in the case of Engine manufacturers, Pratt & Whitney, GE, IAE, SNECMA and Rolls Royce.

Operational Specifications (Op Specs)

Operational Specifications (Op Specs) are the specifications that delineate the scope of an air carrier's operations, including any applicable authorizations and limitations, and maintenance responsibility for U.S.-registered Aircraft. While the Op Specs, with its operations manual and general maintenance manual, is the bible for operating and maintaining an airline's Aircraft, it can be and often is revised.

Airlines revise, add to and delete from the Op Specs and general maintenance manual when changes are made to policies, procedures or intervals. These changes are implemented by the airline's chief inspector in the form of a revision to the general maintenance manual or to the inspection program.

In the U.S., changes are made through a revision process that is acceptable to the FAA and conforms to the established procedure. As an example, an additional Aircraft added to the operator's system will require revisions to the Op Specs before that Aircraft can be operated in passenger service. The Op Specs may also be changed to facilitate planning. This allows the operator to schedule Maintenance Checks at specific times, in order to better utilize hangar space and manpower.

Op Specs will differ in many ways between airlines, such as:

- types of equipment;
- inspection intervals;
- inspection content; and
- terminology.

Passenger Convenience Equipment/In-flight Entertainment Equipment (PCE/IFE)

As it sounds, this is the equipment on Aircraft like televisions, telephones and so on, that provide entertainment or communication outlets for airline passengers. Its relevance is that Passenger Convenience Equipment/In-flight Entertainment Equipment (PCE/IFE) is often financed separately from the Aircraft itself, and may, therefore, not be subject to the Security Interest of a Mortgage or the ownership interests of a lessor.

PMA Parts

Parts Manufacturer Approved (PMA) for any Aircraft asset are Aircraft Parts that are not manufactured (or originally sourced) by the related OEM; the manufacturer of such part is rather an Aviation Authority-approved manufacturer of such Parts. The manufacturers of these parts usually do not have access to the original manufacturers' data so necessarily design their own parts and have to prove (in the U.S.) to the FAA, under FAR 21.303 regulations, that their Parts are as good as the original.

PRACTICE NOTE

Many lessors and other financiers seek to restrict lessees from using PMA Parts in the Aircraft they have leased and financed as such usage is often perceived to diminish the value of such Aircraft, notwithstanding that PMA Parts may be cheaper and/or better than the original OEM Part. This perception is derived from the following facts (some of which are self-perpetuating).

- PMA Parts are not universally accepted by airlines and national Aviation Authorities.
- Installation will harm Aircraft liquidity by making Aircraft less desirable to subsequent buyers and/or operators.
- Lenders do not like PMA Parts.
- OEMs will not support Aircraft with PMA Parts:
  - warranty invalidation;
  - not acceptable for OEM Total Care Programs/Power-by-the-Hour; and
  - OEM secondary market Engine purchase constraints.

There is, however, a growing cautious acceptability for PMA Parts for Aircraft interiors and non-critical Parts.

Quick Engine Change (QEC)

A Quick Engine Change (QEC) kit is a collection of components and accessories such as pumps, generators, reverser, nose cowl, wiring harnesses and fluid lines installed onto a bare Engine to speed the eventual installation of the entire power plant onto an Aircraft. The actual make-up of the QEC kit will usually depend on the type of Aircraft that the Engine will be used on, and may also be different for different Engine positions on the same Aircraft. Accordingly, a QEC must be identified by a listing of all the related applicable parts.

There are three types of QEC kits.

1   Basic – this QEC kit includes all major parts and accessories (such as brackets, wiring harnesses, fluid lines, tubes, ducting and various air and anti-icing values that interface between the Nacelle and Engine) that are necessary in having an engine test run, but does not include the primary nozzle and centerbody.

2   Neutral – this QEC kit includes the Basic kit plus additional parts and accessories necessary for installation on an airframe, although such parts and accessories are not airframe-specific.

3   Full – this QEC kit includes the Neutral kit and parts and accessories specific for the particular airframe, including the primary nozzle and centerbody.

A QEC does not include the thrust reverser or nose cowl.

PRACTICE NOTE

When financing or acquiring an Engine, it is important for valuation purposes to understand whether such Engine is 'bare' or with a QEC.

Registration Certificate

With certain exceptions, no Aircraft may be operated in the U.S. without a Registration Certificate that is issued to its owner by the FAA. The Registration Certificate is also the basis for assigning a U.S. identification number (N Number). Generally, the Registration Certificate remains effective until the Aircraft is sold, exported, destroyed or scrapped. Note that this definition pertains specifically to the U.S. but comparable regulations apply in most other jurisdictions.

Repairable Part

This is a replaceable Part or component, commonly economical to repair, and subject to being rehabilitated to a fully serviceable condition over a period of time less than the life of the flight equipment to which it is related. Examples include engine blades and vanes, some tires, seats and galleys.

Rotable Part

This is a Part that can be economically restored to a serviceable condition and, in the normal course of operations, can be repeatedly rehabilitated to a fully serviceable condition over a period of time approximating the life of the flight equipment to which it is related. Examples include Avionics units, Landing Gears, APUs and major Engine accessories.

Service Bulletin

A Service Bulletin is a document issued by an OEM to notify the owner or operator of an Aircraft (or Engine or other device) of recommended improvements and modifications, substitution of Parts, special inspections/checks, reduction of existing life limits or establishment of first-time life limits and conversion from one Engine model to another. Service Bulletins may or may not be Aviation Authority-approved. Service Bulletins are different in character from ADs in that they are often less critical, and are issued by the OEM. Service Bulletins often provide the instructions for accomplishing an AD, but, under the FAA system, Service Bulletins themselves are not mandatory. Service Bulletins usually recommend the implementation of general improvements. They are sometimes issued to assure that a future AD will not be necessary. They may also suggest an increased inspection cycle of certain areas or items.

Stage [X] Aircraft Noise Standards

This is the categorization of Aircraft noisiness by the establishment of acceptable stages at different Aircraft noise levels. Aircraft are, by nature, noisy, and create noise pollution, especially in areas around airports during take-offs and landings. In the U.S., the FAA has been tasked with implementing noise reduction programs. These programs dictate the level of permitted noise by reference to different permitted levels designated as 'Stages'. Currently, Stage 4 is the most restrictive noise standard for subsonic jet Aircraft which is intended to ensure that the latest available noise reduction technology is incorporated into new Aircraft designs. Aircraft operated in the U.S. must be at least Stage 3. Aircraft Noise Standards at the various Stages are measured at particular decibel levels using specified multi-point criteria.

PRACTICE NOTE

Traders and operators of older Aircraft need to be concerned with the Stage levels of the Aircraft they are purchasing or selling, or using. An Aircraft at an inappropriate Stage will not be able to operate in certain jurisdictions. Hushkitting is a method to improve an Engine's noisiness so as to be able to avoid scrapping the Engine entirely.

Supplemental Type Certificate (STC)

A Supplemental Type Certificate (STC) is issued by the FAA or other Aviation Authority to grant approval for an alteration of a product by a major change in the type design, where such a change is not great enough to require a new application for a Type Certificate. STCs can address a broad range of matters, including passenger seat configurations, avionics and retrofitted Winglets. The STC is kept by the applicant and is then the basis for issuing or retaining Airworthiness Certificates to all Aircraft (or Engines) subsequently modified in the same way. In the case of alterations by the OEM, approval is normally in the form of an amendment to the original Type Certificate, rather than an STC (see FAR 21). See also 'Type Certificate'.

PRACTICE NOTE

An STC is a necessary feature for cargo conversions and Winglet installation (or de-installation).

Type Certificate

This is the OEM-arranged certificate obtained from an Aviation Authority containing the requisite details and approval for the specific design of a particular type of Aircraft Asset. The FAA issues a Type Certificate (ATC) when the applicant (normally the OEM) submits the type design, test reports, and computations and proves to the FAA's satisfaction that the product meets the applicable requirements of the FARs regarding airworthiness, noise and emissions. The Type Certificate is kept at the OEM's facility and is the basis for issuing airworthiness certificates to all Aircraft (or Engines) subsequently manufactured according to the same type design (see FAR 21). Type Certificates may also be issued for products manufactured in foreign countries with which the U.S. has an agreement for the acceptance of these products if the country of origin certifies that the product meets airworthiness, noise and emission standards equivalent to the U.S. standards, and the OEM submits the appropriate supporting technical data.

Winglets

Winglets are structural additions at the end of an Aircraft's wing that provide enhanced aerodynamics to the Aircraft and improvements in fuel economy. Airbus calls them 'sharklets'.

Exhibit 5.3 Boeing 737 MAX Winglet (with scimitar)

Source: Author's own

MAINTENANCE/RETURN CONDITIONS

Introduction

While the Pricing provisions identified in Part 4 pertain to ongoing economic return by Aircraft Financiers, the Maintenance/Return Condition matters covered in Part 6 go to ongoing preservation of the Aircraft Assets as collateral and, ultimately, the value of the asset when it is returned (due to a scheduled expiration of a lease term (in the case of a lessor) or repossession) to the Aircraft Financier. This is a high value proposition, and could involve many millions of dollars on a single Aircraft Asset.[1] In the context of operating lease financing of Aircraft, there is no overstating the importance of these matters insofar as they have a direct correlation to Aircraft values.

Part 6 covers two components: (i) the responsibility of the operator of an Aircraft Asset during the period it operates such asset; and (ii) the responsibility of the operator of an Aircraft Asset on its return to the Aircraft Financier.

Ongoing maintenance

As to the responsibility of an operator during the operational period, this is largely subsumed by its maintenance obligations in respect of an Aircraft Asset. Maintenance is required for three principal reasons.

1 Operational: to keep the Aircraft Asset in a serviceable and reliable condition so as to generate revenue.
2 Value Retention: to maintain the current and future value of the Aircraft Asset by minimizing its physical deterioration throughout its life.
3 Regulatory Requirements: the condition and the maintenance of Aircraft Assets are regulated by the Aviation Authorities of the jurisdiction in which the Aircraft is registered. Such requirements establish standards for repair, periodic overhauls, and alteration by requiring that the owner or operator establish an airworthiness maintenance and inspection program to be carried out by certified individuals qualified to issue an Airworthiness Certificate.

Such maintenance would be performed in accordance with a Maintenance Program. Here are some of the factors influencing Maintenance Program structure.

- External factors:
  - aircraft design and technology;
  - improved inspection methods and techniques;
  - Original Equipment Manufacturer (OEM);

- ○ aviation authority (FAA or its counterpart); and
- ○ maintenance review board.
- Operator considerations:
  - ○ characteristics of route system and flight schedule on which this aircraft type will be operated:
    - – average flight segment time;
    - – average daily flight hour and cycle utilization per aircraft;
    - – aircraft overnighting pattern relative to maintenance base location(s);
    - – maintenance 'windows' inherent in the schedule pattern, if any;
    - – seasonal or other periodic variations in the level of flight activity; and
    - – operating physical environment (altitude, desert, temperature and so on);
  - ○ fleet size:
    - – total number of this type of Aircraft in the fleet; and
    - – number of spare Aircraft included in the total;
  - ○ operator's maintenance philosophy;
  - ○ whether this Aircraft type is a derivative of a type already in the fleet;
  - ○ structure of existing programs for other Aircraft type(s) already in service in the operator's fleet;
  - ○ capabilities and loading of in-house maintenance facilities (if applicable); and
  - ○ where and by whom maintenance will be performed (may vary by Check).

Return

As to the responsibility of an operator on the return of an Aircraft Asset, these are some of the considerations that will be built into the return conditions.

- Minimum return lives and performance/overhaul life (generally engines, Auxiliary Power Unit (APU), components and landing gear).
- FAA and/or EASA compliance.
- No component/part fitted >X% total time of the Airframe.
- Permanent/flush repairs.
- No locally approved modifications.
- Documentation availability.
- Full repaint – redelivery is 'white tail'.
- Same configuration as delivered – no unapproved reconfigurations and so on.
- Airframe check minimum limitations, for example, fresh from C Check.
- Demonstration flight and ground inspection.
- Same Engines (or Engine value) as at delivery.
- Airworthiness Directives (ADs) for which termination is required up to X days after the redelivery date must be terminated on return.

- De-registration and Export Certificate of Airworthiness (C of A).
- Engine and APU Borescope.
- Engine Power Assurance Run.
- Receipt of non-incident/accident letter.

A satisfactory return condition of an Aircraft being returned from a lessee is critical for the Aircraft's owner since that owner will seek to redeploy such Aircraft with a new operator as expeditiously (and seamlessly) as possible. Most potential lessees will require that an Aircraft taken by it on lease be capable of operating at least one full C Check interval (usually 18 to 24 months) before major maintenance is due. Lessees will sometimes accept an Engine with as little as one year operating time remaining until the next anticipated shop visit, but in such cases the lessor will still be responsible for a major portion of the cost of the shop visit when it occurs.

Costs for reconfiguration of galleys, seat layouts and avionics systems vary widely but can range from less than U.S.$100,000 to U.S.$2 million or more even for narrowbody Aircraft. Reconfiguration requirements are negotiated on a deal-by-deal basis, and are influenced by many factors – the urgency of lessee's need for the Aircraft, lessee's credit standing and thus leverage in the negotiations, and of course, the degree of customization of the interior by the previous operator.

Back-to-birth

As applied to Aircraft Records, Back-to-birth records are maintenance records that go back in time so as to trace repair history to when the related Life Limited Part (LLP) was first installed on an Aircraft Asset. Back-to-birth traceability is an absolute imperative for LLPs.

Block Check

A Block Check is a grouping of scheduled maintenance tasks. See 'Check'.

Bore Blend

A Bore Blend is a repair carried out using a borescope and a small grinding tool which is inserted through a borescope port. This allows the technician to grind off small amounts of material in such a way as to remove cracked or torn material and leave the blade with a smooth, contoured profile. It is normal to perform another borescope inspection within 100 Cycles to check on the integrity of the repair performed. This is a method of repair of an Engine that can be accomplished outside an Engine repair facility.

Borescope

A Borescope is an optical device consisting of a rigid or flexible fiber optic tube used to see the innards of, most usually, Engines and APU; that is, Parts not otherwise

Exhibit 6.1 Borescope

Source: Photo: © Joel Silverman

visible by physical inspection. A borescope inspection allows the internal condition of the compressor, combustor or turbine section of the typical Engine to be viewed clearly. Such an inspection will not just highlight defects and problems that could cause failure if left unresolved, but it also helps the customer to better estimate the eventual cost of the Engine repair and the estimated remaining service life. Reasons that often prompt an inspection are reports of Engine surge, Foreign Object Damage (FOD), excessive vibration and high Exhaust Gas Temperature (EGT). Borescope inspections are also part of a scheduled maintenance program, when an Aircraft is going on or coming off lease or being sold.

Any defects discovered by a Borescope that threaten the integrity of the Engine or may cause the Engine to fail to produce power is likely to be a 'no go' item. Blade defects in either a compressor or turbine which could fracture and cause wipe-out of the module are usually not acceptable. These include cracks or tears in compressor blades and cracks, tears or holes in turbine blades. If the strength of the blade is compromised, failure can occur very rapidly.

Calendar Limit

In respect of a particular Part, the Calendar Limit is the period of time the mere passage of which requires the taking of certain action (such as removal or maintenance).

115

Checks (A, B, C and D)

Aircraft maintenance checks are periodic inspections that have to be done on all Aircraft after a certain amount of time or usage. Airlines and other commercial operators of Aircraft follow a continuous inspection program approved by the applicable Aviation Authority as contained in their Maintenance Programs. Each Maintenance Program includes both routine and detailed inspections. Airlines and airworthiness authorities casually refer to the detailed inspections as 'checks', commonly in one of the following categories: 'A' Check, 'B' Check, 'C' Check, or 'D' Check. 'A' and 'B' Checks are lighter checks, while those designated 'C' and 'D' are considered heavier checks.

Transit Check

A Transit Check is accomplished when an Aircraft transits designated stations where the operator has maintenance staff. This is a visual inspection of such things as tire pressure, Engine oil, Landing Gear, wings, Engines, empennage, cabin and flight compartment interior, and review of the logbook. Many airlines accomplish a series of heavier Transit Checks in a larger overnight check called a 'Daily Check'.

A Check

This is a primary examination of the Airframe, Avionics, Engines and accessories to determine the general condition of the Aircraft, which is usually accomplished in one or two days. The A Check is performed approximately every 500–800 Hours or 200–400 Cycles. It needs about 20–100 man-hours and is usually performed overnight at an airport gate or hangar. The actual occurrence of this Check varies by Aircraft type, the Cycle count, or the number of Hours flown since the last Check. This Check may be delayed by an airline if certain predetermined conditions are met.

B Check

This is an intermediate inspection that includes selected operational checks, fluid servicing and some lubrication tasks. This is performed approximately every four to six months. It needs about 150 man-hours and is usually performed within one to three days at an airport hangar. It will include an 'A' Check. This type of check is not common on today's Aircraft.

C Check

This is a detailed inspection of the Airframe, Engines and accessories. It is performed approximately every 15–21 months or a specific amount of actual Hours as defined by the OEM. This maintenance Check is much more extensive than a 'B' Check, as pretty much the whole Aircraft is inspected. This Check puts the Aircraft out of service until it is completed. It also requires more space than 'A' and 'B' Checks – usually a hangar at a maintenance base. The time needed to complete such a check

116

is generally one-to-two weeks and the effort involved can require up to 6,000 man-hours. The schedule of occurrence has many factors and components as has been described, and thus varies by aircraft category and type. Some access panels will be removed and the newly-exposed areas inspected. Heavy lubrication will be done on this inspection. Generally, this check includes a portion of the Corrosion Protection and Control Program (CPCP). Service Bulletin requirements and Aircraft appearance work will be done. It will include 'A' and 'B' Checks. 'C' Check tasks themselves may also be segmented into sub-block checks.

D Check

This is an overhaul which returns the Aircraft to its original condition, insofar as possible. It constitutes a thorough examination of the entire Aircraft. The interior of the Aircraft, and many of the Aircraft's components, are completely removed. Items that cannot be removed are inspected and refurbished in place. It includes structural inspection and CPCP tasks. This is by far the most comprehensive and demanding check for an Aircraft. It is also known as a Heavy Maintenance Visit (HMV). This check occurs approximately every five years. It is a check that, more or less, takes the entire Aircraft apart for inspection and overhaul. Also, if required, the paint may need to be completely removed for further inspection of the fuselage metal skin. Such a check will usually demand around 40,000 man-hours and it can generally take up to two months to complete, depending on the Aircraft and the number of technicians involved. It also requires the most space of all maintenance checks, and as such must be performed at a suitable maintenance base. Given the requirements of this check and the tremendous effort involved in it, it is also the most expensive maintenance check of all, with total costs for a single visit being well within the million dollar (and higher) range. Like 'C' Checks, 'D' Checks may be accomplished in segmented blocks during 'C' Checks.

Because of the nature and the cost of such a check, most airlines – especially those with a large fleet – have to plan 'D' Checks for their Aircraft years in advance. Oftentimes, older Aircraft being phased out of a particular airline's fleet are either stored or scrapped upon reaching their next 'D' Check, due to the high costs involved in comparison to the Aircraft's value. On average, a commercial Aircraft undergoes two to three 'D' Checks before it is retired. Many Maintenance Repair and Overhaul providers (MROs) state that it is virtually impossible to perform a 'D' Check profitably at a shop located within the U.S. As such, only a few of these shops offer 'D' Checks.

Condition-monitored Maintenance

Condition-monitored Maintenance is a primary maintenance process under which data on the whole population of specified items in service is analyzed to indicate whether some allocation of technical resources is required. Not a preventive maintenance process, Condition-monitored Maintenance allows failures to occur, and relies upon analysis of operating experience information to indicate the need for appropriate action.

Corrosion Protection and Control Program (CPCP)

Corrosion Protection and Control Program (CPCP) is the systemic approach for controlling corrosion in an Aircraft's primary structure – the purpose of which is to limit corrosion to a level necessary to maintain airworthiness. This program is part of the Aircraft's Maintenance Program. CPCP tasks are mainly structural inspections and often require extensive work to allow access to the inspection area. If necessary, any discovered corrosion or damage is repaired. Prior to access close, the area is treated with corrosion protection compounds.

Crazing

Crazing is the degradation of Aircraft windows by the inclusion of a web of fine scratches on the window, caused by stresses placed on the acrylic of the window pane. Return conditions often require that the Aircraft windows not be subject to crazing.

Cycle

This is a complete flight sequence including taxi, take-off, flight en route, and landing. In the case of Engines, a Cycle includes starting, acceleration to maximum rated power, deceleration and stopping, see FAR 33.14.

Deferred Maintenance

This is the deferral of maintenance tasks beyond their normal schedule. Maintenance repairs may be deferred by operators for a number of reasons.

- Part Availability – a Part may be unavailable and therefore cannot be replaced immediately.
- Aircraft Scheduling – the Aircraft may be unavailable, or may have insufficient ground time to accomplish the necessary maintenance or repair.
- Cost of Repair – the airline may be avoiding what it considers to be unnecessary expense by indefinitely deferring non-airworthy items. This is a short-sighted policy, but, unfortunately, one which is common among airlines that are struggling financially.

Not all maintenance items can be deferred. Minimum Equipment List (MEL) items can only be deferred for a limited period of time; however, there are many items which can be deferred indefinitely.

Airworthiness Directive Terminating Actions

Many ADs offer two options for accomplishment. For these ADs, one option is to accomplish inspections or some other minor action at specified intervals. The more desirable option from a lessor's view is to accomplish a one-time action, usually a modification or replacement, which terminates the AD and the need for any further action.

The terminating action usually requires more time and money, but pays off over the long term due to overall reduced maintenance requirements.

Interior Items

- Flight station windows – as long as vision is not impaired.
- Seats.
- Passenger air vents and reading lights.
- Overhead bins.
- Passenger cabin windows – outer pane only.
- Lavatories.
- Galleys.
- Carpets.
- Sidewalls.

Skin Surface

Unsightly and drag inducing doubler repairs[2] can be placed over areas of skin damage and permanent corrections can often be deferred indefinitely.

Wings and Empennage

Dents, doubler repairs and leaks in these areas can sometimes be deferred indefinitely, as long as they are within acceptable airworthy limits.

Cargo Bays

Automatic loading devices and non-structural damage need never be repaired as long as fire containment capabilities are not compromised.

Engines

Deterioration of internal components such as combustion liners and airfoils can often be deferred for short periods of time (10 Cycles, 50 Hours and so on) to allow an airline to operate an Aircraft until it reaches a maintenance base where the Engine can be removed. These deferrals are a necessary part of any airline's operation. In the context of crafting return conditions, from a lessor's perspective, return condition language should be carefully crafted to prevent the return of an Engine with such deferrals.

Exhaust Gas Temperature (EGT)

The Exhaust Gas Temperature (EGT) of an Engine is an Engine operating limiter and is used to monitor the mechanical integrity of the Engine's turbines. Properly maintained Engines of a certain type/manufacturer will have EGT within prescribed ranges. Excess EGT will reduce turbine blade life, while low EGT reduces turbine efficiency and thrust.

Exhaust Gas Temperature (EGT) Margin

In a turbine Engine, the Exhaust Gas Temperature Margin is the difference between the EGT limit (or EGT red line) and the Engine's actual EGT when producing maximum thrust at the full rated take-off power setting. The actual EGT should be lower than the EGT limit, and the magnitude of this difference (the EGT Margin) is indicative of the time remaining before normal deterioration of the Engine will require removal for restoration.

Engine Performance Restoration

These are services performed during an Engine Shop Visit in which, at a minimum, the compressor, combustor and high-pressure turbine are exposed and subsequently refurbished. For those refurbished modules, the Hours and Cycles since such restoration are then re-set to zero.

Engine Shop Visit

An Engine removal is classified as a 'shop visit' whenever the subsequent engine maintenance performed prior to reinstallation entails one of the following.

- Separation of pairs of major mating engine flanges (other than solely for shipment); or
- Removal/replacement of a disk, hub, or spool.

Sometimes the definition is specifically tailored, as in some ADs that say, 'For the purpose of this AD, an engine shop visit is defined as input to an engine repair shop where the low pressure turbine module is removed' . . . or 'the front and rear flanges of the combustion case are separated' . . . or 'any major module is separated' . . . or 'the inlet gearbox is exposed'.

Foreign Object Damage (FOD)

Foreign Object Damage (FOD) refers to damage to any portion of the Aircraft (most commonly Engines) caused by impact or ingestion of birds, stones, hail or other debris. See Exhibit 6.2.

Full-time, Full-life

These terms are commonly used by appraisers and Aircraft maintenance personnel to describe the maintenance status of an Aircraft or Engine, or associated LLPs, which assumes the relevant equipment has been serviced and no usage (Hours or Cycles) has occurred.

Half-time, Mid-time, Mid-life, Half-life

These terms are commonly used by appraisers and Aircraft maintenance personnel to describe the maintenance time status of an Aircraft or Engine, or associated LLPs, which assumes that every component or maintenance service which has a prescribed interval that determines its service life, overhaul interval or interval between maintenance services, is at a condition which is one-half of the total interval.

- Half-time and Mid-time pertains to scheduled inspections or overhauls that are repeated at specified intervals of time, with 'mid-time' (or half-time) implying that the status is mid-way through such an interval.
- Half-life and Mid-life pertains to items with mandated life limits (engine disks, for example), and 'mid-life' (or half-life) implies that such items have been in service for one half of their life limits.

PRACTICE NOTE

Overhaul periods are changeable (for example, if an Engine is performing better than originally anticipated, its overhaul period may be extended from 8,000 Hours to 10,000 Hours). If this type of criteria is utilized, from a lessor's perspective, some level of certainty may be sought as to what is that overhaul period.

Exhibit 6.2 Foreign object damage potential
Source: Getty Images: #184047221

Hard Landing

This is an uncontrolled landing of an Aircraft that places excessive stress on the Airframe such that the Airframe may suffer structural damage. A hard landing noted in an Aircraft's records will adversely affect its value.

Hard Time Maintenance

This is a primary maintenance process to ensure continued airworthiness which requires removing the Aircraft Asset from service before a previously specified time in order to perform some required maintenance actions such as inspection or refurbishment.

Hot Section Inspection

Hot Section Inspection (HSI) is the inspection and restoration of the hot section items of an Engine (principally the combustion and turbine sections), usually at a predetermined time/cycle limit. A Hot Section Inspection is not necessarily considered an Engine Shop Visit if no major disassembly or repairs are required.

Hour

Any hour or any fraction of an hour (rounded to two decimal places) measures the time from the time the wheels of an Aircraft leave the ground until the wheels next touch ground.

Maintenance Program

A program, either acceptable to or approved by Aviation Authorities, which defines a logical sequence of maintenance actions to be performed as events or pieces of a whole which, when performed collectively, result in achievement of the desired maintenance standards. The program is typically originated by the OEM and, subject to approval by the relevant Aviation Authority, amended by the operator. The general maintenance manual sets out the operator's continuous airworthiness maintenance program of sufficient scope to maintain the Aircraft in an airworthy condition in accordance with Aviation Authority requirements. An Aircraft and all of its component parts and equipment are required by the applicable Aviation Authorities to be maintained in an airworthy condition by the airline with respect to the maximum operation time limits and cycles set out in its Maintenance Program. These time limits and cycles refer to the various inspections listed earlier, the hard time components and on-condition program.

Parts and components not specifically listed are checked, inspected and overhauled within the same time limitations specified for the component or accessory to which they are related.

- A 'Block Maintenance Program' is one which allocates major structural inspections and/or maintenance tasks into groups, or blocks, which permit convenient, economical and effective accomplishment. A program of recurring 'C' Checks and 'D' Checks may be a block maintenance program. See also 'Phased Maintenance Program' below.

- A 'Continuous Airworthiness Maintenance Program' is a compilation of the individual maintenance and inspection functions utilized by an operator to fulfill its total maintenance needs (see Advisory Circular AC120–16C and FAA publication 8300.9). The authorization to use continuous maintenance programs is documented in the operator's Operations Specifications. The basic elements of a continuous airworthiness maintenance program are:
  - Aircraft inspection;
  - scheduled maintenance;
  - unscheduled maintenance;
  - Engine and appliance repair and overhaul;
  - structural inspection program and Airframe overhaul;
  - required inspection items; and
  - maintenance manuals.
- A 'Phased Maintenance Program' (sometimes called an 'equalized' or 'segmented' program) is one where some of the maintenance effort is apportioned to smaller packages that may be accomplished more frequently than the packages in a block maintenance program. Usually, the objective of this subdivision of effort is to even out the maintenance workload over time and shorten the length of each period of down-time.
- A 'Progressive Maintenance' program is one which provides for the complete inspection of an Aircraft within each 12 calendar months, consistent with the OEM's recommendations and other regulatory requirements.

PRACTICE NOTE

The distinction between 'block' and 'phased' programs is not very clear. Different airlines and different Aviation Authorities have adopted many variations, so these terms do not have unique meanings applicable to all circumstances. For example, the 'C' Check might be divided into phases while the 'D' Check is left intact, or the 'D' Check might also be divided into phases, and the number of phases could be large or small. Moreover, different airlines have adopted different lettering and numbering terminologies to designate their Checks and there may be different terminology based on OEM.

Maintenance Reserves

Maintenance Reserves are periodic (usually monthly) payments by a lessee to a lessor to prefund on a real-time basis the cost to do maintenance on an Aircraft or Engine. Maintenance Reserves are payments often based on the lessee's Hours and/or Cycles of use of Aircraft Assets during the preceding period. The reserves so paid are then available to the lessee to reimburse the lessee for the lessee's costs of maintaining such assets (or paying such costs on behalf of the lessee directly to the maintenance provider). Because Maintenance Reserves are both paid by a lessee and then recoverable by the lessee, they have a dual nature and are regarded very differently by the lessor and the lessee.

From the lessor's perspective, Maintenance Reserves are payments made by its lessee to compensate the lessor for the lessee's use of the lessor's Aircraft and as security for the maintenance of the collateral.

From the lessee's perspective, on the other hand, Maintenance Reserves are just what they are named: reserves paid (or deposited) over time by the lessee to be held by the lessor as a reserve to pay for upcoming maintenance. The lessee views the Maintenance Reserves as nothing more than cash management – the lessee is setting aside its own money so that it has sufficient cash available to pay for the significant periodic maintenance. However, having to provide Maintenance Reserves to a lessor is not just a time value of money consideration for a lessee. The lessee may wish to push-back on having to supply Maintenance Reserves for the following reasons.

- As a credit matter, a lessee may require a lessor to rely on its (the lessee's) credit for performing maintenance.
- The provision of Maintenance Reserves may adversely affect a lessee's cash flow and cash balance position.
- Lessees often overpay Maintenance Reserves and, consequently, the lessor walks away from a Lease with 'extra' compensation.
- A lessee may have trouble accessing its posted Maintenance Reserves in a default situation (when it may still be operating the Aircraft).
- A lessee may have trouble accessing its posted Maintenance Reserves if there is a disagreement on whether the item for reimbursement qualifies for reimbursement.
- A lessee is taking lessor credit risk for reimbursement, as the Maintenance Reserves may be held in co-mingled unrestricted accounts.
- A lessee may end up double-paying for maintenance during any period it is waiting for reimbursement.

Reserved Items

Maintenance Reserves are reserves that can be established against maintenance relating to:

- Airframe heavy 'D' Check(s);
- 'C' Check(s) (or different phases thereof);
- Landing Gear Overhaul (LGO);
- APU overhaul;
- Engine performance restorations; or
- Engine LLPs.

Reserved Amounts Calculations

Maintenance Reserves are typically assessed on the particular item based on Hours of usage times an agreed rate/Hour (or, in the case of Heavy Checks, a monthly rate). The agreed rate may be subject to adjustment based on the actual Hour/Cycle ratio for the relevant item.[3] The agreed rate may also be subject to adjustment based on increases in MRO maintenance costs, inflation indexing and changes in OEM catalogue prices.

Reimbursing the Lessee

Lessees are reimbursed their paid-in Maintenance Reserves for a particular item after that item has undergone the related maintenance/restoration. A lessee must provide full work package and invoices to support a request for reimbursement for Maintenance Reserves, unless the maintenance work was preapproved. Certain maintenance events do not qualify for reimbursement:

- any maintenance or Parts replacement required as the result of, or any cost resulting from, foreign object damage, operational misuse, mishandling, abuse, faulty maintenance, negligence, accidental damage, Airworthiness Directives or service bulletin, modification, addition or alteration, or any cost which is reimbursable from insurance after due diligence or claims against the OEM, supplier or repairer of any Part or part thereof, in respect of the condition or performance of such Part or part thereof, whether based on warranty claims or otherwise;
- replacement, repair or rental of engine line replaceable Parts (regardless of whether or not such Parts must be operational for the repair facility to return an Engine to service);
- labor at premium rates; or
- taxes or shipping and handling charges or the like incurred in connection with any of the foregoing maintenance or overhaul activities or purchase of LLPs.

On Watch

An Engine, APU or other component is On Watch if it requires special monitoring due to performance trends falling outside certain specified ranges.

On-condition Maintenance

On-condition Maintenance is a primary maintenance process having repetitive inspections or tests to determine the condition of units, systems, or portions of structure with regard to continued serviceability. Corrective action is taken when required by the item's condition. For example, a hydraulic component may be tested regularly to determine its internal leakage rate, but refurbishment is required only when the rate exceeds a specified limit. Engines and most Avionics components are on-condition.

One-way Settlement

In the case of an Aircraft or Engine return, this is a payment by the lessee to the lessor if the item does not satisfy return thresholds, but without any payment by the lessor to the lessee if the item exceeds return thresholds.

Overhaul

This is the disassembly, inspection and/or check of an Aircraft, Engine, Part or appliance to an extent necessary to determine, as substantiated by service experience and accepted practices, that it is in satisfactory condition to operate over one complete overhaul period. It includes the replacement, repair, adjustment or refinishing of such parts as required, which, if improperly accomplished would adversely affect the structural strength, performance, flight characteristics or safety of the Aircraft involved.

Power By the Hour (PBH)

Power By the Hour (PBH) is an Engine maintenance-outsourcing concept whereby an MRO (which may be an OEM) provides to an airline/operator heavy engine maintenance, charging the operator based on the time used by an Engine in operations. PBH programs typically either assess costs per Hour for both parts and labor, or for labor only with parts billed at cost. See also 'Total Care Package'.

Power Assurance Run/Maximum Power Assurance

Power Assurance Run/Maximum Power Assurance (MPA) is the running of Engines at maximum thrust to ensure performance vitality, typically in connection with an Engine return, delivery or maintenance Check. Off-wing Engines would be run using a test cell.

Progressive Inspection

This is an inspection that may be used in place of an annual or 100-hour inspection. It has the same scope as an annual inspection, but it may be performed in increments so the Aircraft does not have to be out of service for a lengthy period of time.

Reduced Interval Inspections

Reduced Interval Inspections are the heightened frequency of inspections of specified Aircraft components due to such components' falling outside of normal performance specifications.

Scheduled Maintenance

This is the maintenance performed on Aircraft Parts at defined intervals to retain such Part in a serviceable condition by systematic inspection, detection, replacement of worn out items, adjustment, calibration, cleaning and so on.

Storage

Storage is the physical act of placing an Aircraft Asset out of commission and safeguarding that asset during such period. Aircraft Assets are placed in storage for a variety of reasons. A lessor will store Aircraft Assets while seeking to redeploy the asset with another operator or, failing redeployment, pending Part-Out. An operator will store Aircraft Assets if they are surplus to its needs and it is unable to (sub)lease those assets. A financier will store an Aircraft Asset if it has repossessed an Aircraft Asset and needs time to lease or sell it.

Aircraft Assets are usually (and preferably) stored at specialized facilities, and in dry (desert-type) climates that minimize the corrosive effects of rain and humidity. Storage takes one of the following forms.

- *Flyable Storage.* Flyable storage is the prescribed procedure to maintain a stored Aircraft in operable condition. Next to daily use, this category of storage keeps the Aircraft in the best possible condition. All scheduled preventative maintenance will be performed on Aircraft in flyable storage, and periodic operation of the Aircraft and all systems is required. There is no time limit on flyable storage.
- *Short-term Storage.* Short-term storage is used to store an Aircraft for a period not to exceed 60 days. Aircraft in short-term storage require extensive preservation but very little periodic attention.
- *Intermediate Storage.* Intermediate storage is used to store an Aircraft for a period of 60 to 180 days. Aircraft in intermediate storage require very extensive preservation but minimal periodic attention.
- *Long-term Storage.* Long-term storage is used to store Aircraft for a period in excess of 180 days. Aircraft in long-term storage require very extensive preservation and periodic attention.
- *Bone Yard Storage.* Where Aircraft are placed when they no longer will be flown and are likely subject to Part-Out.

Induction of an Aircraft into a storage program would entail the following procedures.

- Airframe:
  - installation of protective coverings and closing of all external openings (except drains);
  - parking/mooring procedures;
  - installation of safety pins;
  - washing of Aircraft (due to environment, may be repetitive)
  - landing gear strut servicing, lubricating and protection of the landing gear shock absorbers (oleo);
  - tire inflation and rotation;
  - fuel system decontamination;
  - primary and secondary flight control cycling and lubrication;
  - protection of windows;

Exhibit 6.3 Aircraft storage

Source: Corbis Images: #42–29561020

- o   inspection of seats and carpet for moisture/mildew (if stored in humid environments);
- o   preservation of lavatories and water systems; and
- o   opening of closets, cabinets, and interior doors to supply ventilation and to prevent mildew.
- •   Engine/APU:
  - o   procedures to operate the Engine/APU on an established interval; and
  - o   complete preservation of the Engine/APU (shrink-wrap).
- •   Electrical:
  - o   opening/closing of circuit breakers;
  - o   battery servicing/disconnection; and
  - o   removal of batteries from emergency devices such as megaphone, flashlights, power supplies for emergency lights, emergency beacons and so on.
- •   Operational Checks:
  - o   procedures to transition the Aircraft from preservation to a state acceptable for engine operations and operational checks of systems, back to the preserved state; and
  - o   operational checks of hydraulics, electrical, engine, fuel systems and Avionics and so on.

Return of an Aircraft to airworthy condition once taken out of storage would entail the following procedures.

- Audit the current status of the Aircraft to the Maintenance Program and comply with required tasks, including ADs, LLPs, certification maintenance requirements, Avionics databases and so on.
- Conduct other inspections and operational checks, as deemed necessary, based on the amount of time the Aircraft was in storage and the environment to which it was exposed.
- Conduct any operational check flights or test flights prior to return to service.

The type of storage chosen for a particular Aircraft Asset would usually be dictated by the perceived length of time such Aircraft Asset will need storage. During periods of Storage, the Aircraft Asset will need to be insured, and that will be effected through Storage Insurance.

Top-case Repair

This is the repair of the upper half of the high pressure compressor module case. This case can be removed to allow access to the compressor rotor blades. If blades are damaged they can be repaired or replaced. The repair manual normally specifies the maximum number that can be changed in this way. This is a method of repair of an Engine that can be accomplished outside of an Engine repair facility.

Total Care Package

This is a contractual arrangement between an MRO (which may be an OEM) and an airline/operator where such airline's Engines are subject to an outsourced maintenance care arrangement with that MRO. Total Care Packages relieve the airline of maintaining staff and the facilities for conducting heavy Engine maintenance. The MROs conducting such maintenance are oftentimes the Engines' manufacturer. Under these arrangements, the airline would typically make regular payments to the MRO based on Hours and Cycles usage of its Engines (much like a Maintenance Reserve payment). APUs may also be subject to Total Care Packages.

PRACTICE NOTE

In financing transactions where the financier is taking Maintenance Reserves from an airline and that airline's Engines are in a Total Care Package program, the financier will likely want to receive an assignment from the airline of the airline's rights in the Total Care Package since the receipt of such an assignment is the only way the financier can receive the protection equivalent to what it would have received had it directly been receiving Maintenance Reserves in respect of the financed Engines. MROs are often very reluctant to agree to such an assignment, and any such assignment would likely only cover a single

---

shop visit if an Engine had to be repossessed; the MRO will not likely grant any new operator the same terms as were available to the original operator. The issue of assignability of these contractual arrangements is of increasing concern in the lessor community in the light of the increasing role that these types of arrangements are playing for airlines.

Two-way Settlement

In the case of an Aircraft or Engine return, a Two-way Settlement is a contractual arrangement for payment by the lessee to the lessor if the item does not satisfy return thresholds, and payment by the lessor to the lessee if the item exceeds return thresholds.

Notes

1   Under Basel II rules, in order for banks to reduce the equity/risk reserve levels for a loan by reason of the availability of Aircraft collateral, the value of the collateral must be regularly reviewed and agreements put into place to ensure that the collateral is properly maintained.
2   Sheet metal or other materials used to repair cracks or holes in or otherwise stiffen portions of an Aircraft's skin.
3   The lower the ratio, the more expensive it is likely to be because the Aircraft Asset will be undergoing the increased stress of more take-offs and landings per Hour.

LEGAL

Introduction

Similarly to many areas of specialized finance, Aircraft Finance is replete with legal terms and jargon that relate to the many facets of obtaining title in Aircraft Assets, obtaining Security Interests in Aircraft Assets, Perfecting those interests and enforcing all of the above. In this Part 7, we will review the principal terms used in this area.

An important focus point in Part 7 is the legal regime of Cape Town: the international treaty that has revolutionized Aircraft Finance by: (i) simplifying the protocol for registration of leases, mortgages and ownership interests; and (ii) creating some level of certainty for repossession of Aircraft Assets in case of default, in each case, of course, in countries that have adopted Cape Town as their law. The detailed intricacies of the application of Cape Town is beyond the scope of this Handbook, which offers only high-level summaries of Cape Town's most salient features. We would highly commend to practitioners (the aptly named): *Practitioner's Guide to the Cape Town Convention and the Aircraft Protocol*, published by the Legal Advisory Panel of the AWG (available for free on their website: www.awg.aero/assets/docs/Practioner's%20Guide%20Final%20_4V_.pdf). See Appendices B-1a and B-1b.

A final note on the matter of *Legal* issues: the matter of what the law says (as to rights and remedies under contracts, for example) in any particular jurisdiction may not necessarily play out in *practice*. A country may have very fine laws on its books, and may even have adapted Cape Town, but whether these laws are enforced in that country is an entirely separate matter that must be investigated. This is explored in some detail below in 'Country Risk'.

Accession

Accession is the legal doctrine that when an object is incorporated in a larger object, the incorporated object is subsumed by the larger object and loses its individual character for certain legal purposes (see UCC 9–102(a)(1)). For example, when an engine blade is installed in an Engine, the blade becomes part of the Engine and does not need to be individually identified for UCC perfection purposes so long as the Engine itself is properly perfected. The doctrine of accession, however, could become problematic at the point when Engines are installed on an Airframe. In this instance, absent other authority, the Engine would become part of the Aircraft, so that the ownership of the Engine would be dictated by the ownership of the Airframe. Most jurisdictions, however, have legal regimes that recognize individual owner-ship and security interest rights in Engines, whether or not installed on an Airframe.

As well, international treaties such as the Cape Town Convention recognize that Engines do not accede to the Airframes on which they are installed.[1]

PRACTICE NOTE

Not all Parts when installed on an Aircraft accede to such Aircraft; certain parts, such as Landing Gear, Auxiliary Power Unit (APU) and avionics retain their individualized identity. For this reason, financiers should make sure that such Parts are free and clear of Liens when installed on their financed Aircraft since such Liens would survive installation.

Accredited Investor

See 'Appendix L'. An Accredited Investor may participate in Private Placements without the necessity for the related securities to be registered with the SEC.

Aircraft Objects

Aircraft Objects as per Cape Town are: (i) airframes that are type-certified to transport at least eight persons including crew or goods in excess of 2,750 kilograms; (ii) engines having at least 1,750 pounds of thrust if jet propulsion and at least 550 rated take-off shaft horsepower if turbine-powered or piston-powered; and (iii) helicopters that are type-certified to transport at least five persons including crew or goods in excess of 450 kilograms. Each of the foregoing includes all installed, incorporated or attached accessories, Parts and equipment (in the case of Airframes, other than Engines) and all data, manuals and records relating thereto. Engines under the Cape Town Convention are treated separately from airframes since they are highly valuable independent units and are increasingly dealt with and financed separately. The Protocol specifically provides that ownership of or an interest in any such Engine is not affected by its installation on or removal from an Airframe.

Assignments of Contracts

This is the general proposition of assigning the rights in a contract from one person (usually the airline) to another (usually the financier). This Legal category is intended to capture the general concept of the assignment of rights in Purchase Agreements, Warranty Agreement and Total Care Agreements, and the difficulty in obtaining the agreement of the obligors thereunder (usually an Original Equipment Manufacturer (OEM)) to such an assignment. This issue comes up in the following contexts.

- Pre-delivery Payments (PDPs) – where the airline seeks to collaterally assign its interest in its Purchase Agreement with the related manufacturer to the PDP financier.

- Warranties – where the Aircraft Financier seeks to obtain the benefits of Warranties provided to an airline in the event the financier repossesses the financed Aircraft.
- Total Care Packages/Power By the Hour (PBH) – where the Aircraft Financier seeks to obtain the benefits of Total Care Package arrangements provided to an airline in the event the financier repossesses the financed Aircraft.

As noted elsewhere in association with the referenced provisions, OEMs, who are usually the obligors under such arrangements, are typically reluctant to agree to such assignments and, accordingly, require significant negotiation to achieve results acceptable to the Aircraft Financiers.

Aircraft Sector Understanding (ASU) (2011)

The Aircraft Sector Understanding (2011) (ASU)[2] is an internationally-agreed protocol that seeks 'to provide a framework for the predictable, consistent and transparent use of officially supported export credits' and to maintain a level playing field across the global aviation industry among manufacturers, airlines, and Export Credit Agencies (ECAs). The ASU, importantly, is a non-binding 'gentlemen's agreement' among the participating countries: Australia, Brazil, Canada, the EU, Japan, New Zealand, Norway, South Korea, Switzerland and the U.S.

The ASU attempts to bring the economic terms of ECA Financings more in line with commercial market lending conditions and to minimize the support of the ECAs as a factor in the choice by buyers among competing aircraft. Significantly, the ASU makes no distinction among the various types and categories of aircraft models. The terms and conditions stipulated under the ASU apply to all new civil aircraft for any ECA financing. The ASU emphasizes the risk profile of buyers/borrowers and requires that the ECAs classify all buyers/borrowers into one of eight risk categories, reflecting the buyer/borrower's senior unsecured credit ratings. The risk categories are determined by the ECAs and are recorded and maintained by the Organization for Economic Co-operation and Development (OECD) Secretariat. These categories are valid for a maximum period of 12 months commencing from the date recorded by the OECD Secretariat and are binding on the buyers/borrowers at all stages of a transaction. In addition, ECA Financings must be structured as eligible asset-backed transactions. An eligible asset-backed transaction must include the following components: (i) a first-priority Security Interest on the applicable Aircraft and Engines; (ii) if such transaction is a lease structure, an assignment of and/or first-priority Security Interest in the related Lease payments; and (iii) if allowed under applicable law, Cross-default and Cross-collateralization of all Aircraft and Engines owned by the same parties under such transaction.

The ASU establishes a maximum repayment term of 12 years, but this may be extended to 15 years on an exceptional basis provided a 35 percent surcharge is added to the minimum premium rates. The amortization profile in respect of the ECA-supported debt may either be Mortgage Style Amortization (equal payments of principal and interest) or 'straight-line' (equal principal payments, with interest payable on a declining basis); provided, in either case, repayment is made on a quarterly basis. The ASU also allows for repayment on a semi-annual basis provided a 15 percent surcharge is added to the minimum premium rates (MPRs).

The ASU requires that the ECAs charge buyers/borrowers an MPR based on a percentage of the amount officially supported by the ECAs. The MPR may be paid either up front or over the life of the transaction and is based on a 12-year repayment term. The MPR corresponds to the risk category of each buyer/borrower and is determined using a complex equation involving: (i) the minimum risk-based rates, which correspond to the buyer/borrower's risk category (and are set annually using a four-year moving average of the annual Moody's 'Loss Given Default' for first lien senior secured bank loans); and (ii) a market-reflective surcharge, which serves as reflection of broader market conditions (which is based on, among other things, the borrower's risk category and Moody's 'Median Credit Spreads' with an Average Life of seven years). Although somewhat complex in calculation, the new premium rates are intended to balance ECA pricing with commercial market pricing at any point in time. The pricing mechanism also uses publicly available measures (for transparency) that can be easily accessed by bankers in order to determine spread volatility.

In addition to setting these ground rules for the provision by ECAs of guarantees, the ASU permits ECAs to provide direct financing support for both fixed and floating financings of all types of Aircraft. In this regard, the ASU provides guidance on the minimum level of interest rates that may be charged to borrowers.

Finally, the ASU sets a limit on the maximum level of support that may be provided to Aircraft buyers. This maximum is set at 85 percent of the net purchase price for the Aircraft, plus the up-front premium.

The ASU largely superseded the Large Aircraft Sector Understanding (LASU), except that certain Aircraft were grandfathered under the LASU arrangements.

The ASU allows for up to a 10 percentage reduction in the MPR if the obligor is in a qualifying Contracting State under Cape Town; this is the so-called 'Cape Town Discount'. A country qualifies for this discount if it fulfills certain conditions set forth in the ASU (at Appendix II, Articles 35 and 38). As at the date of this writing, the following countries are entitled to this discount: Angola, Canada, Ethiopia, Fiji, Indonesia, Kazakhstan, Kenya, Luxembourg, Malaysia, Mongolia, New Zealand, Nigeria, Norway, Oman, Pakistan, Panama, Rwanda, Senegal, Singapore, Tajikistan and Turkey.

Aviation Authority

This is the FAA (in the U.S.), EASA (in the EU), NavCanada (in Canada), CASA (in Australia) or other aviation authority in any jurisdiction that: (i) regulates Aircraft operation and maintenance in such jurisdiction; and/or (ii) regulates matters concerning Aircraft Registration, Leases and Security Interests. EASA does not replace the national aviation authorities within the EU jurisdictions; such national authorities remain charged with enforcement of safety standards and the issuance of Airworthiness Certificates and certificates of registration.

Bargain Purchase Option

This is an option given to a lessee to purchase a leased Aircraft Asset at lease expiry for a price which is significantly lower than the expected fair market value of that Aircraft Asset at such expiry.

Blue Sky

This refers to an English law case, *Blue Sky One Limited & O'rs v. Mahan Air & Ano'r [2010] EWHC 631 (Comm)*. This case reaffirmed the principle that the validity of an English law mortgage on an Aircraft is to be determined by the *lex situs* of the Aircraft; the law of the place where the Aircraft is situated at closing. Accordingly, a valid security interest is created under English law without additional requirements only when an Aircraft is located in England at the time of closing or where the location of an aircraft is unknown. In summary, the requirements are as follows.

- If an Aircraft is outside England at closing, an English mortgage must be valid under the law of the jurisdiction where the Aircraft is located in order to be effective without considering such jurisdiction's conflict of law rules.
- If an Aircraft is over international waters at closing, best practice is to ensure the mortgage is valid under the law of the jurisdiction where the Aircraft is registered to ensure the mortgage is effective (without consideration of such jurisdiction's conflict of law rules).[3]

These new requirements have cost, risk and timing implications for transactions using an English law mortgage. A best case scenario resolution addressing the new requirements is that local counsel in the jurisdiction where an Aircraft is located or registered will be able to give a clean opinion confirming that the English law mortgage is valid under local law. At worst, local counsel will give an opinion containing assumptions or exclusions that push the risk of a mortgage being invalid back to the parties, or will not be able to give an opinion at all – potentially because the English law mortgage will not, in fact, be effective under local law (as was the case in Blue Sky One).

PRACTICE NOTE

The issues with Blue Sky One can be sidestepped by having an Aircraft mortgage governed by laws other than English law. New York law is an alternative to consider. A second solution is to rely solely on a mortgage governed by the law of the jurisdiction where the aircraft is located or registered at closing. This will be less desirable if local rules on enforcement are not as familiar or as effective as the laws of a 'money center' jurisdiction like New York. Taking only a local mortgage may also necessitate local counsel and local courts becoming more involved in the enforcement process, potentially reducing certainty and increasing enforcement risk for lenders.

It is worth noting that, if the debtor is located in a country that has adopted the Cape Town Convention, then the parties arguably have a broader choice for the mortgage's governing law; the *lex situs* rules of Blue Sky should not apply (so long as they have not agreed to English as the governing law, and/or are not adjudicating the matter in English courts). The Cape Town Convention provides that, so long as the relevant contracting state has made the election under Article XXX(1), the transaction parties are free to choose the governing law of their agreements.[1] In this case, a New York law mortgage still would be a sensible choice, as this would give the parties the choice of law protections afforded by both the Cape Town Convention and New York law.[2] The parties utilizing a New York mortgage for these purposes should be aware of the need to file a UCC-1 financing statement in an applicable jurisdiction for perfection purposes.

[1] See Article VIII of the Protocol.

[2] Gee, C, 'Choice of law After England's Blue Sky One case', Vedder Price Newsletter/Bulletin June 2011. Available at www.vedderprice.com/Choice-of-Law-After-Englands-Blue-Sky-One-Case-06–30–2011/.

Cabotage

Cabotage is the transport of goods or passengers by Aircraft between two points in the same country where the Aircraft is not Registered in that country. Most countries do not permit non-locally Registered Aircraft to engage in Cabotage.

Cape Town

Cape Town is the generic reference to the international treaty adopted on 16 November 2001, at the conclusion of a diplomatic conference held in Cape Town, South Africa, where 53 countries from around the world supported the adoption of two documents, namely the Convention on International Interests in Mobile Equipment (the 'Convention') and an associated Protocol to the Convention on Matters Specific to Aircraft Equipment (the 'Protocol'). Since the adoption of the Convention, along with the Protocol (herein collectively referred to as the 'Cape Town Convention'), over 70 countries have become Contracting States. Central to the purpose of the Cape Town Convention is the enhancement and harmonization of private laws in respect of the financing, lease and sale of mobile equipment. The Cape Town Convention is intended to give parties involved in such transactions greater confidence and predictability, principally through the establishment of a uniform set of rules guiding the constitution, protection, prioritization and enforcement of certain rights in Aircraft and Engines. It alters the rules governing Aircraft Object sales, leases and financing on a jurisdiction-by-jurisdiction basis by establishing a new international framework and providing for the creation of an International Registry to be supervised by the International Civil Aviation Organization (ICAO). The intent of the Convention is to establish primacy as regards matters within its scope relating to the creation, enforcement, perfection and priority of interests in Aircraft Objects. As such, to the extent applicable, it supersedes the Geneva Convention.

At a very high level, the Cape Town Convention has two primary areas of coverage. The first relates to the establishment of an International Registry that serves as a central database to record, among others, mortgage and lease interests (that is, International Interests) in the airframes and engines constituting Aircraft Objects. This central database allows parties to fairly easily record on the public record, and perfect, their interests in aircraft assets and allows interested persons to review what encumbrances may be on these assets.

The second area of coverage is to provide a body of law that will govern the exercise of remedies in respect of Aircraft Objects by a creditor in the face of a default by a debtor, including the deregistration of an Aircraft by a creditor from an offending debtor's jurisdiction by an instrument commonly referred to as deregistration power of attorney (or, in Cape Town lingo, an IDERA). The Convention and the Protocol provide remedies upon default with respect to Aircraft Objects that may be exercised by lessors, conditional sellers and secured parties in respect of International Interests in their favor and by assignees of International Interests, all in their roles as creditors. The adoption of the Cape Town Convention in a particular jurisdiction, then, creates greater certainty that, upon default, creditors can swiftly, but in a commercially reasonable manner, exercise their remedies to repossess, de-register and extract, if applicable, and sell or otherwise realize upon Aircraft Objects. As to this second area of coverage, one of the more significant achievements of Cape Town (where the Contracting State has opted for Alternative A), is that Aircraft Financiers will have the benefit of Section 1110-type arrangements addressing matters of certainty in their ability to repossess Aircraft.[4]

An aircraft financier contemplating a financing in a particular jurisdiction, then, might first be interested to know whether that jurisdiction is a party to the Cape Town Convention as it assesses the legal risks associated with that jurisdiction. However, the inquiry does not stop with that. Even if that jurisdiction is a party to Cape Town, the financier will want to explore certain variations that may be applicable to the two areas of coverage specified above. As to matters relating to the International Registry 'perfection' regime (which, admittedly, is more of a mechanical matter), the financier will want to understand whether there are local procedures for the perfection of International Interests in aircraft assets that go beyond the simple matter of recording interests on the International Registry. In the United States, for example, in addition to making the appropriate registrations on the International Registry, filings must also be made with the FAA in Oklahoma City.

As to variations to the remedial regime outlined above, as part of the second area of coverage, a critical assessment is whether the particular jurisdiction has opted for 'Alternative A' or 'Alternative B', with Alternative A providing a framework that provides that the debtor must cure defaults within a specified time frame and continue to perform under the financing documents if it wants to retain possession of the financed Aircraft Object or return that Aircraft Object within such time frame. Alternative B, on the other hand, puts the matter of the return of Aircraft Objects in the hands of local courts (and, therefore, without any hard dates for establishing return of Aircraft Objects). See Appendix E-2 for a summary of Alternative A or B elections, and specified time frames, which summary was prepared by the Aviation Working Group. In addition, inquiry should be made of other local laws that may affect exercise of rights and remedies under Cape Town (as well as other local laws concerning liens that may rank superior under local laws to a registered Cape Town International Interest) and to 'rule of law' (i.e., corruption and political risk) therein.

The Aviation Working Group has compiled a Summary of National Implementation (go to AWG.aero) that summarizes the key implementation provisions of national authorities in their adoption of Cape Town.

Exhibit 7.1 may be a useful tool to ascertain whether Cape Town is available.

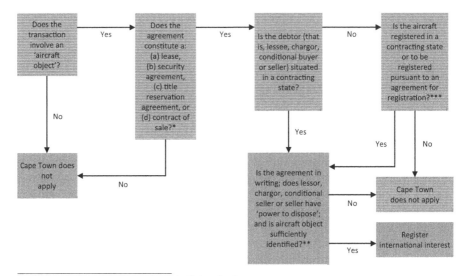

* This flow chart does not address assignments of International interests or associated rights.

** For security agreement, also must be able to determine secured obligations.

*** Applies only to airframe.

Exhibit 7.1 Chart to ascertain whether Cape Town is available

Source: Vedder Price PC

Chattel Paper Original

This is the tangible counterpart of a Finance Lease or Operating Lease designated as the 'original' for the purposes of perfecting therein by physical possession under the UCC. Under UCC § 9–102, 'chattel paper' includes a 'record ... that evidence(s) both a monetary obligation and a security interest in specific goods [such as a Finance Lease] [and] a lease of specific goods [such as an Operating Lease]'. While chattel paper can be perfected by filing (UCC § 9–312(a)), it may also be perfected by possession (UCC § 9–313(a)), and the person who perfects by possession may have a priority perfection status as against those who perfect by filing (UCC § 9–330).

Under U.S. law, the rules on perfection in an Operating or Finance Lease relating to Aircraft assets (for example, the collateral assignment thereof in a Back-leveraging transaction) is largely superseded by Cape Town and FAA regulations. Accordingly, the chattel paper rules governing perfection do not apply (See UCC § 9–201(c)). Nevertheless, it is customary in Aircraft Financings to designate a Chattel Paper Original of a Lease, and to have the secured party hold that in pledge.

Chicago Convention

This is the international treaty designated as the Convention on International Civil Aviation. This Convention was signed on 7 December 1944 in Chicago, Illinois. The Chicago Convention sets certain standards and practices of civil aviation among the member states as applies to civil Aircraft. The principal standard set by the Chicago Convention is the concept whereby it is recognized that every contracting nation has complete and exclusive sovereignty over the airspace above its territory and, therefore, each such nation, subject to the terms of the Chicago Convention, is entitled to regulate air services and cabotage within its borders. According to the Chicago Convention, Aircraft have the nationality of the 'State in which they are registered' and cannot be validly registered 'in more than one State, but its registration may be changed from one State to another' so long as it is in compliance with the law and regulations of the contracting state in which the Aircraft is registered. The Chicago Convention also requires every Aircraft of a contracting nation engaged in international navigation to carry on board, among other things, its certificate of registration and certificate of airworthiness.

Contract of Sale

As per Cape Town, this is a contract for the sale of an Aircraft Object by a seller to a buyer (but which is not one of the three agreements otherwise constituting an International Interest). Contracts of Sale are recorded on the International Registry, and, therefore, show the transition of ownership of Aircraft Objects from one owner to another.

Contracting State

As per Cape Town, this is a country that has ratified or acceded to the Cape Town Convention.[5] Appendix E-1 contains a list of Contracting States as at the time of writing, as well as a summary of the particular Declarations made.

PRACTICE NOTE

'Ratification' of Cape Town by a country does not necessarily mean that it has the force of law in such country. For example, the Treaty was purportedly in force in Brazil on 1 March 2012, but only with the issuance of the Presidential Decree (number 8008) published on 16 May 2013 did it become effective. Accordingly, close attention should be given by Aircraft Financiers to the actual state of the law in each particular jurisdiction.

Country Risk

This is the risk assessment of whether: (i) a particular country has decent laws for the benefit of Aircraft Financiers (such as whether that country has adopted Cape Town); and (ii) whether that country will enforce the laws on its books. The matter of enforcement spans considerations of corruption, court processes and bureaucratic intransigence.[6] Certain organizations rank countries by the levels of corruption therein (for example, Transparency International's 'Corruption by Country' guide at www.transparency.org/country) and economic freedom (which evaluates, among other things, freedom from corruption and rule-of-law; see the Heritage Foundation's Country rankings in its Index of Economic Freedom at www.heritage.org/index/ranking). The rankings may prove to be a useful guideline as to Country Risk assessments.

Department of Transportation (DOT)

The Department of Transportation (DOT) is a U.S. Executive Branch department.

European Aviation Safety Agency (EASA)

The European Aviation Safety Agency (EASA) is a regulatory body in the EU having governmental oversight authority of aviation comparable to that of the U.S.'s FAA. However, unlike the FAA, EASA has no enforcement authority, which authority is kept by the national aviation authorities in the EU. EASA is primarily tasked with ensuring the harmonization of standards in the EU states.

Employee Retirement Income Security Act (ERISA)

This is the Employee Retirement Income Security Act of 1974 (ERISA) and the regulations promulgated and rulings issued. ERISA is a U.S. federal statute that governs U.S. pension plans (among other matters) (29 USC § 18). Its relevance in Aircraft Finance is two-fold. First, in reviewing the credit of U.S. airlines, ERISA rules are pertinent in the assessment of when such an airline has an unfunded pension plan liability, which can be a significant liability. Second, ERISA rules govern the ability of pension plans to invest in securities. Since a violation of those rules may create a 'prohibited transaction', parties arranging Aircraft Finance financings are very careful to ensure that any U.S. pension plan monies that are utilized are invested in compliance with such rules.

PRACTICE NOTE

Nothing – not Increased Costs, tax indemnities, jurisdictional clauses – glazes the eyes of market participants in negotiation sessions as much as ERISA discussions. Fortunately, there are professionals who are expert in this particular field and who oversee (and, hopefully, take over) these discussions.

Eurocontrol/Airport charges

These are airport charges, landing fees and other navigation services charged/assessed by airport/navigation authorities to operators based on services rendered. These charges may be assessed against a particular Aircraft (whether or not the charges were incurred by the current operator of the Aircraft) and on any other Aircraft operated by the defaulting operator. These charges assessed against any particular Aircraft allow the authority to detain that Aircraft and sell it to pay for those unpaid amounts insofar as the charges usually have a Super Priority Lien status.

The broad scope of this provision means that a financier's Aircraft may be detained to satisfy charges incurred not only by its own lessee/borrower, but by a previous operator, or charges that do not relate to the financed aircraft at all, but have been incurred in respect of other aircraft in the operator's fleet.

This statutory right of detention exists under the UK's Civil Aviation Act 1982, Eurocontrol and the Government of Canada's NavCanada.

In the light of these rules, Aircraft Financiers developed the practice of requiring the airline being financed to authorize these authorities to provide periodic statements to the financiers of any outstanding charges. With an airline's express consent, these authorities will send a lessor/financier an electronic file of that airline's monthly statement of account. This keeps the lessor/financier up to date on the general situation and could serve as an early warning of potential problems.

Federal Aviation Administration (FAA)

The Federal Aviation Administration (FAA) is a U.S. government agency established under the Federal Aviation Act of 1958 (49 USC § 106). Under such Act, the FAA became responsible for all of the following.

- Regulating air commerce to promote its development and safety and to meet national defense requirements.
- Controlling the use of navigable airspace in the U.S. and regulating both civil and military operations in that airspace in the interest of safety and efficiency.
- Promoting and developing civil Aeronautics, which is the science of dealing with the operation of civil, or non-military, aircraft.
- Consolidating research and development with respect to air navigation facilities.
- Installing and operating air navigation facilities.
- Developing and operating a common system of air traffic control and navigation for civil and military aircraft.
- Developing and implementing programs and regulations to control aircraft noise, sonic booms, and other environmental effects of civil aviation.

A component agency of the DOT ever since the Department of Transportation Act was passed in 1967 (49 USC § 1651), the FAA engages in a variety of activities to fulfil its responsibilities. One vital activity is safety regulation. The FAA issues and enforces rules, regulations, and minimum standards relating to the manufacture, operation, and maintenance of Aircraft. In the interest of safety, the FAA also rates and certifies people working on Aircraft, and certifies airports that serve air carriers. The agency performs flight inspections of air navigation facilities in the U.S. and, as required, abroad.

The FAA also registers Aircraft and records documents related to the title or interest in Aircraft, Engines and Spare Parts.

Federal Aviation Administration (FAA) Bill of Sale

The FAA's AC Form 8050–2 aircraft bill of sale is a form of bill of sale required to be filed with the FAA when title to a U.S.-Registered Aircraft changes (see 'Appendix K').

Federal Aviation Regulations

Federal Aviation Regulations (FAR) are maintenance requirements for Aircraft Assets in the U.S. and are contained in the FAA's Federal Aviation Regulations. Four of these FARs concern the types of Aircraft and operations that may be pertinent to Aircraft Financing transactions.

1   FAR Part 91 – contains flight operation rules, noise requirements, equipment requirements and general regulations for all operators.
2   FAR Part 121 – applies to domestic and flag carriers operating transport category aircraft in commercial service.
3   FAR Part 129 – defines rules for U.S. Registered Aircraft with foreign operators, and foreign airlines flying into the U.S.
4   FAR Part 145 – contains requirements and procedures for certifying and operating Maintenance, Repair and Overhaul providers (MROs).

Full Warranty Bill of Sale

This is a bill of sale from the seller of an Aircraft Asset by which title is transferred from the seller to a designated buyer and the seller warrants and guarantees that it has good title to such asset, and will defend the buyer's interest as such.

Geneva Convention

The International Recognition of Rights in Aircraft is an international treaty signed in Geneva on 19 June 1948. Under this Convention each contracting state undertakes to recognize security interests and certain other property rights in aircraft which have been created under the laws of the state of registration of the Aircraft (provided that the state of registration is also a contracting state) and which rights are recorded in a public register in that country. The rights recognized are: (i) rights of property in Aircraft; (ii) rights to acquire Aircraft coupled with possession; (iii) rights to possession of Aircraft under leases of six months or more; and (iv) mortgages and other similar security interests in Aircraft. While any contracting state may recognize rights in aircraft other than those specified in the Geneva Convention, the rights specified in the Geneva Convention should have priority. The priority of the rights specified in the Geneva Convention is determined in accordance with the laws of the state of registration.

While the Geneva Convention successfully addresses the rules of priority over property rights in aircraft, it does not provide any valuable enforcement remedies in the case of a default by a debtor or in insolvency proceedings against a debtor, since these areas are exclusively dependent upon the applicable national law of the particular jurisdiction.

> **PRACTICE NOTE**
>
> The Geneva Convention has been largely superseded by Cape Town.

Granting Clause

This is the provision in any agreement (such as a Mortgage or Security Agreement) that purports to grant a Security Interest in the specified collateral. The Granting Clause is a very important component of a Mortgage or Security Agreement insofar as it serves as the basis for obtaining a Security Interest in such specified collateral.

Hell or Highwater

This is a term used in relation to Leases that describes the requirement of a lessee to pay the lease rentals come what may, including its inability to use or operate related Aircraft Assets or any right of setoff the lessee may have against the lessor. This is a critical feature for the financing of Operating Leases on a Back-leveraged basis insofar as it assures the financiers that the lessee will make its payment of rent (which is necessary to pay Debt Service) regardless of any circumstance.

In the case of a dispute with a lessor, the lessee is entitled to initiate a separate action against the lessor (in lieu of setting off against rental payments).

> **PRACTICE NOTE**
>
> While Hell or Highwater provisions in Operating Leases are rather standard and customary, some airlines try to qualify their unconditional rental payment obligation by having an exception which allows them to abate their rental payments if the lessor has interfered with the lessee's Quiet Enjoyment rights. While it is obvious why a lessee would want such a qualification for its Hell or Highwater obligations, it is a rather disconcerting provision for financiers, and addresses a risk that may have never materialized in over 50 years of Aircraft Finance.

Inchoate Lien

This is a Lien on an Aircraft Asset that has not yet attached. Taxing authorities, for example, may have Liens that are assertable against a debtor for unpaid taxes, but a Lien may not actually attach unless and until such taxes are past due.

International Interest

As per Cape Town, this is an interest created by or subject to security agreements, lease agreements and title reservation agreement relating to an Aircraft Object, both current and prospective. These interests may be recorded on the International Registry by reference to the manufacturer's name, the manufacturer's generic model and the manufacturer's serial number with respect to such Aircraft Object. These interests are subject only to certain declared local priorities arising by law (not contract), otherwise such interests are accorded priority based upon the order of registration. The Protocol extends certain provisions of the Convention to outright sales, enabling buyers to avail themselves of the registration facilities and priority provisions. Failure to register an International Interest would render such International Interest junior to competing registered interests (even if the earlier unregistered interest was known to the holder of the registered interest at the time of registration). Similarly, the purchaser of an Aircraft Object would take its interest in such equipment subject to all interests of record on the International Registry.

Outside certain insolvency scenarios, registration of an International Interest is not necessary to protect the creditor against its own debtor, so the fact that a chargee or lessor fails to register its International Interest should not in any way affect such party's rights against its chargor or lessee. An International Interest would be effective in insolvency proceedings against a debtor so long as it is registered with the International Registry prior to the commencement of such proceedings (this would be so even if the International Interest would otherwise be void for want of compliance with local law perfection requirements).

For an International Interest to be valid, it must relate to an Aircraft Object and be:

- granted by a chargor under a security agreement;
- vested in a person who is a conditional seller under a title reservation agreement; and
- vested in a person who is a lessor under a leasing agreement.[7]

International Registry

As per Cape Town, this is the central registry for the purposes of registering applicable interests in Aircraft Objects in accordance with the Cape Town Convention. The International Registry is an electronic web-based system that is located in Dublin, Ireland, and is operated by the Registrar, Aviareto Limited. The International Registry, established by the Supervisory Authority under the Cape

Town Convention, is the facility for effecting and searching registrations created pursuant to the Cape Town Convention. It provides a mechanism to determine the priority of registrations made against a specific aircraft object. The Protocol requires the centralized functions of the International Registry to be operated and administered seven days a week on a 24 hour basis.

The system is 'object specific' (that is, asset-based), meaning registrations are performed against an Aircraft Object. In turn, searches are made against Aircraft Objects and (for purposes of determining priority) cannot be made against debtors or other parties to a transaction. In an asset-based registry such as the International Registry, the most critical task is to properly identify the Aircraft Object in question. The registration system is intended to be wholly automated and operative 24 hours a day, seven days a week. It is intended that the International Registry may be searched at any time to determine the existence of interests affected by or related to Aircraft Objects.

Large Aircraft Sector Understanding (LASU)

Large Aircraft Sector Understanding (LASU) is the gentlemen's agreement in place among certain Aircraft-producing countries prior to ASU as related to export financing by their respective ECAs. This initial arrangement governing the extension of credit or guarantees in connection with the ECA financing of civil aircraft was concluded in 1986 between the U.S. and the European countries involved in the manufacture of Airbus aircraft (the UK, Germany, France and Spain, the 'Airbus Countries'). The LASU provided guidelines relating to the extension of credit or guarantees in connection with the financing of large civil aircraft. Among other things, LASU included a home-country rule, whereby U.S. Ex-Im would not support Boeing deliveries into the Airbus Countries, and the ECAs of the Airbus Countries would not support Airbus deliveries into the U.S. This was intended by OECD member countries to be a gentlemen's agreement that would create a uniform standard for ECA financing.

LASU has been superseded by ASU, although certain in or pre-production Aircraft have maintained 'grandfathered' status under LASU.

Lender Lien

This is a Lien on a leased Aircraft Asset caused by, or arising through, a mortgagee (or related lender) of such asset.

PRACTICE NOTE

A Lender Lien on an asset is not legally possible, since the mortgagee/lender does not own the asset. It is a silly concept that has, unfortunately, worked its way into lease and loan documentation.

Lessor Liability

Lessor Liability is liability risks arising to owners (and financiers) of an Aircraft Asset (most likely, an Aircraft) arising by virtue of their ownership (or mortgagee) interest therein. The crash of an Aircraft can have catastrophic results, not just for the poor souls on board the Aircraft, but also for folks on the ground.[8] As well, a crash may cause substantial property damage. Lessor Liability for these casualties and losses due to the Aircraft Asset in which a lessor has an interest is a frightening prospect for lessors and others with interests therein. These risks are accentuated by the fact that Lessors may be viewed as 'deep pockets' from which to obtain sizeable settlements. Theories of liability, such as Strict Liability, can keep lessors awake at night. There are three mitigants to this risk.

1   Insurance – the lessee/operator will be required to keep the Aircraft Asset properly insured for liability claims at levels satisfactory to the lessor (and many lessors carry Contingent Insurance coverage in respect of such risks).
2   Statutory – statutes, such as 49 USC §44112 under U.S. law (discussed in Strict Liability), should have a prophylactic effect.[9]
3   Structural – the placement of Aircraft Assets into Special Purpose Vehicles (SPVs) (assuming corporate formalities are observed) should protect the lessor from personal liability for third-party claims.

Lessor Lien

This is a Lien on a leased Aircraft Asset caused by, or arising through, the owner of the asset. While lessees are required to keep their leased assets free and clear of liens, they are not responsible for removing Lessor Liens. Lessor Liens come in two categories: permitted and non-permitted.

1   'Permitted' Lessor Liens would include any acceptable Back-leveraged Operating Lease financing and certain Inchoate Liens and would also include Liens caused by lessees (and otherwise permitted under the transaction documents).
2   All other Liens would not be permissible and would normally require immediate removal by the lessor.

Lien

This is a mortgage, pledge, charge, Security Interest or encumbrance of any kind on or in respect of any property.

Mechanic's Lien

A Mechanic's Lien is a statutory Lien that secures payment for labor or materials supplied in improving, repairing or maintaining an asset. Depending on the jurisdiction, Mechanic's Liens may have a Super-Priority Lien status on the Aircraft

Asset subject to maintenance. This is especially true if the MRO (mechanic) is in physical possession of the Aircraft Asset. Mechanic's Liens are usually Permitted Liens since they occur automatically.

Mortgage Convention

See 'Geneva Convention'.

Non-discrimination

As relating to any financed Aircraft Asset, this is the agreement by a lessee, servicer or borrower to treat such financed asset no differently than it treats its other (non-financed) comparable assets.

PRACTICE NOTE

Non-discrimination provisions are commonplace. They are most frequently seen in the context of maintenance obligations by the operator. While rather easy for an operator to abide by – the Aircraft coming in for maintenance are not distinguished by financing status – there may be some tricky elements to consider for Aircraft in the last months of a lease term, when the operator may want to manage the Aircraft's condition to its return condition obligations.

Novation

This is the substitution of one party with another in a contract, typically used in the context of the sale of an Aircraft Asset under lease where the new lessor is substituted for the old lessor under the related lease. This is largely an English law construct. In the U.S., one would normally expect to see an assignment and assumption agreement among the parties to achieve a similar result. However, in contrast to an assignment, all parties to the contract (including the lessee) must be a party to the Novation agreement.

Office of Foreign Assets Control (OFAC)

The Office of Foreign Assets Control (OFAC) is an office in the U.S. Department of Treasury. OFAC administers and enforces economic and trade sanctions based on U.S. foreign policy and national security goals against targeted foreign states, organizations, and individuals. In enforcing economic sanctions, OFAC acts to prevent 'prohibited transactions', which are described by OFAC as trade or financial transactions and other dealings in which U.S. persons may not engage unless authorized by OFAC or expressly exempted by statute. OFAC administers and enforces economic sanctions programs against countries, businesses or groups of individuals, using the blocking of assets and trade restrictions to accomplish foreign policy and national security goals. In the realm of Aircraft Finance, because of the international character in the trading of Aircraft Assets, attention needs to be paid to ensure that transactions are not effected in contravention of OFAC (and other comparable regulations issued by, among others, the relevant jurisdiction, UN and EU).

Patriot Act

The U.S. Patriot Act of 2001 is an Act of the U.S. Congress that was signed into law by President George W Bush on 26 October 2001. The Act, as a response to the terrorist attacks of September 11, among other things, placed greater burdens on financial institutions to 'know your customer' in connection with financial transactions in order to facilitate the prevention, detention and prosecution of international money laundering and the frequency of terrorism. The Act tightened the record keeping requirements for financial institutions, making them record the aggregate amounts of transactions processed from areas of the world where money laundering is a concern to the U.S. government. It also made institutions put into place reasonable steps to identify beneficial owners of bank accounts and those who are authorized to use or route funds through payable-through accounts.

Perfection

In relation to any particular asset, Perfection is the act of causing a Security Interest in such asset to be effective as against third parties so as to provide the person holding such Security Interest a priority right recognized by applicable law. Perfection usually requires the taking of some action, but can also be automatic. Part 2 summarizes the perfection methodologies for the Aircraft Assets and other collateral utilized in Aircraft Finance. While Perfection crystallizes a financier's rights against third parties, such financier's rights against other persons with Perfected interests will be determined by Priority.

Permitted Lien

A Permitted Lien is a Lien on a mortgaged or leased Aircraft Asset that is permitted; that is, the borrower or lessee is not required to remove it. Permitted Liens include:

- Mechanic's Liens that, typically, are not past due (or not past due for more than a specified number of days);
- Judgment Liens that, typically, are not past due (or not past due for more than a specified number of days);
- Inchoate Liens, such as for taxes;
- Liens permitted by the transaction documents (such as permitted Leases/subleases, pooling arrangements); and
- Permitted Back-leveraging Lease transactions.

Priority

This is the matter of the rights of a creditor relating to others to have its claim paid from collateral in which it has a Perfected Security Interest as compared with other persons who might have an interest in such collateral. Priority is often dictated by the relative timing of Perfection as among persons with interests in collateral. However, certain methods of Perfection (such as possession) may have Priority rights over other methods of Perfection. As well, certain types of Liens may have Super Priority Lien status by operation of law.

Priority Search

As per Cape Town, this is a search of the International Registry to identify existing registrations against a specific Aircraft Object. These searches would normally be conducted prior to closing to identify any registered loans that must be terminated and after a closing to confirm the release of any pre-existing registration and to evidence the registration of new registrations, thus establishing the intended priorities under the Cape Town Convention. The 'priority search certificate' provided by the International Registry is a reflection of the official records of the International Registry with regard to an Aircraft Object. The priority search certificate sets out the information relating to any registrations against a particular Aircraft Object, or it will confirm that no such registrations have been made with regard to such Aircraft Object. Any registrations with respect to an Aircraft Object will be listed in chronological order on the priority search certificate. Although the priority search certificate specifies the type of interest registered with respect to an Aircraft Object, it will not state whether such interest was registered as an International Interest or a Prospective International Interest. Exhibit 7.2 shows a sample Priority Search certificate.

Private Placement

See introduction to Part 3.

Prospective International Interest

As per Cape Town, this is an interest in an Aircraft Object that is intended to be created as an International Interest upon the occurrence of a stated future event (which may include the debtor's acquisition of an interest in the Aircraft Object or registration of the Airframe in a Contracting State). If the stated event occurs, then an interest initially registered as a Prospective International Interest will become an International Interest and it will be treated as registered from the time of registration of the Prospective International Interest, provided that such registration was still current immediately before the International Interest was constituted under Article VII of the Protocol. No additional registration is required when the International Interest comes into being (for example, when the documents are signed and the transaction is completed (assuming this to be the stated event)).

Priority Search Certificate

Issued by

The International Registry for International Interests
In Mobile Equipment (Aircraft Equipment)

This certificate was created on 01 Dec 2010 at 17:19:13 GMT

Certificate Number: 13153448

Requested by:	Kelly Wood of Crowe & Dunlevy, P.C.
Beneficiary of Priority Search Certificate:	Crowe & Dunlevy

Search Criteria

Manufacturer:	BOMBARDIER
Model designation:	CRJ-700
Manufacturer's serial number:	10317

Search Results

The International Registry system is designed to use percentages with a maximum of six decimal places when recording fractional and partial interests in aircraft objects e.g. 12.123456%. Please consider that certain fractions cannot be fully represented within six decimal places. Percentages shown are of the full aircraft object.

Date/time	Details of Interest	
	Registration	
01DEC2010	Type.	Contract of Sale
17:07:23 GMT	File number:	676534
	Fractional or Partial Interest:	100.000000%
	Seller:	
	Bombardier Inc.	Contact details: francis.lecomte@aero.bombardier.com
	Buyer:	
	AMERICAN EAGLE AIRLINES, INC.	Contact details: ROSE.GACILOS@AA.COM

This document has been digitally signed by the Registrar and the signature has been filed.

Exhibit 7.2 Priority Search Certificate sample

Source: Author's own

Protocol

See 'Cape Town'.

Professional User Entity (PUE)

As per Cape Town, a Professional User Entity (PUE) is a company or other grouping of persons providing professional services to a Transacting User Entity (TUE). A PUE is typically a law company or other company that assists TUEs in making registrations on the International Registry when authorized to do so. A prospective PUE must establish an account with the International Registry in order to act in such capacity, and must also appoint an administrator who will have sole authorization to submit and consent to registrations on behalf of the TUE it represents and to authorize other employees of the PUE to submit and consent to registrations on behalf of such TUE. Similar to a TUE, the PUE, once established, must act through individuals.

Qualified Institutional Buyer (QIB)

A Qualified Institutional Buyer (QIB) is an investor that may participate in a Rule 144A offering without the necessity for the related securities to be registered with the SEC (see 'Appendix M').

Quiet Enjoyment

This is the right of a lessee or mortgagor to use, enjoy and possess leased or mortgaged Aircraft Assets without interference by the owner/lessor or mortgagee. Lessees and mortgagors (rightly so) insist on Quiet Enjoyment rights. Quiet Enjoyment rights, of course, fall away in case of an event of default by the lessee or mortgagor under the related Lease or Mortgage. However, the default by a mortgagor in a Back-leveraged Lease will not allow a lender to disturb the related lessee's quiet enjoyment unless that lease is subject and subordinate to the mortgage financing. Article XVI of the Protocol provides for the comparable 'quiet possession' rights of an operator.

Quitclaim Bill of Sale

This is a bill of sale from the seller of an Aircraft (or other asset) by which title is transferred from the seller to a designated buyer and the seller warrants and guarantees that it is conveying only such title as it received originally from the person who originally sold that asset to that seller. Putting it differently, the seller is telling the buyer that it is selling whatever interests it may have in the Aircraft, and is not stepping up to warrant that it has good title to the Aircraft (as it would in a Full Warranty Bill of Sale).

Registration (Aircraft)

Registration[10] is the legal domiciling of an Aircraft in a particular jurisdiction. An Aircraft's registration is a unique alphanumeric string that identifies a civil Aircraft, in similar fashion to a license plate on a vehicle. In accordance with the Chicago Convention, all aircraft must be registered with a national aviation authority and they must carry proof of this registration in the form of a legal document called a Certificate of Registration at all times when in operation. Most countries also require the Aircraft's Registration to be imprinted on a permanent fireproof plate mounted on the fuselage for the purposes of post-fire/post-crash aircraft accident investigation. Because Aircraft typically display their registration numbers on the aft fuselage just forward of the tail, in earlier times, more often on the tail itself, the registration number is often referred to as the 'tail number'.

PRACTICE NOTE

In the U.S., an Aircraft's registration must be renewed every three years.

Re-registration (Aircraft)

This is the changing of an Aircraft's Registration from one jurisdiction to another.

PRACTICE NOTE

From an Aircraft Financier's perspective, the Re-Registration of an Aircraft from one jurisdiction to another is of great importance as an oversight matter. There are a number of critical considerations that need to be addressed on such an occasion (which considerations, of course, would be the same as a *de novo* transaction).

- Legal issues: is the jurisdiction a Cape Town Contracting State? If not, is there a body of local law that would allow enforcement of legal rights, like repossession? Is a governmental taking of the Aviation Asset protected by just compensation laws?
- Political Risk issues: assuming a favorable outcome to the Legal Issues analysis, are contractual rights enforceable in the new jurisdiction? What is the timing for court processes and repossession?
- Perfection and Priority issues: will the ownership and Security Interests in the related Aircraft Asset be recognized; is the Perfection and Priority at least as good as it was?
- Doing Business Issues: are there increased liability issues for the Finance Parties (for example, Lessor Liability)? Are there taxes that have to be paid? Is there a need to qualify to do business?

- Documentation and Filing issues: has everything been done on the public record (filings, registration and so on) to protect the interests of the Finance Parties? A de-registration power of attorney obtained?
- Insurance Issues: is new Insurance in place? Are the insurers internationally recognized – or is Re-insurance necessary?
- Tax issues: is there withholding on rent payments? Import, Registration or export taxes?
- Metal issues: is the Aircraft Asset properly certified as airworthy by the Aviation Authority? Are maintenance standards as good as FAA/EASA?
- Other issues: are foreign exchange permits necessary? Import/export permits necessary? Engine accession?

Rule 144A

See introduction to Part 3.

Securities and Exchange Commission (SEC)

The Securities and Exchange Commission (SEC) is the U.S. governmental authority charged with the supervision and regulation of securities and securities markets.

Securities Act

See introduction to Part 3.

Security Interest

This is an interest in an asset created by agreement or by operation of law to secure performance of an obligation (see UCC § 1–201(37)). A Security Interest attaches in an asset when it becomes enforceable against a debtor, and such Security Interest in that asset occurs only if:

- value has been given;
- the debtor has rights in the collateral or the power to transfer rights in the collateral to a secured party;
- one of the following conditions is satisfied:
  - the debtor has authenticated a security agreement that provides a description of the collateral;
  - the collateral is not a certificated security and is the possession of the secured party under Section 9 313 pursuant to the debtor's security agreement;
  - the collateral is a certificated security in registered form and the security certificate has been delivered to the secured party; and
  - the collateral is deposit accounts, electronic chattel paper investment property, or letter-of-credit rights, and the secured party has control.

UCC § 9–203(b). See 'Blue sky' for English law-related issues. A Security Interest is usually granted pursuant to a Granting Clause.

PRACTICE NOTE

Security Interests are, absent Perfection, only enforceable against the related debtor. They are enforceable against third parties only upon Perfection.

Strict Liability

This is a legal doctrine that, in the context of Lessor Liability, provides that the owner of an Aircraft Asset is liable for damages associated with its asset, regardless of fault. The mere fact of ownership brings on such liability.

Subject and Subordinate

This is the subordination of the right of the user of an Aircraft Asset to another financing (a lease or mortgage) which would allow the lessor or mortgagee under such financing to repossess from such user such asset due to, typically, the default by the head lessee or mortgagor under the terms of the financing. Subject and subordinate provisions may make (sub)leasing a surplus Aircraft Asset difficult as the new operator would be taking the credit risk of its lessor.

Subrogation

This is the legal doctrine whereby one person takes over the rights or remedies of another against a third party. Rights of Subrogation can arise two different ways: (i) either automatically as a matter of law; or (ii) by agreement as part of a contract. Subrogation by contract most commonly arises in contracts of insurance. Subrogation as a matter of law is an equitable doctrine, and forms part of a wider body of law known as unjust enrichment. The two most common areas where Subrogation is relevant are insurance and sureties. In each case, the basic premise is that where one person (that is, typically an insurer or a guarantor) makes a payment on an obligation which, in law, is the primary responsibility of another party, then the person making the payment is subrogated to the claims of the person to whom they made the payment with respect to any claims or remedies which are exercisable against the primarily responsible party (see 'Waiver of Subrogation' in Part 10 for Aircraft Finance context).

Super Priority Lien

This is a Lien that will take Priority over an Aircraft Financier's interest in an Aircraft Asset, notwithstanding that the Aircraft Financier's interest was placed on record first by the standard Perfection methodologies (outlined in Part 2). Super

Priority Liens may include Aircraft/Navigation Charges and Mechanic's Liens, as well as (depending on the jurisdiction) crew wages, taxes and other governmental charges.

True Sale

This is a sale of an asset that is not subject to re-characterization as a secured loan.

The various factors evidencing a true sale are not found in any single source but rather have developed over time through case law, scholarly writings and industry practice. Not all such factors need be present in a given transfer for it to be characterized as a true sale. The following is a list of the principal true sale factors.

- *Recourse.* The absence of recourse by the transferee to the transferor for non-payment of the transferred asset.
- *Intent.* The parties' intention to accomplish a True Sale, rather than a loan.
- *Identification of the Transferred Assets; Administration as a Sale.* The asset transferred in a true sale must be identified with specificity, and if a party other than the transferee is servicing the accounts receivable, the collections should be segregated in a special collection account rather than commingled with the servicer's other funds. Notification of the transfer to the account debtors will also favor true sale.
- *Payment of Fair Value.* Payment of less than fair value for an asset could be evidence that something other than a true sale was intended.
- *Irrevocability.* In a true sale, the risks and benefits of ownership must pass to the transferee upon closing, and those risks and benefits cannot then be reallocated. Lack of irrevocability may be evidenced, among other things, by an agreement that the transferee will receive a specified rate of return on its investment when in fact fluctuations in the dates on which accounts receivable are paid mean that a specified rate of return cannot be guaranteed. Lack of irrevocability may also be evidenced by an agreement to terminate a transaction at a given time and to re-convey any unpaid accounts to the seller in exchange for the outstanding balance of the accounts.

Re-characterization changes a sale of an asset into a loan secured by the ostensibly transferred asset. Thus, instead of the transferee simply continuing to collect the transferred accounts upon the transferor's bankruptcy, the transferee is placed in the position of a pre-petition lender to the now bankrupt transferor with a security interest (duly perfected one would hope) in the ostensibly transferred asset and their pre- and post-petition proceeds.

In bankruptcy (the venue in which a re-characterization challenge most likely would arise), such proceeds constitute cash collateral that the transferee can use to fund its cash needs, subject to court approval and Adequate Protection of the transferee's/lender's interest in the accounts.

PRACTICE NOTE

True Sale issues and concerns arise in the contexts of Securitizations and Warehouses of Aircraft Assets where the originator of the assets conveys them to the SPV. It is critical for the Securitization and Warehouse transaction that that conveyance constitutes a True Sale.

Transacting User Entity (TUE)

As per Cape Town, a Transacting User Entity (TUE) is one of the parties to an Aircraft Financing who is obtaining or granting an International Interest. In order to 'register' notice of an interest with the International Registry, each party to the interest must first establish an account with the International Registry as a transacting user entity. Registrations of interests in Aircraft Objects are then accomplished electronically (that is, data entry onto the website of the International Registry and consent given by way of digital signatures utilizing Public Key Infrastructure (PKI) technology and Digital Certificates). The primary entity which makes registrations on the International Registry is known as a TUE, which is a legal entity, natural person, or more than one of the foregoing acting jointly, intending to be a named party in one or more registrations.

Uniform Commercial Code (UCC) (generally)

The Uniform Commercial Code (UCC) is the body of law in all 50 states of the U.S. (and the District of Columbia) governing commercial transactions that is more or less consistently adopted (that is, uniform) across all such jurisdictions. In the realm of Aircraft Finance, Article 9 of the UCC is the portion that is of the greatest interest, insofar as it pertains to secured transactions; the granting of security interests and the perfection in and the enforcement on collateral.

UCC-1

This is the instrument filed with the Secretary of State in the jurisdiction in which the debtor is organized (and if not U.S.-organized, in the District of Columbia) to Perfect a Security Interest in the case of assets that are susceptible to Perfection by such a filing. UCC-1s expire on the fifth anniversary of their recordation date. Exhibit 7.3 is a sample UCC-1. The cost of filing a UCC-1 is a combination of the filing fee and the fee charged by the service company performing the filing. Such charges typically range from U.S.$80 to U.S.$150 per filing.

PRACTICE NOTE

While the primary assets usually dealt with Aircraft Financing, Aircraft Assets, are not usually susceptible to Perfection by a UCC-1 filing, there may be critical

---

UCC FINANCING STATEMENT
FOLLOW INSTRUCTIONS

A. NAME & PHONE OF CONTACT AT FILER (optional)

B. E-MAIL CONTACT AT FILER (optional)

C. SEND ACKNOWLEDGMENT TO: (Name and Address)

| Print | | Reset |

THE ABOVE SPACE IS FOR FILING OFFICE USE ONLY

1. DEBTOR'S NAME: Provide only one Debtor name (1a or 1b) (use exact, full name; do not omit, modify, or abbreviate any part of the Debtor's name); if any part of the Individual Debtor's name will not fit in line 1b, leave all of item 1 blank, check here ☐ and provide the Individual Debtor information in item 10 of the Financing Statement Addendum (Form UCC1Ad)

1a. ORGANIZATION'S NAME				
OR 1b. INDIVIDUAL'S SURNAME	FIRST PERSONAL NAME	ADDITIONAL NAME(S)/INITIAL(S)	SUFFIX	
1c. MAILING ADDRESS	CITY	STATE	POSTAL CODE	COUNTRY

2. DEBTOR'S NAME: Provide only one Debtor name (2a or 2b) (use exact, full name; do not omit, modify, or abbreviate any part of the Debtor's name); if any part of the Individual Debtor's name will not fit in line 2b, leave all of item 2 blank, check here ☐ and provide the Individual Debtor information in item 10 of the Financing Statement Addendum (Form UCC1Ad)

2a. ORGANIZATION'S NAME				
OR 2b. INDIVIDUAL'S SURNAME	FIRST PERSONAL NAME	ADDITIONAL NAME(S)/INITIAL(S)	SUFFIX	
2c. MAILING ADDRESS	CITY	STATE	POSTAL CODE	COUNTRY

3. SECURED PARTY'S NAME (or NAME of ASSIGNEE of ASSIGNOR SECURED PARTY): Provide only one Secured Party name (3a or 3b)

3a. ORGANIZATION'S NAME				
OR 3b. INDIVIDUAL'S SURNAME	FIRST PERSONAL NAME	ADDITIONAL NAME(S)/INITIAL(S)	SUFFIX	
3c. MAILING ADDRESS	CITY	STATE	POSTAL CODE	COUNTRY

4. COLLATERAL: This financing statement covers the following collateral:

5. Check only if applicable and check only one box: Collateral is ☐ held in a Trust (see UCC1Ad, item 17 and Instructions) ☐ being administered by a Decedent's Personal Representative

6a. Check only if applicable and check only one box:　　　　　　　　　　　　　6b. Check only if applicable and check only one box:
☐ Public-Finance Transaction ☐ Manufactured-Home Transaction ☐ A Debtor is a Transmitting Utility ☐ Agricultural Lien ☐ Non-UCC Filing

7. ALTERNATIVE DESIGNATION (if applicable): ☐ Lessee/Lessor ☐ Consignee/Consignor ☐ Seller/Buyer ☐ Bailee/Bailor ☐ Licensee/Licensor

8. OPTIONAL FILER REFERENCE DATA:

FILING OFFICE COPY — UCC FINANCING STATEMENT (Form UCC1) (Rev. 04/20/11)　　International Association of Commercial Administrators (IACA)

Exhibit 7.3 Sample UCC-1 Financing Statement

Source: Author's own

components of a financing that are, such as credit support documents and Warranties. Institutions benefitting from UCC-1 filings should have a tickler system that is sufficient to remind them to renew the UCC-1s prior to their expiry, which is five years from original filing date (or renewal).

Notes

1 The Cape Town Convention served to clarify certain contrary interpretations of the Mortgage Convention (by, for example, Denmark and the Netherlands), which concluded that Engines did accede to the Airframes on which they were installed.

2 Gerber, D, 'The 2011 aircraft sector understanding: calming the turbulent skies', *The Air & Space Lawyer 24(1)*, 2011.

3 This best practice has not been tested in the English courts so is not free from doubt.

4 Goode, Professor Sir R, *Official Commentary on the Convention on International Interests in Mobile Equipment and Protocol thereto on Matters Specific to Aircraft Equipment*, 3rd edition, 2013.

5 For updated information and status concerning country ratification, visit the International Institute for the Unification of Private Law website at www.unidroit.org/english/implement/i-2001-convention.pdf.

6 For example, on certain risks associated with India (and that may be applicable in other jurisdictions). Several banks and operating lessors have had bad experiences in India recently in the retrieval of Aircraft from a failed Indian carrier. Even though India is a Contracting State to the Cape Town Convention, local authorities would not allow banks and lessors to de-register their Aircraft or, in those cases where de-registration was permitted, the airport authorities demanded millions of dollars in unpaid dues from the banks and lessors as a condition to allowing the Aircraft to fly (even within India). Needless to say, India now appears more regularly on the list of Excluded Countries.

7 Article 7 of the Cape Town Convention details some further formal requirements for such an interest to be properly 'constituted', such as: (i) the agreement must be in writing; (ii) the grantor must have power to dispose of the subject property; (iii) the property must be properly identifiable; and (iv) there must be a related security agreement identifying the obligations that are secured.

8 As the horrific events of 9/11 showed.

9 Although, even with such statutory protection, a number of courts try to find reasons for such statutes not to apply.

10 This term, of course, has a verb-orientated meaning too: the act of recording an interest by filing or taking of some other action with a government authority. This alternative meaning is not what is address here (and needs no elaboration).

PART 8

BANKRUPTCY/WORKOUT/REMEDIES

Introduction

If a financier wants to engage in Aircraft Finance, he must be prepared for his airline debtor to file for bankruptcy at some time during the term of the related financing. Bankruptcy is an inevitable looming presence in Aircraft Finance insofar as airlines are so regularly entering its realm. The seeming inevitability of bankruptcy may be due to the cyclical nature of the airline business, coupled with its very high fixed asset costs, which makes pricing (air travel tickets) stability/discipline much harder to achieve.[1]

"Ladies and gentlemen, is there a bankruptcy attorney on board?"

Exhibit 8.1

Source: The Cartoon Bank, Condé Nast: TCB-31055

As an Aircraft Financier faces a bankrupt or defaulting debtor-airline, it will need to entertain its options (unless the airline has exercised: (i) Section 1110(a) (or equivalent) rights, in which case the financier is at the mercy of the airline's ability to continue to perform; or (ii) Section 1110(c) (or equivalent) rights, in which case it has no option except to repossess; see 'Section 1110'.

- Renegotiate the financing of its Aircraft Assets operated by the defaulting airline-debtor.
- Exercise foreclosure remedies on such Aircraft Assets.
- Repossess such Aircraft Assets.

Renegotiation

Renegotiation of an Aircraft Asset financing agreement – a lease or loan – invariably means the making of concessions by the financier to the debtor. Concessions are made by the financier to avoid having to take back the related financed Aircraft Asset. The unwillingness to repossess an Aircraft Asset may be due to any of the following reasons.

- The Aircraft Financier has assessed that the value it can realize on a returned asset (or there is an acceptable level of uncertainty as to the value it can realize on such asset), whether by sale or by lease, is less than it can achieve by keeping the asset with the current debtor.[2]
- The Aircraft Financier is simply not prepared or equipped to take back the asset (while there are certainly service providers that can do the actual work, the financier may not have the stomach (or authority) to undertake the responsibility).
- The Aircraft Financier determines that the debtor can be rehabilitated with the proper (and shared) concessions, which outweigh the costs and trouble of a repossession.
- The concessions are only temporary, and are to be paid back in the future.
- The Aircraft Financier has other agenda items with the debtor, and in making concessions, has other business opportunities with the debtor that it can cash in.

When a debtor is facing financial difficulties, a single Aircraft Financier with significant exposure to that debtor may find that it is the debtor, rather than the financier, who can call the shots. This may arise because a liquidation of the debtor would create too large a loss for the financier, or the return of so many of a given aircraft type would adversely affect the market for that aircraft type.

Foreclosure

In addition to contractual and statutory considerations (discussed elsewhere in Part 8), there are several additional considerations for when, or if, a creditor should pursue its foreclosure remedies.

- In the first instance, where both: (i) an equity cushion exists; and (ii) the interest rates are above market, the creditor rationally may elect to delay – or forbear from – exercising its foreclosure remedies. In these situations, the creditor can continue to collect its contractual rate of interest and wait until the equity exercises its redemption rights. Conversely, where the contractual rates of interest are low and there is an equity cushion, the creditor may have an incentive to exercise remedies as soon as practicable to provide incentive for the equity to exercise its redemption rights quickly.

- A creditor's collateral is often comprised of two types of assets: (i) the Aircraft Asset – often described as the 'metal' collateral; and (ii) the claims and other rights against the debtor arising under the Aircraft Finance documentation (which rights may be collaterally assigned in the case of a Back-leveraged Lease financing). Because the market values for both the metal and claims often fluctuate widely over time, the creditor's view regarding these values and the future changes in collateral value often may affect whether the creditor exercises foreclosure remedies expeditiously (to ensure control over the deficiency claims before any anticipated 'pop' in value) or in a more deliberate manner.

- In addition, if the creditor desires to go through with a foreclosure, however conducted, there are several measures that are recommended to be put into place prior to the time the foreclosure is completed so that the creditors are prepared in case they become the owner of the financed Aircraft Asset through a Credit Bid. The creditor should:
  - if deemed appropriate, hire a qualified servicer with an agreement on servicing responsibilities (including as to scope of duties and compensation; see Part 12, 'Management/Servicing Agreement') to take care of the asset (whether on-lease monitoring (if on-lease at the time of foreclosure) or physical possession (if off-lease));
  - establish a corporate structure for a new aircraft-owning entity ('New Owner') with agreed upon voting and management provisions; creditors will need to be satisfied that the structure for holding the financed Aircraft Asset best addresses any regulatory, internal governance and public relations issues;
  - form the New Owner so that it has tax attributes/characteristics acceptable to the creditors;
  - ascertain the optimal location for effecting the formal transfer of title to the Aircraft Asset so as to minimize tax and other costs of transfer; and
  - implement a capital (debt/equity) structure for the New Owner that is acceptable to the creditors.

- If the creditor receives possession of an off-lease Aircraft, then the creditor will need to be prepared to take physical possession of the Aircraft, and, therefore, have in place the storage, maintenance and insurance facilities necessary to preserve properly the Aircraft Asset and related records.

161

• If the creditor wants to sell, lease or otherwise dispose of an Aircraft Asset while it is on-lease (whether to sell it currently with the lease attached or to sell it on a forward basis when the Aircraft is scheduled to be returned), the creditor will need to make post-foreclosure remarketing arrangements; see Part 12, 'Remarketing Agreement'.

Bankruptcy experiences have been especially acute in the U.S. As a measure to protect Aircraft Financiers in the face of that risk, Section 1110 of the Bankruptcy Code (discussed elsewhere) contains special protections for those financiers. In the bankruptcy context, the question of whether a lease is a true lease or a financing may also have a significant bearing on outcomes (see Part 3, 'Finance/Capital Lease').

Repossession

Repossession of an Aircraft Asset may be consensual, in which case the debtor hands over the keys (so to speak). This handover may be orderly, where the Aircraft Asset is returned at an agreed location with all Aircraft Records, or it may be by way of abandonment, with a 'come and get it' disorderly repossession. In the case of an insolvent debtor, the local law will dictate the conditions under which repossession may be effected. The two most prominent regimes for repossession in such circumstances are Section 1110 of the U.S. Bankruptcy Code and Alternative A under Cape Town (each of which is described in some detail below). Insofar as these two regimes are so important, a comparison of the two might be beneficial. Exhibit 8.2 summarizes those differences.[3]

In addition, to the foregoing avenues for repossession, repossession may also be by surprise by:

• capturing an Aircraft Asset while it is on the ground in a favorable jurisdiction (like London or New York); likely at an airport on its regular routing;[4] or
• sending in a 'repo team' to grab the asset and ferry-fly it to a friendly jurisdiction.

The flow charts that follow display the decision-tree dynamics that creditors often face in default situations. Exhibit 8.3 illustrates the general preparations for the exercise of remedies, Exhibit 8.4 shows the decision-tree for obtaining possession or control of an Aircraft and Exhibit 8.5 shows the decision-tree for foreclosure and title transfer options.

Adequate Protection

This is the protection permitted by a bankruptcy court, as an exception to the Automatic Stay, to a creditor of a bankrupt entity to address the deterioration of collateral or its diminution in value.

Exhibit 8.2 Comparison of remedies on insolvency/bankruptcy

Cape Town Article XI (ALT A) and Section 1110		
Issue category	*Cape Town Alternative A*	*Section 1110*
Applies to whom?	• Any debtor.	• Only to certificated air carriers (U.S. airlines) but not other debtors.
Applies to what equipment?	• Limited to Aircraft Objects (airframes and engines), including installed equipment. • Includes 'all data, manuals and records relating to' the Aircraft Object.	• Covers airframes, engines, propellers, appliances, spare parts. • Includes only those records which are required by the underlying documents to be returned by debtor in connection with the surrender or return of the equipment.
Applies to what financing?	• Applies to any 'international interest', that is a lease agreement, security agreement, and title reservation agreement (see 'International Interests'). • An international interest be registered or otherwise effective under applicable law.	• Only applies to a lease, security agreement or conditional sale contract (for equipment first placed in service prior to 22 Oct, 1994, a security agreement only qualifies if it constitutes a purchase money equipment security interest). • Section 1110 does not require that the applicable interest be perfected (1110 rights can be available, for example, in respect of a security agreement even if the security interest created thereby is not perfected).
Applies where?	1 *This Article applies only where a Contracting State that is the primary insolvency jurisdiction has made a declaration pursuant to Article XXX(3).* • 'Primary insolvency jurisdiction' is where the center of a debtor's main interest is situated, with a rebuttable presumption that it is the place of incorporation.	• The U.S.
Applies when?	2 *Upon the occurrence of an insolvency-related event . . .* • Insolvency related event is defined as: (i) the commencement of insolvency proceedings; or (ii) the declared intention to suspend or the actual suspension of payments by a debtor where a creditor may not commence proceedings or exercise Convention remedies by law or State action. • Allows for trigger and commencement of waiting period when creditor might otherwise be prohibited from exercising remedies.	• Only applicable in a Chapter 11 reorganization proceeding – would not apply in a Chapter 7 liquidation. Triggers off commencement of bankruptcy proceedings.

Exhibit 8.2 continued

Cape Town Article XI (ALT A) and Section 1110		
Issue category	*Cape Town Alternative A*	*Section 1110*
Waiting period:	3 *Upon the occurrence of an insolvency-related event, the insolvency administrator or the debtor, as applicable, shall, subject to paragraph 7, give possession of the aircraft object to the creditor no later than the earlier of: (a) the end of the waiting period; and . . .* • Turn-over of the collateral is mandatory.	• After 'waiting period', right of a lessor, secured party or conditional vendor to take possession of (and obligation of debtor to surrender and return) such equipment in compliance with a security agreement, lease or conditional sale contract and such parties' rights to enforce its remedies is not limited or otherwise affected by any other provision of the Code or any power of the court (including the automatic stay). • Only gives relief from the automatic stay – the agreement and applicable law must provide applicable remedy.
	4 *Upon the occurrence of an insolvency-related event, the insolvency administrator or the debtor, as applicable, shall, subject to paragraph 7, give possession of the aircraft object to the creditor no later than the earlier of: (b) the date on which the creditor would be entitled to possession of the aircraft object if this Article did not apply.* • Does not limit shorter period under applicable law.	• Practically, there would be no ability to achieve possession of the equipment prior to expiration of the waiting period, even if the term of the underlying instrument expired.
	5 *For the purposes of this Article, the 'waiting period' shall be the period specified in a declaration of the Contracting State which is the primary insolvency jurisdiction.* • Emerging standard for waiting period: 60 days. • A declaration could say that the waiting period is X days or such longer period as agreed. No Contracting State has done so (to date).	• Waiting period is 60 days from the commencement of the bankruptcy case (subject to extension by agreement of the parties).
Adequate protection:	6 *Unless and until the creditor is given the opportunity to take possession under paragraph 2: (a) the insolvency administrator or the debtor, as applicable, shall preserve the aircraft object and maintain it and its value in accordance with the agreement; and (b) the creditor shall be entitled to apply for any other forms of interim relief available under the applicable law.*	• Financier can seek adequate protection for the use of the aircraft collateral and leased equipment. • Requires motion and court order to implement.

Exhibit 8.2 continued

Cape Town Article XI (ALT A) and Section 1110

Issue category	Cape Town Alternative A	Section 1110
	• Automatic obligation to preserve aircraft object and maintain it and its value in accordance with agreement.	• No comparable provision in Code except to the extent ordered by the court after motion of the financier.
	7 *Sub-paragraph (a) of the preceding paragraph shall not preclude the use of the aircraft object under arrangements designed to preserve the aircraft object and maintain it and its value.*	
Ability of debtor to avoid repossession:	8 *The insolvency administrator or the debtor, as applicable, may retain possession of the aircraft object where, by the time specified in paragraph 2, it has cured all defaults other than a default constituted by the opening of insolvency proceedings and has agreed to perform all future obligations under the agreement . . .*	• The trustee must get court approval and agree to perform all obligations of the debtor under such agreement and cure, before the expiration of the 60-day period, any default, other than a default of a kind specified in section 365(b)(2), under such security agreement, lease, or conditional sale contract that occurred prior to the petition date.
	• Covers any default other than default constituted by opening of insolvency proceeding.	• Substantially the same.
	9 *A second waiting period shall not apply in respect of a default in the performance of such future obligations.*	• No second waiting period applies in the event of a post-waiting period default.
Supplemental remedies:	10 *With regard to the remedies in Article IX(1): (a) they shall be made available by the registry authority and the administrative authorities in a Contracting State, as applicable, no later than five working days after the date on which the creditor notifies such authorities that it is entitled to procure those remedies in accordance with the Convention; and . . .*	• No comparable provision. Section 1110 merely provides relief from the automatic stay. The FAA requirements for de-registration are not modified or altered.
	• Article IX(1) remedies are de-registration and export. When creditor is 'entitled to procure' the remedies depends on other declarations, particularly under Article XIII (IDERA).	
	11 *With regard to the remedies in Article IX(1): (b) the applicable authorities shall expeditiously co-operate with and assist the creditor in the exercise of such remedies in conformity with the applicable aviation safety laws and regulations.*	• No comparable provision.

Exhibit 8.2 continued

Cape Town Article XI (ALT A) and Section 1110		
Issue category	*Cape Town Alternative A*	*Section 1110*
	12 *No exercise of remedies permitted by the Convention or this Protocol may be prevented or delayed after the date specified in paragraph 2.*	• No comparable provision. Remedies would still need to comply with the other requirements of applicable law (for example, Uniform Commercial Code).
	13 *No obligations of the debtor under the agreement may be modified without the consent of the creditor.*	• Consent of financier to modifications is required.
	14 *Nothing in the preceding paragraph shall be construed to affect the authority, if any, of the insolvency administrator under the applicable law to terminate the agreement.*	• No comparable provision, but likely the same result.
Priority:	15 *No rights or interests, except for non-consensual rights or interests of a category covered by a declaration pursuant to Article 39(1), shall have priority in insolvency proceedings over registered interests.* • Only those rights or interests covered by declaration under 39(1) have priority in insolvency proceedings. If no declaration is made, then such non-consensual rights or interests would not have priority, notwithstanding contrary local law.	• Priority is determined by underlying law (Section 1110 only provides relief from the automatic stay but any exercise of rights would be subject to prior claims allowed by applicable law). • Example: if under state law a mechanics lien or air navigation lien has priority over a previously perfected security interest, then the exercise of 1110 rights are subject to such liens.
Remaining uncertainties:	• Section 1110 has been tested in the courts and has been the subject of considerable litigation over the years, which has given practitioners greater certainty as to its implementation. The Cape Town Convention has not been tested by the courts. • Political risk issues (that is, will courts enforce/corruption/delays and so on) in certain countries.	• Interplay with 365(d)(5) – For leases, 365(d)(5) requires all obligations be performed after 60th day (subject to the 'equities of the case'). If Section 1110 is applicable, would the equities of the case require obligations to be performed if financier had right to seek return of the equipment? • Does Section 1110 require the payment of default interest as a cure payment?

Source: Dean Gerber

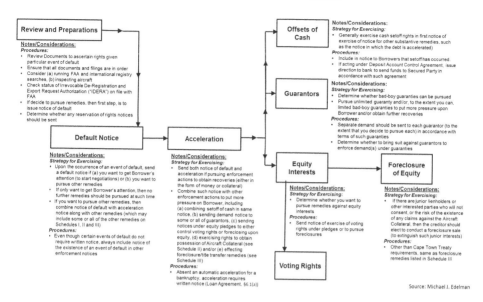

Exhibit 8.3 General preparations and exercise of remedies

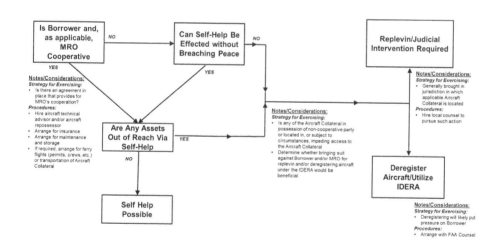

Exhibit 8.4 Actions to obtain possession or control of aircraft collateral

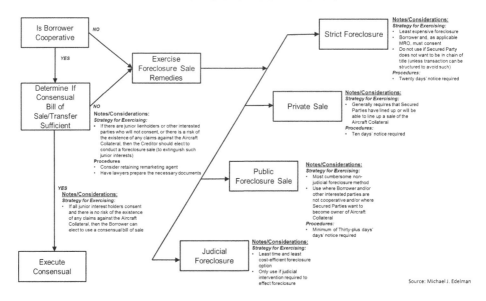

Exhibit 8.5 Foreclosure and title transfer options

Alternative A (Cape Town)

This is an option as to which a Contracting State may elect upon acceding to Cape Town, to be found under Article 23 of the Protocol (Remedies on Insolvency), which provides Section 1110-like rights and remedies for Aircraft Financiers in respect of Aircraft Objects where the debtor is the subject of an insolvency-related event. From an Aircraft Financier's perspective, Alternative A is preferred over Alternative B because it requires the debtor, no later than the earlier of: (i) the end of the waiting period (the period comparable to the Section 1110 (60 Day) Period) specified by the Contracting State that is the primary insolvency jurisdiction and that has adopted Alternative A; or (ii) the date on which the creditor would otherwise be entitled to possession if the Convention and Aircraft Protocol did not apply, either: (i) to give possession of the Aircraft Object to the creditor; or (ii) cure all defaults (other than a default constituted by the opening of insolvency proceedings) and agree to perform all future obligations under the related financing agreement. Unless and until the creditor is given the opportunity to take possession of the Aircraft Object, the insolvency administrator or debtor must preserve the Aircraft Object and maintain it and its value in accordance with the related financing agreement and the creditor is entitled to apply for any other forms of interim relief available under applicable law. In addition, the remedies of de-registration and export of the related Aircraft Object are required to be made available on an expedited basis by the Aircraft registry authority and administrative authorities of a Contracting State which opts into Alternative A. Alternative A adds a special provision that only those non-consensual rights or interests covered by a declaration under Article 39(1) of the Cape Town Convention have a Super Priority Lien status over registered interests in insolvency proceedings. To date, most Contracting States have opted for Alternative A.

Alternative B (Cape Town)

This is an option that a Contracting State may elect upon acceding to Cape Town, found under Article 23 of the Protocol (Remedies on Insolvency). This option is much less useful to creditors than Alternative A. Alternative B provides that there is a time specified in the declaration after which the insolvent debtor, upon request of the creditor, must give notice that it will either: (i) cure all defaults other than a default constituted by the insolvency proceedings and agree to perform all future obligations under the financing agreements; or (ii) give the creditor the opportunity to take possession of the Aircraft Object in accordance with applicable law. If the insolvent debtor does not give such notice, or if the debtor notifies the creditor that it will give the creditor the opportunity to take possession of the Aircraft Object but fails to do so, it is then within the discretion of the local court in the relevant insolvency jurisdiction to decide whether or not to permit the creditor to take possession of the related Aircraft Object and, if so, to decide upon the terms and conditions to be applicable to such taking of possession. To date, Mexico is the only Contracting State to opt for this Alternative B.

PRACTICE NOTE

As one industry observer has put it, any Contracting State that elects for Alternative B has rendered the entire default/remedy benefit of Cape Town moot in that jurisdiction.

(Automatic) Stay

This is the automatic injunction that arises when a debtor files for Chapter 11 pursuant to Section 362 of the U.S. Bankruptcy Code. This injunction prohibits creditors from taking action against the debtor. Prohibited actions would include demanding repayment of debt, foreclosing on a Lien and demanding repossession of leased or mortgaged Aircraft Assets. There are exceptions to this injunction, including Section 1110 and Adequate Protection.

Cannibalizing

This is the act of removing from a financier's financed Aircraft certain Parts that are then installed on an operator's other Aircraft, either *without* replacing that Part or replacing that Part with an inferior Part. This is not, unfortunately, an uncommon occurrence with operators who are in a fiscal pinch, and can leave the financier with an incomplete asset that cannot be repossessed other than *in situ*. Removal without replacement may also leave other Parts exposed to the elements and further diminish the value of the asset.[5]

> **PRACTICE NOTE**
>
> The obvious way to tackle this risk from the financier's perspective is to monitor any of its assets that are Aircraft on Ground (AOG) as an operator enters into distress.

Certificate of Repossession

This is a certificate that allows a financier or other person who had repossessed a U.S. registered Aircraft to register that Aircraft in its (or a designee's) name under U.S. Aircraft registration regulations. U.S. federal regulations specifically address a financier's right to register an Aircraft following its repossession in connection with the exercise of remedies (as it would be unlikely for a financier to have received a bill of sale or other evidence of ownership sufficient to abide by the normal registration provisions). In such a case, registration may be accomplished by the compliance by the repossessing party of the following conditions:

- such party must submit a certificate of repossession on FAA AC Form 8050 4 signed by the applicant stating that the subject Aircraft was repossessed or seized in accordance with the applicable security agreement and local law (see Appendix G);
- such party must provide the FAA with a certificate copy of the applicable security agreement (unless it is already on record with the FAA); and
- when repossession results from foreclosure proceeding in which the Aircraft is sold, such party must provide a bill of sale signed by the sheriff, auctioneer or other authorized person conducting the sale, which bill of sale states that the sale was made in accordance with the applicable local law.

If the applicant is the purchaser of an Aircraft at a judicial sale or foreclosure proceeding or in connection with any sale of an Aircraft to satisfy a lien or other charge, the FAA will require the purchaser to submit the items specified in the third bullet above. If title is adjudicated in a court of law, an applicant may submit a certificate copy of the court's decision and the FAA will waive the evidence of ownership requirements.

Chapter 11

This is the portion of the U.S. Bankruptcy Code that allows for the rehabilitation of debtors through their reorganization.[6] This is in contrast to a liquidation of the debtor under Chapter 7 of the Bankruptcy Code (which would be the likely result if there is no chance that the business can be salvaged).

Claw Back

In the context of a U.S. bankruptcy proceeding of a borrower/buyer under a Pre-delivery Payment (PDP), Claw Back is the situation in which the borrower/buyer (the airline or lessor) wishes to retrieve its PDP. In U.S. bankruptcy cases, security deposits are considered to be a type of cash collateral paid by the buyer under a Purchase Agreement (who is then considered and referred to as the 'purchaser-debtor'), and can be returned to it subject to a ruling by the court. As such, an airline or lessor, as the purchaser-debtor, has the legal right to request that the Original Equipment Manufacturer (OEM) re-pay any cash collateral received by the OEM. However, the airline/lessor requesting bankruptcy court approval also has the burden of proving that each entity with an interest in the collateral (that is, the lender and the manufacturer) is adequately protected. Although there are no reported bankruptcy cases that address the use of cash collateral comprised of PDPs under an Aircraft Purchase Agreement, the analysis regarding whether a bankruptcy court would allow PDPs to be used by a purchaser-debtor is the same as for other types of security deposits. The first prerequisite is for the bankruptcy court to view the PDPs as a security deposit. If or when that characterization is made, details of the 'equity cushion' will be decided. Here, a bankruptcy court would examine whether the interests of the manufacturer and the lender – the two parties having an interest in the PDPs – are being protected.

The legal basis for a Claw Back in a PDP faces numerous hurdles. Manufacturers will use precise and considered language in their purchase agreement to minimize the chances that the PDPs are characterized as security deposits; rather, the manufacturer uses express contractual terms stating that the PDPs, once paid, are the property of the manufacturer as compensation for the cost of the construction of the aircraft. Under the term of most Aircraft Purchase Agreements, the PDPs are described as absolute and unconditional payments – as such only the manufacturer has an interest in the PDPs, not the purchaser. Based upon the express terms of the Purchase Agreement, a purchaser-debtor may have difficulty either characterizing the PDPs as a security deposit or other interest in which it has any reversionary interest. If the manufacturer's interpretation of the agreement is upheld, the Claw Back of the PDPs should not occur.[7]

PRACTICE NOTE

OEMs typically refuse to take Claw Back risk, shifting that risk onto the PDP financier. In light of the legal analysis as to the unlikely application of Claw Back to PDP transactions, PDP financiers are willing to take this risk. As well, Claw Back risk would only be applicable where the purchaser-debtor is in a jurisdiction where the bankruptcy laws apply rules that entitle debtors to a return of security deposits. In addition, PDP financiers have employed special purpose vehicles in some of their financings to further mitigate the risk of a Claw Back.

Consolidation

This is the merger of the bankruptcy cases pending against two or more debtors, resulting in the multiple bankruptcy estates being consolidated, sharing assets and liabilities. In the context of Aircraft Finance, Consolidation concerns arise in the structuring of Back-leveraged Operating Lease financings (whether a one-off, a Warehouse or a Securitization). In these financings, it is a goal of the persons structuring the transaction to keep the Special Purpose Vehicle (SPV) owning the Aircraft Asset separate from any related (that is, owning, sponsoring or originating) operating lessor such that the bankruptcy of such operating lessor would not allow the SPV to be Consolidated in that operating lessor's bankruptcy (see Part 3, 'Bankruptcy Remote' and 'Special Purpose Vehicle').

Credit Bid

In a Public Foreclosure Sale, this is the use by the creditors of the debt and other amounts owing to them as the purchase price for collateral subject to the Foreclosure Sale. In other words, the creditors may use the debt owing to them as currency with which to bid for the Aircraft Asset subject to such a Foreclosure.

De-registration

De-registration is the removal of an Aircraft from the registry of a particular country. In the exercise by an Aircraft Financier of remedies, repossession of an Aircraft may not be enough; the Aircraft may need to be eligible for export from its then-registered jurisdiction. In other words, in order to have the freedom to re-deploy an Aircraft that has been repossessed, the Aircraft Financier may need to have such Aircraft registered in a different jurisdiction. Insofar as Aircraft registration in any country is largely a matter controlled by the Aviation Authority, procedures must be followed in order to allow for the de-registration and export of the Aircraft. A major hurdle to De-registration might be the need for the airline/debtor to co-operate with such procedure. In the case of a deadbeat or otherwise recalcitrant airline, De-registration may be difficult or impossible. To avoid the need for debtor co-operation, certain instruments may be signed at the time of entering into an Aircraft Financing that preauthorize the Aircraft Financier to De-register the subject Aircraft. These instruments are known as De-registration Powers of Attorney. The De-registration Powers of Attorney used in Contracting States are IDERAs. In the U.S., the practice is not to use a De-registration Power of Attorney, but rather to use a Certificate of Repossession.

PRACTICE NOTE

It is incumbent on an Aircraft Financier to ascertain whether a particular jurisdiction will recognize a De-registration Power of Attorney so as to assess its rights on a repossession.

De-registration Power of Attorney

This is a power of attorney granted by an airline (or, in an owner registry jurisdiction, lessor) to an Aircraft Financier that will enable the Aircraft Financier to De-register a financed Aircraft. De-registration Powers of Attorney take the form of an IDERA in Contracting States.

Debtor in Possession Financing (DIP)

This is new debt financing obtained by a debtor subject to a Chapter 11 (Debtor in Possession (DIP)). The DIP Financing has a priority position over existing debt, equity and other claims. Airlines subject to Chapter 11 often seek DIP Financing for financing during the course of its bankruptcy (including to build a 'war chest') and there is usually no shortage of financiers who may be willing to supply such financing in the light of the preferential treatment such financiers receive in the event of a liquidation. Aircraft Financiers who may be 'underwater' in their pre-petition financings with these airlines often offer DIP Financing to the affected airline and seek to bootstrap these pre-petition financings into the DIP to improve their position.

Deficiency Claim

This is the amount still owed by a bankrupt debtor to a creditor when the collateral securing the debt owed to such creditor is sold at a foreclosure sale (or Marked-to-Market).

PRACTICE NOTE

In U.S. airline bankruptcies especially, Deficiency Claims take on a life of their own. Insofar as a reconstituted airline emerging from U.S. Chapter 11 will be owned (predominantly) by its unsecured creditors (which includes holders of Deficiency Claims), Deficiency Claims may have great value since they will be converted into the equity of the emerging airline. Deficiency Claims may trade widely in value at any point in time or as the bankrupt airline wends its way through the Chapter 11 process.[1] Hedge Funds and investment banks take a keen interest in assembling and trading these assets. A creditor pursues the claim for its Deficiency Claim by filing a 'proof of claim' with the applicable bankruptcy court by the date (the 'bar date') established in the related bankruptcy.

[1]   In the American Airline bankruptcy for example (which commenced November 2011), Deficiency Claims traded, initially, at 20 cents on the dollar and rose to over 90 cents on the dollar.

Financial Accommodation

This is the granting by a creditor to a debtor of concessions or incrementally more credit to assist the debtor in meeting its obligations, enhancing its financial position and/or avoiding bankruptcy. Under the U.S. Bankruptcy Code, Financial Accommodations are not subject to assumption by the debtor upon a bankruptcy filing by that debtor (see Bankruptcy Code § 365(c)(2)(12)). Accordingly, if a lessor grants Financial Accommodations to a lessee as part of a pre-bankruptcy work-out, these accommodations are not assumable (that is, allowed to be used) by the lessee if the lessee were nevertheless to file for bankruptcy.

Foreclosure

In the U.S., there are three types of non-judicial foreclosures pursuant to which creditors may seek to eliminate an equity's interests in collateral that serves as security for debt:[8] (i) a strict foreclosure under UCC § 9–620; (ii) a commercially reasonable private foreclosure sale under UCC § 9–610; and (iii) a commercially reasonable public foreclosure sale under UCC § 9–610. Importantly, creditors must not only keep in mind the foregoing legal options for effecting foreclosure, they must be sure to comport with any contractual arrangements they are party to which might control the exercise, and timing for the exercise of remedies, as well as Equity Squeeze provisions.

Strict Foreclosure (UCC § 9–620)

In most secured transactions, the secured party may be able to obtain title to its collateral in full or partial satisfaction of its debt – which procedure is generally called a Strict Foreclosure. Strict Foreclosures are generally simple, quick and inexpensive (relative to other types of foreclosure). Unless there is an express written agreement from the debtor, no Deficiency Claim (vis-à-vis the borrower) survives in a Strict Foreclosure. The statutory notice period to effect a Strict Foreclosure is 20 days unless either: (i) additional time is required under the terms of the security agreement; or (ii) the equity agrees (after receipt of the Strict Foreclosure notice) to a shorter notice period (that is, the equity consents to the Strict Foreclosure).[9] Once such notice is given, if the equity (or other party in interest) does not object to the Strict Foreclosure prior to the end of the applicable notice period, then the creditors may effect the Strict Foreclosure. If the equity (or any other party with an interest in the collateral) objects to a Strict Foreclosure proposal from a secured lender within the applicable notice period, the secured lender cannot utilize a Strict Foreclosure but, instead, must dispose of the collateral pursuant to a commercially reasonable private or public foreclosure sale under UCC § 9–610. Importantly, a debtor cannot pre-agree (prior to a default) to the utilization of a Strict Foreclosure. Only after the debtor has defaulted in its obligations (such that the creditors are entitled to exercise remedies) may the creditor issue its notice to effect a Strict Foreclosure to which the debtor may or may not, at such time, object.

While Strict Foreclosure might appear to be 'binary' in nature (that is, the obligor (here, the equity) can agree to the Strict Foreclosure by not objecting or can disagree to the Strict Foreclosure by objecting within the 20-day statutory period), there are actually a number of variations on how Strict Foreclosures can be effected. The options range from: (i) sending out a Strict Foreclosure notice, followed by a confirmatory simple foreclosure bill of sale after the notice period elapses; to (ii) detailed, multi-party agreements pursuant to which the equity agrees to turn over its ownership interest in the collateral (including any Deficiency Claims) to the secured party upon terms and consideration detailed in such agreements. Sometimes the existing equity agrees to enter into such arrangements without additional consideration (in effect, honoring its secured loan arrangements). In other situations, the secured party may agree to provide some relatively minor consideration to the equity so that the equity agrees to co-operate (that is, the secured creditor buys the equity's co-operation so that it does not object to the strict foreclosure). Examples of such purchased co-operation range from: (i) an agreement to pay some minor amount of legal costs; to (ii) detailed profit sharing arrangements. These arrangements are known as 'Consensual Strict Foreclosures'. A secured creditor may decide to provide additional consideration to the equity to obtain a Consensual Strict Foreclosure to: (i) avoid the (not insubstantial) costs of a commercially reasonable public foreclosure sale; (ii) secure the equity's co-operation in 'handing over the keys', which, in the context of Aircraft Finance, might be valuable given the complexity of Aircraft Asset turnovers; and (iii) provide incentives for the equity to execute transfer documents, which will clear title issues and, accordingly, facilitate future transfers. In any Strict Foreclosure, the creditors can either elect to become the direct owner of the collateral or select a nominee (usually an affiliated entity under common control) to become the owner.

Where no objection is made by the equity (or any other parties with an interest in the collateral), effecting the transfer of ownership utilizing the creditors' Strict Foreclosure rights is usually the optimal (and, certainly, most cost-effective) means for transferring title to Aircraft Assets and other collateral to the secured party and/or its nominee. There are several factors, however, that may weigh against completing a Consensual Strict Foreclosure and, instead, may warrant the use of either a private foreclosure sale or a public foreclosure sale under UCC § 9–610. These include:

- the equity or a junior lienholder refusing to consent to the Strict Foreclosure arrangements;
- both: (i) the creditors finding a third-party, arms-length purchaser to whom the creditors want to sell the Aircraft Assets and/or other collateral; *and* (ii) the creditors determining that they want to have the collateral sold directly to such purchaser and do not want to be an intermediary owner in the chain of title; in such situations, a private foreclosure sale may be the optimal means for utilizing a secured creditors' foreclosure powers to transfer Aircraft Assets and/or other collateral directly from the original equity to the purchaser; and
- any other tax, regulatory, contractual impediment, internal governance and public relations issues leading the creditors to avoid becoming the direct or indirect owner of Aircraft Assets and/or other collateral.

If any of these factors exist, then creditors may elect to exercise their foreclosure remedies under UCC § 9–610 to effect a commercially reasonable private or public foreclosure sale of the collateral.

Commercially Reasonable Foreclosure Sale (UCC § 9–610)

The statutory alternatives to Strict Foreclosure are either a Private Foreclosure Sale or a Public Foreclosure Sale, both of which are governed by UCC § 9–610. In either case, every aspect of the disposition must be effected in a commercially reasonable manner.[10]

Private Foreclosure Sales

A commercially reasonable private disposition of an Aircraft Asset can potentially be utilized where the creditors have identified an arms-length, third-party purchaser. UCC § 9–610 specifically allows Private Foreclosure Sales to an arms-length, third-party purchaser regardless of whether there is an established market for the collateral being sold. Under UCC § 9–610, Private Foreclosure Sales cannot be utilized to transfer ownership of the collateral from the equity to the creditors by means of a Credit Bid: Private Foreclosure Sales can only be used to transfer collateral from the mortgagor to the secured creditor if there is an established market with widely distributed, standard price quotations. This type of market likely does not exist for the 'metal' Aircraft Asset collateral.[11] Because of this restriction, Private Foreclosure Sales are not utilized for Aircraft Assets where the secured parties are seeking to retain any part of the collateral or any residual interests in such collateral.

While a Private Foreclosure Sale may not be suitable for the 'metal' collateral in an Aircraft Financing, Private Foreclosure Sales may be utilized to transfer ownership of Deficiency Claims to the creditor if these claims satisfy the market requirements referred to above. In most bankruptcies of U.S. airlines, a robust market for Deficiency Claims typically develops, thereby creating situations where Private Foreclosure Sales may be utilized.

For both private and public foreclosure sales under UCC § 9–610, the notice of foreclosure sale must satisfy each of the following requirements for notices to the equity and other parties with a potential interest in collateral (that is, holders of junior liens):[12] (i) satisfy the notice to the equity and other interested parties as mandated under the NY UCC (which is 10 calendar days' notice); (ii) satisfy the Cape Town Convention's requirement that prior written notice to interested parties of a proposed sale of 10 or more 'working days' since Aircraft Assets (and Deficiency Claims) collateral are collateral and interests of the type governed by the Cape Town Convention; and (iii) satisfy any longer notice period to the equity and other interested parties as provided under the contractual financing documents. The notice of the private foreclosure sale must specify the date after which such sale may occur (subject to the foregoing required notice periods) and the assets included in the private sale. Similar to strict foreclosures, the affected equity and junior lien holders can agree to waive or shorten these notice periods (which waiver must be effected after the occurrence of the default giving rise to the foreclosure sale).

Public Foreclosure Sales

Notice requirements to the debtor and other parties in interest for effecting a commercially reasonable public sale are dependent upon the facts and circumstances involved in the sale. In addition to the requirements for notices to parties with an interest in collateral as detailed in the private sale summary above, which requirements apply to Public Foreclosure Sales, the notice for a Public Foreclosure Sale must include the date, time and place of the public sale and the assets included in such sale. Furthermore, in a Public Foreclosure Sale, the creditors will need to provide notice to the public in a manner that is commercially reasonable, and otherwise conduct the sale in a commercially reasonable manner. (Such notice to potential bidders is not needed in a Private Foreclosure Sale as the purchaser has already been identified.) In other words, a commercially reasonable amount of notice must be provided to give sufficient time for prospective bidders to participate at a public sale of sophisticated assets the likes of which are Aircraft Assets. The notice to the public of such sale would similarly specify, among other matters: (i) the date, time and place of the auction; (ii) a description of the Aircraft Asset and other collateral (including the related Lease and the claims under the Lease) offered for sale; (iii) whether the Aircraft Asset and other collateral may be purchased as a whole and/or in partial lots; (iv) certain rules on bidding and sales terms; (v) that other sales terms are available upon request; (vi) whether all of the collateral under the related Mortgage is being sold at the foreclosure sale or whether certain collateral is not being sold (and will remain subject to the creditors' security interests); and (vii) bidder qualifications (minimum cash on hand, deposit requirements and so on). The notice to the public of the auction, including subject matter and pertinent details, will need to be disseminated in a manner that reaches out to potential interested bidders in a commercially reasonable manner. Accordingly, publication of the notice would be necessary in trade publications, as well as in general circulation press.

The timing of a public foreclosure auction to be specified in the various notices will be dictated by the amount of time the creditors determine is commercially reasonable for the type of assets subject to sale. This determination will take into account the complexity of the collateral and number of Aircraft Assets to be sold, which complexity will likely expand the time horizon to between 30 and 45 days from the date publication of the sale is first presented to the public.

As part of the public presentation (and prior to 'launch'), the creditors will need to have available to potential bidders, to the extent reasonably available, detailed lists of the Aircraft Assets and associated Leases, if any, and pertinent summaries of maintenance status of the Aircraft Assets, so that potential bidders will have sufficient information on which to base their bids. Generally, a foreclosing lender will require that prospective bidders execute commercially reasonable confidentiality agreements prior to disseminating proprietary business information to such parties. After executing a confidentiality agreement, this detailed information may be produced in one or more of the following formats: (i) summaries, to share with first round interested parties (to the extent that these summaries do not contain confidential information, this may be disseminated without the need for a confidentiality agreement); and (ii) detailed compendium, to share with those parties that express a further interest after reviewing the summaries. As part of the detailed materials, the creditors may present appraised values of the Aircraft collateral obtained from one or more reputable appraisers. The foregoing information may be included with, and be part of, such marketing materials as the creditors think appropriate so as to best conduct a commercially reasonable auction sale. In recent foreclosure sales, such information has been made available through electronic due diligence rooms/websites.

In addition to the proper advertising of the auction to achieve commercial reasonableness, the creditors are well advised to reach out to those persons who might be possible bidders to ensure that those persons have been properly apprised of the upcoming sale. In the light of the complexity of the collateral, the number of Aircraft available and the special nature of the equipment, it is often worthwhile for the creditors to hire a preforeclosure marketing agent to assist in the above, especially if the creditors come to the conclusion that they do not have the necessary expertise, staff and market presence to conduct the aforementioned tasks in a commercially reasonable manner. A marketing agent, then, could: (i) identify the appropriate periodicals for publication of the auction information, and handle such publication; (ii) identify and contact potential bidders; (iii) co-ordinate with counsel to address confidentiality agreements and other issues; and (iv) prepare marketing materials. To be sure, the hiring of a well-respected marketing agent could help deflect any concerns that the mortgagor of the property subject to foreclosure would claim that the creditors were not equipped to, and did not, conduct a commercially reasonable auction sale.

Also, in advance of the public foreclosure auction, ground rules for bidding at the auction would need to be established, and bidders would need to be made aware of those rules.[13]

At the foreclosure sale, the creditors may reject non-conforming bids. Further, the creditors may Credit Bid; if the creditors are the highest and/or best offer, then the creditors will become the owners of the auctioned Aircraft Assets and other collateral. Of course, at any private or public foreclosure sale under UCC § 9–610, if the highest and best bid is submitted by a third party purchaser, the foreclosure sale can be made directly to such winning bidder (and ownership would not need to pass through the creditors).

Fraudulent Conveyance

A Fraudulent Conveyance is a transfer made with actual intent to hinder, delay or defraud any creditor. There are two types of fraudulent transfer – *actual fraud* and *constructive fraud*. *Actual fraud* typically involves a debtor who as part of an asset protection scheme donates his assets, usually to an 'insider', and leaves himself nothing to pay his creditors. *Constructive fraud* does not relate to fraudulent intent, but rather to the underlying economics of the transaction, if less than a reasonably equivalent value was realized at a time when the debtor was in a distressed financial condition.

PRACTICE NOTE

In a U.S. Bankruptcy, Fraudulent Conveyances are subject to unwind, with the fraudulently conveyed asset restored to the debtor. Parties to a Securitization will want to ensure that no Fraudulent Conveyance occurred in the

---

contribution of assets to the SPV issuer; this risk is largely addressed through the receipt of appraisals of the financed Aircraft Assets and comparing the appraisal results with the consideration received by the originator. Similar concepts may be applicable in other jurisdictions.

Non-consolidation

This is the prevention of Consolidation. This would be a typical goal for a Bankruptcy Remote SPV as relates to: (i) in the case of a Securitization, the originator; and (ii) in the case of an Aircraft Asset financing, the related lessor. In many Aircraft Financings, issuer (or sponsor) counsel is asked to provide a legal opinion (a 'non-consolidation opinion') to the effect that the risk of Consolidation of the issuer and the originator/lessor is remote.

Preferential Transfer

These are payments by a debtor to a creditor made within a short time before the filing of the debtor's bankruptcy. Specifically, Section 547 of the U.S. Bankruptcy Code sets out six elements of a preferential transfer under U.S. federal bankruptcy law. These elements are:

1   the existence of a transfer;
2   the transfer is for the benefit of the creditor;
3   the transfer must be on account of a prior existing debt;
4   the debtor must have been insolvent at the time of the transfer;
5   the transfer must be 90 days prior to the filing of the debtor's bankruptcy (or one year if the creditor is an 'insider' as defined by bankruptcy law); and
6   the transfer must have enabled the creditor to receive more than the creditor would have otherwise received if the case was a Chapter 7 bankruptcy (liquidation), the transfer had not been made, and the creditor received payment of its debt to the extent provided under Chapter 7.

Preferential Transfers are recoverable by a bankrupt debtor for deposit in its bankruptcy estate. A defense against a preference claim is that the transfer was an 'ordinary course' transfer. There are three components to this defense that the creditor must prove in order to successfully assert it. One, the debt that the transfer satisfied must have been one that typically existed between the creditor and the debtor. Two, the transfer must have been paid in a time and manner that is consistent with other prior payments made by the debtor to the creditor. Three, the transfer must be consistent with other transfers that are standard in the creditor's or debtor's industry.

Other defenses include the 'new value' defense, which rewards creditors that continued to do business with the debtor during the final weeks prior to the debtor's bankruptcy filing; the 'contemporaneous exchange' defense; and the defense for transfers in situations in which: (i) the creditor is granted a Purchase Money Security Interest 90 days before the bankruptcy filing to enable the debtor to receive inventory; and (ii) the creditor perfects the security interest on or before 20 days after the debtor receives the property.

Public Foreclosure

See 'Foreclosure'.

Purchase Money Security Interest (PMSI)

Purchase Money Security Interest (PMSI) is a Security Interest in favor of a lender that is created in an asset at the time the buyer of that asset uses the lender's loan proceeds to make the purchase (see UCC§ 9–107).

This term has particular relevance in the context of Section 1110 for mortgaged Aircraft and Engines first placed in service prior to 22 October 1994, where, for such equipment, the mortgage financing must be a 'purchase-money equipment security interest' in order to qualify for Section 1110 protection, Section 1110(d)(2). Insofar as 'purchase-money equipment security interest' is not defined, practitioners come to understand that term by reference to PMSI interpretations.

Repossession

This is the act of physically gaining possession of a financed Aircraft Asset. There are significant costs associated with Repossession:

- legal;
- insurance;
- temporary parking;
- storage;
- ferry flights;
- maintenance and refurbishment;
- records retrieval, assembly, review and retention; and
- remarketing.

PRACTICE NOTE

Not only is actual physical repossession necessary to realize on collateral, but the ability to remove the Aircraft Asset from the jurisdiction as well, by means of, among other things, de-registering the Aircraft.

Reservation of Rights

This is a communication to a debtor by a creditor that the creditor is not waiving any of its rights, notwithstanding its current inaction (or accommodation) in the face of a default. In the face of a default by a debtor, a creditor may not be taking immediate action to enforce its rights (for example, accelerating the debt, repossessing the collateral, foreclosing on security) for any of the following reasons.

- The creditor and debtor are trying to work out a restructuring, and that may take time.
- The creditor does not have internal approval to act.
- The creditor has not come to an agreement either internally (credit approval) or with co-creditors on the appropriate action to take.
- The creditor is a 'deer in headlights' and unable to make a move.
- The creditor is taking a wait-and-see attitude to see if favorable developments occur (such as a white knight rescue, a capital infusion and so on).
- The creditor is not too fussed by the default and is willing to let the transaction continue.
- The creditor needs time to assemble the resources to take remedial action (see 'Repossession').

The Reservation of Rights addresses the concern of the creditor that it may be perceived by the debtor (and the law) to be waiving its rights if it does not act on its rights in a timely manner; the Reservation of Rights serves as a notice that the creditor is not waiving (but reserving) its rights vis-á-vis the debtor in the light of specified defaults.

Section 1110

Section 1110 of the Bankruptcy Code is a statutory provision that provides that the right of Aircraft Financiers (lessors, conditional vendors and holders of security interests) with respect to any commercial Aircraft, Engines and Rotable Parts (collectively, 'Section 1110 Equipment') operated by U.S. commercial airlines to take possession of such Section 1110 Equipment in compliance with provisions of the relevant lease, conditional sale contract or security agreement that relates to the financing of such Section 1110 Equipment is not affected by:

- the Automatic Stay provision of the Bankruptcy Code;
- the provision of the Bankruptcy Code allowing the trustee in reorganization or the debtor-in-possession to use, sell or lease property of the debtor;
- the confirmation of a plan by the bankruptcy court; or
- any power of the bankruptcy court to enjoin a repossession.

Section 1110 provides, however, that the right of the Aircraft Financier to take possession of Section 1110 Equipment in the event of a bankruptcy default may not be exercised for 60 days following the date of commencement of the reorganization proceedings (unless specifically permitted by the bankruptcy court) and may not be exercised at all if, within such 60-day period, the airline (acting through trustee in reorganization or the debtor-in-possession) agrees to perform (and, in fact, continues to perform) its obligations that become due on or after such date and cures all existing defaults (other than defaults resulting solely from the financial condition, bankruptcy, insolvency or reorganization of the debtor).[14] Accordingly, Aircraft Financiers find Section 1110 of the Bankruptcy Code to be beneficial because if their airline lessee/borrower goes bankrupt, rather than having to wait until the bankruptcy plan for the airline is worked out (which can take many months, if not years) so as to realize on their Section 1110 Equipment collateral, the financiers are able to have such collateral returned to them and dispose of it following the 60th day of the bankruptcy unless the airline agrees to perform on the Lease or the mortgage *and* becomes (and stays) current on its payment and other obligations.

Section 1110 provides for three options with respect to Section 1110 Equipment. These options are spelt out in Section 1110(a), Section 1110(b) and Section 1110(c), as set out below.

Section 1110(a)

Section 1110(a) of the Bankruptcy Code is exercised at the option of the debtor (with Bankruptcy Court approval) to elect to keep possession of the Section 1110 Equipment, but in doing so, it must cure all past (curable) defaults and agree to continue performing on its mortgage or lease financing in respect of such equipment.

PRACTICE NOTE

The debtor would do this if the related Section 1110 Equipment is 'core fleet', is at advantageous financing costs and/or is owned/financed by a creditor with whom it is cutting other deals. A Section 1110(a) election is not irrevocable for a debtor; its election is *not* an assumption of the related financing. The debtor is permitted at a later date to change its mind either to: (i) seek to negotiate a Section 1110(b) agreement with the Aircraft Financier; or (ii) stop performing, in which case Section 1110(c) may be applicable.[1]

[1] This might occur if the debtor's fleet plans changes over the life of the bankruptcy.

Section 1110(b)

Section 1110(b) of the Bankruptcy Code is an option that must be mutually agreed to by both the debtor (with Bankruptcy Court approval) and the Aircraft Financier whereby the debtor agrees to keep possession of the Section 1110 Equipment on revised financing terms.

PRACTICE NOTE

The debtor would do this if the Section 1110 Equipment is non-core, but useful, equipment for its fleet, is at disadvantageous financing costs and/or is

---

such that the debtor learns that the financiers do not want to risk taking the asset back. The Aircraft Financier would agree to this if allowing the debtor to keep the Section 1110 Equipment is the best use of the equipment, there are dim prospects for the sale or re-lease of the equipment and/or the Aircraft Financier is not prepared to take the asset back. For example, if a bankrupt airline has 100 Boeing 757 aircraft in its fleet, is paying above-market rent for these older aircraft and has a post-emergence fleet plan calling for only 50 Boeing 757s, then the airline can play hardball with the Aircraft Financiers mortgaging and/or leasing this equipment. In fact, the airline can conduct a 'reverse auction' to determine which Leased Aircraft to keep (taking into account, of course, where these Aircraft are in their maintenance cycle (see Part 4, generally).

Section 1110(c)

Section 1110(c) of the Bankruptcy Code is exercised (in a back-handed sort of way) at the option of the debtor whereby, if the debtor does not elect to keep possession under Section 1110(a) and does not cut a deal under Section 1110(b), the Aircraft Financier can demand immediate 'surrender and return' of the related Section 1110 Equipment. This phrase provides the Aircraft Financier with some level of protection against the debtor's mere abandonment of the Section 1110 Equipment (see 'Alternative A (Cape Town)').

PRACTICE NOTE

The debtor would do this if it has no need for the Section 1110 Equipment or cannot reach a 1110(b) deal on the Section 1110 Equipment with the related Aircraft Financier.

Exhibit 8.6 may be a useful tool to ascertain whether Section 1110 is available.

Section 1110 (60 day) Period

In the case of any bankrupt U.S. certificated airline, this is the 60-day period commencing on the date of its bankruptcy filing during which the airline must either: (i) elect (with bankruptcy court approval) to agree to perform (and cure existing defaults) on its lease or mortgage for Section 1110 Equipment (under Section 1110(a)); or (ii) come to an agreement with its lessor or mortgagor pursuant to Section 1110(b), or run the risk that its lessor or mortgagee will demand return of such equipment.

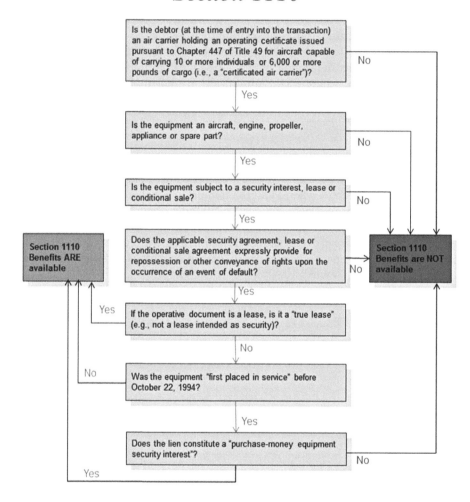

Section 1110

Is the debtor (at the time of entry into the transaction) an air carrier holding an operating certificate issued pursuant to Chapter 447 of Title 49 for aircraft capable of carrying 10 or more individuals or 6,000 or more pounds of cargo (i.e., a "certificated air carrier")? — No

Yes ↓

Is the equipment an aircraft, engine, propeller, appliance or spare part? — No

Yes ↓

Is the equipment subject to a security interest, lease or conditional sale? — No

Yes ↓

Does the applicable security agreement, lease or conditional sale agreement expressly provide for repossession or other conveyance of rights upon the occurrence of an event of default? — No

Section 1110 Benefits ARE available

Section 1110 Benefits are NOT available

Yes ↓

If the operative document is a lease, is it a "true lease" (e.g., not a lease intended as security)? — Yes

No ↓

Was the equipment "first placed in service" before October 22, 1994? — No

Yes ↓

Does the lien constitute a "purchase-money equipment security interest"? — No / Yes

© 2010 Vedder Price P.C.

222 North LaSalle Street
Chicago, Illinois 60601
312-609-7500
Fax: 312-609-5005

1633 Broadway, 47th Floor
New York, New York 10019
212-407-7700
Fax: 212-407-7799

875 15th Street NW, Suite 725
Washington, D.C. 20005
202-312-3320
Fax: 202-312-3322

Exhibit 8.6 Section 1110 Decision Tree
Source: Vedder Price PC

> **PRACTICE NOTE**
>
> Under Cape Town, *Alternative A* of the Remedies on Insolvency (Article 23 of the Protocol) does not specify the length of the waiting period for the debtor to allow the creditor to repossess an Aircraft Object – which is a number of days period that Contracting States must specify when they accede to Cape Town (if they have chosen *Alternative A*).

Standstill

This is a period of time during which a creditor agrees not to exercise remedies against a defaulting debtor. Standstills typically buy time for a debtor (or a junior creditor) to work out its problems (for example, in a Back-leveraged Operating Lease, the lessor may be trying to cut a new deal with the lessee, and, therefore, may ask the back-leveraging lender to hold off exercising remedies for a period of time). A creditor may agree to do this for any number of reasons:

- it is a 'nice' creditor;[15]
- it thinks the problem will be solved within the designated time frame;
- it may receive monetary or other (for example, undertaking to remarket) incentive to agree; or
- it may have other relationships with the debtor (or junior creditor).

Strict Foreclosure

See 'Foreclosure'.

UCC (Remedies)

The remedies of a secured party prescribed under UCC §9–601 *et seq* (see 'Appendix C-2.B'). Mortgages governed by the law of New York (or other U.S. states) typically provide that the secured party has all UCC (Remedies) available to it.

Notes

1 The marginal cost for an extra passenger is so very low.
2 Taking into account all of the costs associated with repossession and remarketing.
3 Thanks to my colleague Dean Gerber for this chart.
4 In New York, for example, a financier can obtain a court order on an *ex parte* basis (that is, without advising the debtor) to seize an Aircraft at the Gate with the assistance of the sheriff. The financier would have to put up a sizeable bond to do this since the action is done on a surprise (*ex parte*) basis, and the court needs to protect the debtor in case the financier was wrong in its action.
5 In one instance, an operator (without replacement) the cockpit windshield on the parked Aircraft of a financier, wreaking havoc on the avionics in the cockpit.
6 USC §101 *et seq*.
7 Gee, CA, 'Aircraft pre-delivery payment financing transactions', *The Journal of Structured Finance*, Fall 2009, pp. 11–21.

8 There is also another method for effecting a foreclosure sale – a judicial foreclosure. Judicial fore-closures for personal property in New York are brought under NY Lien Law §206 and NY UCC §9–601(a)(2). Given the higher costs and longer time periods required, most secured lenders prefer to pursue non-judicial remedies unless either: (i) judicial assistance is needed to effectuate the foreclosure sale; or (ii) it is known that the borrower will seek to stop, and has a reasonable basis to stop, the non-judicial foreclosure through judicial intervention.

9 NY UCC §9–620(d).

10 The main differences between a public and private foreclosure sale is that the purchaser of the collateral has already been identified in a private foreclosure sale transaction. Accordingly, in a private foreclosure sale, after the prospective purchaser is found, all aspects of the foreclosure sale relating to finding a prospective purchaser can be bypassed – that is, there is no need to publish notices or provide additional notice periods reasonably necessary for finding and identifying prospective bidders for the collateral in a public foreclosure auction. Of course, commercially reasonable steps must have been utilized to find such private foreclosure sale purchaser.

11 NY UCC §9–610(c)(2).

12 The Equity and any holders of junior liens known to you, as parties in interest, should be included in any notice of remedies or notice of sale of the aircraft collateral.

13 For example, 'cash only' bids for the entire 'lot' of collateral (Aircraft *plus* Deficiency Claims) may be stipulated.

14 In the case of a secured loan financing (versus a lease), if the Aircraft Asset was first placed into service prior to 22 October 1994, the secured loan financing must be a 'purchase-money equipment' transaction in order to qualify for Section 1110 protection (see 'Purchase Money Security Interest').

15 Yes, these do exist.

PART 9

INTERCREDITOR

Introduction

In structured financings generally, transactions will often have multiple layers of debt Tranches and may include an equity Tranche. While these different investor classes often have plenty to worry about vis-à-vis the debtor/obligor of the transaction, consideration must also be given to the relationships among the creditor classes themselves. Structured transactions in Aircraft Finance, accordingly, must similarly deal with these intercreditor issues.

In the light of the large number of Aircraft Financing transactions structured as Leases, and the propensity for operating lessors to seek to Back-leverage their interests, intercreditor matters are an inherent aspect of Aircraft Finance. What is more, many of the structures, such as Enhanced Equipment Trust Certificates (EETCs) and Securitizations, employ multiple Tranches of debt, thereby offering more opportunities for inter-creditor issues to arise. Part 9 delves into the terminology employed in this area.

Before examining that terminology, a few general observations need to be made on this topic. First, there is no 'right' or 'wrong' result in the matter of intercreditor discussions and outcomes. Each intercreditor issue is a zero-sum game. Rights are sliced and diced in many different ways, and results will vary widely from deal to deal, even deals that are similarly structured.

Second, as opposed to EETCs and the old U.S. Leveraged Lease market where there is or was a semblance of a market 'standard' with little deviation from deal to deal, there is not much of a standard in a typical structured Aircraft Finance transaction.[1]

Third, because of this high level of uncertainty of outcomes, it is recommended that intercreditor terms be negotiated among the principals in a transaction at the Term Sheet stage. The terms set out below should provide guidance on what topics should be addressed by such a term sheet.

Fourth, since there is no right or wrong way for intercreditor matters to be divided up, it is a wonder that any intercreditor agreements are reached. The success in reaching final allocations has much to do with the sophisticated level of industry participants negotiating these matters.

Fifth, there are situations in which one intercreditor party has greater leverage over the other in the matter of intercreditor negotiations. This would typically occur if the financing offered by one party would be significantly more difficult to replace than the other. In these instances, the debtor can always threaten to boot-out the more fungible party from a financing, thereby making that party more amenable to the other party's position (see 'Breakage Cap').

Sixth, if the parties reach loggerheads – they cannot come to agree on terms – the debtor (who is everyone's client) is often brought in to bang heads and, not infrequently, impose a settlement on the warring parties.

Finally, intercreditor issues can arise in many different contexts. In addition to the obvious senior lender vs. junior lender situation and the Back-leveraged lender vs. lessor situation, intercreditor issues will need to be addressed in, for example, a financing where a Swap Provider is utilized. In that instance, the Swap Provider – who is washing the fixed interest/Rent received from the borrower into a floating rate for the benefit of bank participants – will need its credit exposure secured by the collateral and will insist on certain voting rights.

Intercreditor issues can be broken down into four broad categories.

Let us review a few of these categories in a little detail, identifying the questions that arise:

1   Voting Rights/Control Rights – Who gets to vote on changes to deal terms? Who gets to exercise remedies in a default situation? What percentage of any class's vote is required for action?

2   Buyout Rights – Does one class of creditor have the right to buy out another class in a default situation? If so, at what price?

3   Cure Rights – Does one class of creditor have the right to make a payment to cure a default – which cure would then forestall the ability to exercise remedies by the most senior/controlling class? If so, how often – consecutively? Cumulatively?

4   Waterfalls – Who gets paid from incoming cash flow and collateral liquidation in what order of priority?

The charts that follow categorize these, and other, of the intercreditor issues commonly faced. The charts address these issues in three scenarios. The first, Exhibit 9.1, pertains to the borrower/lessor vs. lender in the context of a Back-leveraged Operating Lease. The second, Exhibit 9.2, pertains to the senior lender vs. subordinated lender scenario. The third, Exhibit 9.3, sets forth the intercreditor landscape in an EETC as between the holders of the senior PTCs vs. the junior PTCs.

Asset Subordination

As between a senior financier and a junior financier, this is the recovery by the senior financier of its claims against a debtor from pledged Aircraft Asset collateral in priority to recoveries by the junior financier. This is in contrast to 'Credit Subordination'.

Breakage Cap

In a Back-leveraged Operating Lease (or U.S. Leveraged Lease), a Breakage Cap is the imposition by the equity investor/lessor of a cap on the amount the related lenders can charge for Hedge Breakage losses ahead of the equity investor/lessor's

Exhibit 9.1 Back-leveraged Op Lease (lessor vs lender)

Provision	Provision terms	Related issues	Comments
A Voting Rights			
1 Amendments, modifications and waivers	*Lease:* Non-recourse: both Lessor and Lenders must approve. Recourse: No Lender approval except: • [Economics] • Core Lease	Except: • Administrative matters? • Return condition (no balloon)? • Other?	'Shared rights' Will depend on Lessor credit
2 Acceleration/foreclosure/ repossession	*Lease:* Non-recourse: Lender controls to exclusion of Lessor Recourse: Lessor controls to exclusion of Lender *Mortgage:* Standard, except for Equity Squeeze	Allow remarketing period? Remarketing period constraints?	UCC Protections 'commercially reasonable manner' (§§9–610) Equity Squeeze protection: if lender is foreclosing on lessor, must concurrently dispossess lessee of aircraft
B Buy-out right	Lessor has right to prepay		
C Cure rights	With monthly debt service, Junior Class can cure (i) ___ consecutive and ___ cumulative debt payments to Senior Class and (ii) non-payment defaults up to [$500,000] in aggregate during any 12-month period. Cure rights to be effected within [five] (in the case of (i)) or [ten] (in the case of (ii)) business days of Event of Default. Junior Class subrogated upon payment so long as no further payment default.	***Issues:*** (i) Number of consecutive & cumulative cures. (ii) Are Junior cure rights cumulative or additive to cure rights afforded Equity? (iii) How much time to cure? (iv) $ Limits (non-Rent)	Need to consider in light of Junior Lender cure rights

D Waterfall	*Standard:* First Lender, then, Lessor	
E Prepayment	NA	See Rent Trap below
F Remarketing	**Maturity Date Matters** Lessor controls? Also: • Event of default vs • Lease maturity Also: Balloon Risk (Non-recourse) Situations • Walk-Away Right timing • Irrevocability of election	Balloon-risk (non-recourse); advance notice of walk-away; pre-funding costs? Special Bill of Sale? *Issue:* (1)Length (2)Obligations • Debt Service • repossession • storage/maintenance/ insurance • remarketing (3)Recourse vs Non-recourse of obligations/$Up-front (4)Timing to elect
		Different treatment for recourse Lessors vs non-recourse
G Rent Trap	Rent otherwise to be distributed to Lessor trapped for up to [180] days if Event of default continuing.	
H Sharing the spoils	NA	
I Last look/Right of First Refusal	NA	UCC Safeguards
J Credit subordination	NA	
K Events of default	Non-recourse: Lease event of default Recourse: Lease event of default?	DSCR test? Lessor financial tests?
L Maintenance reserves & security deposits	Pledged? Trigger events?	Credit worthiness of Lessor
M Other provisions	(i) Inspection rights	Using-up allotment

Exhibit 9.2 Mortgage financing (senior lender vs junior lender)

Provision	Provision terms	Related issues	Comments
A Voting Rights			Structural Issue: One Lien vs Two?
1 Amendments, modifications and waivers	Both Majority Senior Class and Majority Junior Class must approve; changes of economic terms require approval of all affected lenders.		[Increasing senior debt amount or senior debt interest rate requires consent of Junior Class.]
2 Acceleration/foreclosure/ repossession	Majority Senior Class controls to the exclusion of the Junior Class; [provided that if the Senior Class shall not have commenced the exercise of significant remedies within [180] days of the occurrence of an Event of Default, the Junior Class may direct the exercise of remedies unless the Majority Senior Class shall seek to (and does) exercise significant remedies, in which case the Majority Junior Class rights in this regard shall fall away.]	*Issues:* (i) 'Fish-or-cut bait' provision. (ii) If so, number of days. (iii) If so, snap back to Senior Class.	No UCC protections [cf EETC] 'Controlling Party'
	Senior Class required to give Junior Class __ day prior 'enforcement notice' before exercising any significant remedies.		Gives Junior Class opportunity to exercise buy-out right.
B Buy-out Right	Junior Class can buy-out Senior Class any time after the 10th business day following an Event of Default at (x) par plus accrued interest plus swap and LIBOR breakage.	*Issues:* (i) Swap breakage in (x) subject to a cap? (ii) accrued interest at standard or past due rate? (iii) Standstill?	*Comment:* (ii) Never saw at standard rate. (iii) Credit issue.

C Cure Rights	With monthly debt service, Junior Class can cure (i) ___ consecutive and ___ cumulative debt payments to Senior Class and (ii) non-payment defaults up to [$500,000] in aggregate during any 12-month period. Cure rights to be effected within [five] (in the case of (i)) or [ten] (in the case of (ii)) business days of Event of Default. Junior Class subrogated upon payment so long as no further payment default.	**Issues:** (i) Number of consecutive & cumulative cures. (ii) Are Junior cure rights cumulative or additive to cure rights afforded Equity? (iii) How much time to cure? (iv) $ Limits (non-Rent)	*Need to consider in light of equity cure rights (e.g., number of days).* *Need to wait to see if will cure*
D Waterfall	In respect of the liquidation of the aircraft and other collateral, the proceeds of such liquidation shall be applied: First, to recovery costs (and other standard super-priority costs); Second, to principal, interest, LIBOR and swap breakage and other amounts owing to Senior Lenders in respect of the Senior Loan; Third, to principal, interest, LIBOR and swap breakage and other amounts owing to Junior Lenders in respect of the Junior Loan; and Fourth, to the Borrower.	***Issue:** Is Senior Loan swap breakage senior in priority to Junior Loan p & i? Some alternatives –* *(i) Senior Loan swap breakage senior in priority to Junior Loan p & i (as reflected on left).* *(ii) ___% of Senior Loan swap breakage senior in priority to Junior Loan p & i; balance is payable after Junior Loan p & i paid.* *(iii) All Senior Loan swap breakage subordinated to Junior Loan p & i.* *Issue: Should Junior Interest rank ahead of Senior principal?* ***Issue:** Rent Trap?*	*'Breakage Cap'* cf EETC
E Prepayment	No Junior Class prepayment prior to prepayment in full of Senior Class.		
F Remarketing	Senior Class controls or Junior Class?	*Same as Lessor vs Lender* *If Junior Class controls, Senior Class must accept sale that covers Senior Class p, i, swap and LIBOR breakage and other amounts owing to them.*	

Exhibit 9.2 continued

Provision	Provision terms	Related issues	Comments
G Rent Trap	Rent otherwise to be distributed to Junior Class to service p & i trapped for up to [180] days if Event of Default continuing.		
H Sharing the Spoils	Allocation of 'upside' between Senior and Junior Lenders		
I Last look/Right of First Refusal	Junior Class to have a right of first refusal on Aircraft sales orchestrated by Senior Class.	*Timing.* *public sales vs. private sales*	*Chilling of sale concern. Not very common.* *Junior Lenders do not have UCC Safeguards*
J Credit Subordination	Junior Class agrees to pay over consideration received in bankruptcy		*Relevant only for transactions with a real 'credit'.*
K Events of Default	Senior x-default to Junior Debt? Vice-versa?		
L Maintenance Reserves & Security Deposits	NA		
M Other provisions	(i) Junior Class has no right to bid-in debt (ii) Inspection Rights	*Id.*	

Exhibit 9.3 EETC Intercreditor Term sheet summary

Provision	Provision terms
A Voting rights	
1 Amendments, modifications and waivers	If no indenture default, each trustee must approve (in each case, as directed by certificate holders holding a majority in interest of the relevant trust). During an indenture default, the Controlling Party (see below) must approve. Changes of economic (and certain other) terms require approval of each affected certificate holder. Liquidity provider has certain very limited voting rights.
2 Foreclosure/ repossession	Remedial actions are controlled by the 'Controlling Party': • The trustee of the senior trust (directed by certificate holders holding a majority in interest of the trust) until payment of final distributions to the holders of such class of certificates; • thereafter, the trustee of the junior trust (by certificate holders holding a majority in interest of the trust); and • in certain circumstances, the liquidity provider with the largest amount owed to it.
3 Limitation on sales	During the nine month period after the earlier of (i) an airline bankruptcy event and (ii) the acceleration of the equipment notes under any indenture, the Controlling Party may not sell any aircraft or the accelerated equipment notes if the net proceeds of such sale would be less than (A) in the case of an aircraft, 75% of such aircraft's appraised current market value and (B) in the case of such equipment notes, 85% of the related aircraft's appraised current market value.
4 Limitation of leases	During an indenture default, the lease of an aircraft as part of the exercise of remedies can be effected at the direction of the Controlling Party if effected in a commercially reasonable manner within the meaning of the Uniform Commercial Code.
B Buy-out rights	If the airline is in bankruptcy and (A) the 60-day period has expired and airline has not agreed to perform (or has agreed to perform, but failure to cure a default under an indenture) or (B) the airline has abandoned any aircraft), each junior certificate holder will have the right to purchase all but not less than all of the senior certificates at par plus accrued interest.
C Cure rights	None
D Waterfall	All payments in respect of the equipment notes (including of Security liquidation and claims against the airline) will be applied in the following order: 1 Reimbursement of certain administration expenses incurred by the subordination agent, any trustee, any liquidity provider or any certificate holder, subject to a cap; 2 Certain liquidity provider expenses; 3 Interest on liquidity obligations; 4 Liquidity obligations; 5 Reimbursement or payment of certain fees, taxes, charges and other amounts incurred by the subordination agent, each trustee and each certificate holder; 6 Interest on the senior certificates; 7 Adjusted interest on the junior certificates; 8 Expected distributions (principal) on the senior certificates; 9 Interest on the junior certificates; 10 Expected distributions (principal) on the junior certificates.

recovery of its investment in the financed Aircraft Asset following the foreclosure on such asset. This cap is typically calculated as a percentage of the outstanding principal amount of the debt at the time the default waterfall is applicable.[2] These caps would also apply in an equity's buyout price pursuant to its Buyout Right.[3] Breakage Caps may also be applicable in the context of protecting the position of a junior/subordinated lender in a senior/sub structure.

PRACTICE NOTE

Breakage Caps of 2 percent to 5 percent were not uncommon in the (old) days of U.S. Leveraged Leases. The equity investors were able to procure such caps because leveraged-lease equity was far more scarce than leveraged-lease debt. These caps are not commonplace today.

Buyout Rights

This is the right of a junior investor (whether an equity investor or junior lender) to buy out the next most senior Tranche. Because the most senior lender is typically granted status as Controlling Party in a default situation, the related junior investors may seek to wrest control of the deal from that lender. Control may be wrested by buying out the most senior lender's loan, thus stepping into that lender's shoes.

The reasons for wanting to obtain control are fairly obvious. The junior investor, for example, may have a very different viewpoint or impetus from the Controlling Party (senior lender) for maximizing the value of the financed Aircraft Assets or may find a strategic benefit from holding the equipment, unencumbered, in its negotiations with a defaulting credit.

Historically, an important reason that the buyout rights found their way into U.S. Leveraged Lease documentation (along with the Cure Rights discussed below) was the frighteningly large economic hit equity investors would be required to take in a leveraged lease transaction that unwound prematurely. These hits could be attributed to the tax recapture relating to the investment tax credit (ITC) taken in connection with the equity investor's original acquisition of the financed equipment and the current recognition of debt as income upon a foreclosure. The severity of the punitive effects from a foreclosure were such that equity investors required the buyout right to allow them to protect their economic position.

There are two primary buyout issues.

1   When, and for how long, can the Buyout Right be exercised?
2   At what price is the Buyout Right exercisable?

Buyout Rights are almost always triggered by an acceleration of the senior debt. They are also triggered by either a fully ripened 'event of default' or a fully ripened 'event of default' that has continued for 180 days. While many senior lenders may be pleased to be bought out in a default scenario, that is not always the case insofar as they may, notwithstanding the default, still like the financing transaction and the yield they are earning. In fact, the senior lenders as a class may be entirely safe in their exposure so that they are just as happy to let the transaction 'ride'. Some Buyout Rights shut down after a period of 180 days from the date they are first exercisable. senior lenders may like the fact that they can stop worrying, at some point, that their deal may be taken away from them, or even that their efforts in dealing with the pending defaults will remain theirs to deal with.

Buyout Right 'price' is almost always to include principal plus the accrued interest[4] owing to the senior lender being bought out. It is also most likely to include other amounts owing to the senior lenders like expenses and, for Floating Rate Lenders, LIBOR Breakage. The Buyout Right price will also need to address as a component thereof the particular costs associated with the prepayment of a Fixed Rate Loan if that is what is being bought out. In case of a senior lender entitled to a Make-Whole, we would often see as follows: (i) no Make-Whole at all if the Buyout Right was triggered on an acceleration; and (ii) a Make-Whole is payable if the buy-out occurs during the first 180 days following the event of default, but none after. In case of a senior lender entitled to Hedge Breakage Loss coverage, we would expect to see full coverage of that loss amount regardless of when that buy-out occurs; however, in the old days of U.S. Leveraged Leases, it was not unusual to see a Breakage Cap.

In addition to the matter of paying Hedge Breakage Losses, an equity or junior lender may seek to have a lender pay over to it any Hedge breakage gains arising by virtue of the termination of any related Hedge transaction.

Controlling Party

This is the financing party in a transaction that is empowered to dictate (to the exclusion of the other parties) remedies against a debtor and rights in collateral. The Controlling Party concept is most pertinent, and typically arises, only in a default context in the exercise of remedies (or workout). Prior to default, most rights (and certainly economic rights) are shared as provided in Shared Rights. In most instances, it is the most senior Tranche of financing parties that would control. For the reason addressed in Buyout Rights, the junior Tranches/equity may have an interest in pulling such control rights away from that senior Tranche by exercising its Buyout Rights (if available). In an EETC, the Controlling Party shifts from the holders of the most senior Tranche to the Liquidity Provider after 18 months if the Liquidity Provider's obligations have not been repaid to it by then.

PRACTICE NOTE

Equity investors/lessors are protected from a Controlling Party not aggressively marketing collateral by virtue of the protections inherent in the UCC's 'commercially reasonable standard' in the event of foreclosure/liquidation (see Part 8, 'UCC (Remedies)'). However, there are no equivalent legal protections available to junior debt holders. Unless there is a Minimum Sale Price for the

---

collateral being foreclosed upon, the junior Tranche holders are at risk of a lethargic foreclosure sale process which may yield the senior debt Tranches with sufficient proceeds but leave the junior debt Tranches 'short' in asset recovery. This may precipitate the exercise of a Buyout Right by the junior Tranches.

Credit Subordination

As between a senior financier Tranche and a junior financier Tranche, this is the recovery by the senior financier Tranche of their claims against a debtor from payments made by the *debtor* in priority to recoveries by the junior financier. This would include a priority interest in any Deficiency Claims from the debtor.

Cure Rights

This is the right of a junior investor to make payments or perform obligations on behalf of a debtor, the due performance of which, from the perspective of the Controlling Party (or the party that would otherwise have become the Controlling Party), prevents that party from declaring a default or exercising remedies in respect of that failed payment or performance; such payment or performance effectively 'cures' the related default for this limited purpose. Of course, the debtor is still in default. Cure Rights may be a temporary solution to avoid the (great) expense of exercising a Buyout Right.

Cure Rights are available in respect of both defaults in the payment by the debtor of basic rent, as well as other defaults curable by the payment of money (such as payments of supplemental rent or the procurement of insurance). Cure Rights are not unlimited – senior lenders (Controlling Party aspirants) are willing to sit on their hands for only so long. Limitations are framed in terms of the number of con-secutive and cumulative cures that may be effected in the case of curing scheduled payments (Debt Service or rent), and the maximum amount of money that may be expended in any 12-month period, in the case of non-scheduled payment cures.

Non-scheduled payment cures are limited to cures that may be effected by the payment of money for two reasons. First, the expenditure on cures is quantifiable and thus subject to policing and limitation. Second, this limitation effectively prohibits the junior investor's ability to have hands-on interaction with the financed Aircraft Assets (which, from the senior lenders' viewpoint, may be too close to its remedial rights in respect of those assets).

Deep Subordination

This is subordination that provides for no distribution whatsoever to the junior Tranches unless the senior Tranches have been paid in full in cash. Deep subordination would also provide few of the Intercreditor Rights that may otherwise be available to the junior Tranches. As one industry observer[5] put it: 'Below Whale S**t' (or BWS).

Equity Squeeze/Squeeze Protection

In a Back-leveraged Operating Lease, this is the divestiture by foreclosure of the equity investor's (lessor's) ownership interest in the Aircraft Asset subject to such financing following a default by the lessee under the related lease without the foreclosing lenders' dispossessing such lessee from such asset. In most Back-leveraged Leases, an event of default under the related Lease is a default under the Mortgage Financing leg of that transaction, since the lessee is the credit and sole source of payment under the transaction documents.[6] If a Buyout Right is not exercised, or Cure Rights (if any) are used up, then the lessor is at risk of Foreclosure if a lease event of default has occurred. Foreclosure will not only divest the equity investor of its ownership stake in the financed Aircraft Asset, but it may also trigger tax benefit (depreciation) recapture and a current recognition of income on the related debt.

Absent the Equity Squeeze Protections described below, the lender, in exercising its right of foreclosure under the Mortgage Financing, would not need to take any remedial action against the lessee. In other words, the lender would be able to take advantage of the lessee's default to squeeze out the equity by Foreclosure under the Mortgage Financing and either allow the lessee to continue on with its default or work out some side deal with the lessee that would allow the lessee to continue possession of the leased Aircraft Asset.

So as to avoid this result, Equity Squeeze Protection is often demanded by the equity investor. The nature of this protection is most usually characterized by requiring the lender, following a lease event of default, and if not stayed by the lessee's bankruptcy, to exercise one or more of the significant (repossessory-type) remedies under the related lease – and to do so concurrently with the exercise of any foreclosure remedies under the Mortgage. This type of Equity Squeeze Protection, then, allows the lender to squeeze out the equity only if it 'means business' with the lessee.

If the lease event of default that triggers the indenture event of default is the lessee's bankruptcy, the Equity Squeeze Protection is further modified by not permitting the lender to squeeze out the equity, as it otherwise could during the period that the parties are stayed by the bankruptcy law (11 USC Sec. 365) from exercising any remedies against the lessee. This block period is usually subject to some ultimate end date after which the lender can effect a squeeze for a period of time commencing on the date of the related bankruptcy event.

In Back-leveraged Leases of Aircraft Assets having the benefit of Section 1110, it is quite standard for such a period to be the 60-day period under such section of the Bankruptcy Code during which the lessee may either reject the lease (and return the Aircraft Asset subject to the leverage financing) or agree to perform under the lease (and cure all defaults (Section 1110(a)) or, if earlier, at such time as the lender is entitled to repossess the applicable Aircraft Asset.

PRACTICE NOTE

There are a number of strategies employed by debt investors to avoid Equity Squeeze Protection limitations.

Excluded Rights

These are certain contractual rights benefiting a junior investor that are obligations of a debtor in which the senior creditor does not obtain a Security Interest and with which the junior creditor or equity investor may deal with debtor to the exclusion of the senior lender. Excluded Rights most typically cover Liability Insurances and indemnities personal to the junior creditor.

Fish-or-Cut-Bait

This is a contractual provision that provides that a junior financier has the right to step in to become the Controlling Party if the senior financier that is otherwise the Controlling Party fails to exercise remedies within a specified period of time. Also known as 's**t or get off the pot' and 'snooze and lose'.

PRACTICE NOTE

Many Fish-or-cut-bait clauses have a snap-back provision that provides that if control has devolved to the junior financier, it would be restored to the senior lender if the senior lender assert its right to exercise (and in fact, does) exercise remedies.

Minimum Sale Price

This is a price level agreed by financing parties as the minimum price at which the Controlling Party will sell collateral. This is often benchmarked to certain appraised market values at the time of sale. Minimum Sale Price protection for junior lenders may be important since the junior lenders do not have UCC-related 'commercially reasonable' foreclosure protections (see Part 8, 'UCC (Remedies)') that are protective of the equity investor by law. While junior lenders may obtain protection from the threat and consequences of foreclosure by the senior/Controlling Party by bootstrapping on the protections available to the equity, that may not always be the case. First, the Aircraft Asset may be so far underwater from the equity investor's perspective that the equity investor does not care to enforce this right (and may have agreed to a Strict Foreclosure). Second, the asset to be sold may not be one that is owned by the equity investor. For example, in an EETC, one of the rights of the Controlling Party is the right to sell the *equipment notes* placed in the pass through trusts (see Part 3, 'Enhanced Equipment Trust Certificate'; Exhibit 3.8), that is, the IOU of the equity investor (which is not one of equity investor's assets).

The Minimum Sale Price is a relatively recent addition in EETC structures. Previously, no such protection existed for the junior Tranche lenders. The current formulation is, with respect to any Aircraft or the equipment notes issued in respect of such Aircraft, at any time, in the case of the sale of an Aircraft, 75 percent, or in the case of the sale of related equipment notes, 85 percent, of the appraised market value of such Aircraft; provided that such limitation falls away nine months after the earlier of: (i) the acceleration of the equipment notes; and (ii) the bankruptcy of the airline issuer.

Rent Trap

A Rent Trap is the blocking of distributions that would otherwise flow through a Waterfall to junior investors and/or equity investors during a period when a default by the debtor is in effect or is impending. Senior lenders often seek to block the distribution of the free cash to junior investors and/or equity investors at times when the debtor is in, or on the precipice of default since remedies may soon be exercised.

This free cash is not usually tied up forever. It is often released on the earliest to occur of: (i) the curing of the event of default (in which case the money is released down to the next level on the Waterfall); (ii) the lapse of 185 days with no remedial action having been taken; and (iii) the acceleration of the debt or exercise of remedies (in which case the money is applied in accordance with the default Waterfall).

Shared Rights

This is the allocation of rights as among different Tranches and, if applicable, the equity investor, to amend, waive, provide consents and otherwise take action under the transaction documentation (a 'Direction Matter') with the debtor prior to the time the Controlling Party takes control. In a Mortgage Financing there may be different Tranches of debt. The holders of each Tranche must come to an agreement as to how the security trustee or agent acting for all the debt is to be directed in its interaction with the debtor. The first item to be agreed is intra-Tranche – what percentage of the debt of that particular Tranche needs to cast a vote for (or against) a Direction Matter ('Required Lenders'); a simple majority or two-thirds would be a typical direction percentage.[7] As among the different Tranches, absent a default (when the Controlling Party would take control) one would typically see that each Tranche (based on its Required Lender voting) would need to approve any amendments or provide consents. As well, each Tranche would typically be protected from any changes in the economics of its Tranche-economics, regardless of voting by other Tranches, including the Controlling Party; and, that voting on a per-Tranche basis would typically require unanimous agreement of all holders of that Tranche (that is, it is not a matter for Required Lenders).[8] Notwithstanding that a Controlling Party may not be permitted to amend the transaction economics of the junior investors, a Controlling Party may have the unilateral right to re-lease an Aircraft Asset in the face of default, and such lease's economics may, by its economic terms, force the economics of the Tranches to change.

In addition, the junior lender-Tranches often require approval rights for changes to the more senior Tranches economics in order to limit, for example, the amount of debt that gets paid in priority to their claims under the Waterfall.

In a Back-leveraged Operating Lease financing, the foregoing would apply if the related debt is Tranched, and brings in the additional complexity of the equity investor's voting rights. The Granting Clause of any Mortgage or Security Agreement assigns all of the related equity's rights under the related Lease to the lender. Without modification (that is, inclusion of a Shared Rights provision), the lender would be entitled to unilaterally amend, waive, modify, and grant consents under the Lease, without any input from the equity.

To address this issue, market practice has been for the lenders in Back Leveraged Operating Leases to provide to the equity investors/lessors two categories of rights in respect of certain provisions of the related Leases that are tantamount to a claw back of the (outright) assignment under the Granting Clause.

1   Matters under the Leases where the equity investor has unilateral control.
2   Matters under the Leases where the lenders and the equity investors have joint control (that is, agreement of both is required).

Falling in the first category are matters relating to the equity's economics in the transaction: rental adjustments (subject always to Debt Service constraints), return conditions (other than in connection with a lease event of default), renewal terms, and end of term buy-out options. Also included in this category is the right of the equity investor to exercise inspection rights (which right is also independently maintained by the lender); to sue the lessee to seek specific performance of the lessee's covenants in respect of insuring, maintaining, possessing, and operating the equipment; to maintain separate insurance in the financed equipment; to sue the lessee in respect of Excluded Rights; and to exercise Cure Rights. In addition, the equity investor is given the right to receive all notices, opinions, reports and other communications, documentation, and information generated by the lessee under the Lease.

The second category encompasses waivers, modifications, and amendments generally under the Lease, with the important exclusion that the exercise of remedies by the lender under the Lease is always within the lender's exclusive control and the further caveat that joint control over approvals and consents under the Lease are only so long as no lease event of default is continuing. The idea is that until the equity is squeezed out of a deal, it should have a say in the deal that it negotiated at the outset.

In transactions with publicly issued securities, decision-making authority may be devolved to a servicer or security/loan trustee for matters other than the most basic economic-related or, at a minimum, that would not have a 'material adverse impact' on the holders; this latter construction is the standard used in EETCs.

Structural Subordination

Structural Subordination is the subordination of claims by reason of the placement of the junior claim at a structural level, such as a parent corporation/LLC, that is one or more legal-structure steps removed from the senior claim (and further removed from the financed Aircraft Asset). The junior claim, accordingly, only has rights to the extent of a distribution or dividend from its subsidiary (see Exhibit 9.4).

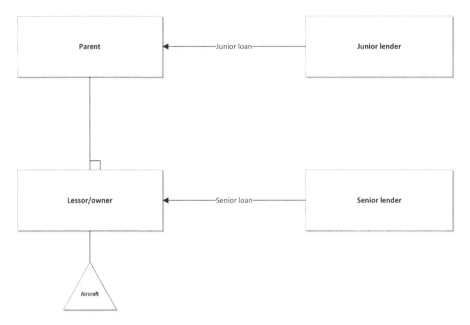

Exhibit 9.4 Structured Subordination

Source: Author's own

Subordination

See 'Asset Subordination', 'Credit Subordination' and 'Structural Subordination'.

PRACTICE NOTE

Subordination is yet another 'loaded term' that has no real meaning unless the terms are fully spelt out.

Waterfall

This is the contractual provision in an Intercreditor Agreement or Mortgage that specifies the order of priority for payment of obligations owing to the Financing Parties from proceeds of collateral and payments by a debtor, which reflects any Credit Subordination and Asset Subordination as among the Tranches and equity investor or borrower. Documentation Waterfall provisions are often split into multiple variations to deal with different scenarios. It is not uncommon in Aircraft Finance transactions to see at least three variants.

1   Regular cash flow supporting Debt Service.
2   Events of Loss/Prepayments.
3   Default/Remedial situations, and the liquidation of Aircraft Assets.

Waterfalls will implement the Asset Subordination and Credit Subordination terms.

Waterfall provisions can get exceedingly complex with the inclusion of incrementally more credit enhancement parties. Two come to mind:

- Hedge Counterparties who are taking transactional risk (who will insist on *pari passu* treatment with regularly scheduled interest payments in respect of ongoing cash flows, and *pari passu* treatment for breakage losses with principal); and
- Liquidity Providers who will require super priority treatment.

In addition, in structured financings where a Servicer is engaged, management/servicing fees (often expressed as a percentage of collections) are usually placed near the top of the Waterfall before debt service.

The discussion on Waterfalls would not be complete without observing the fact that, notwithstanding any other parties' rights to the most senior payment position in the Waterfall, the lawyers[9] always get paid first.

PRACTICE NOTE

In considering Debt Services Coverage Ratio (DSCR) or other debt service coverage tests, it is critical to take into account all costs that come ahead of debt service, such as servicing fees.

Notes

1 Scheinberg, R, 'A modest proposal: standardization of debt/equity points', *Journal of Equipment Lease Financing*, Spring 1997.
2 Although a better methodology would be to have the cap as a percentage of the original principal amount.
3 Scheinberg, R, 'A modest proposal: standardization of debt/equity points, *Journal of Equipment Lease Financing*, Spring 1997; Scheinberg, R and Grier, J, 'Interest rate swaps in traditional leveraged lease transactions', *Air Finance Journal*, November 1991.
4 While it is rather less common, some junior investors are able to negotiate that the accrued interest is payable at the regular interest rate, not the Default Rate.
5 Michael J Edelman.
6 If there is recourse to a lessor in such a transaction, which has its own credit base, then the matter of this Crossdefault is not a certainty but a matter of negotiation.
7 Debtors would typically favor the lower percentage requirement. Among lenders in a particular Tranche, the distribution of holdings as among holders of that Tranche may be a determining factor in the final decision as to the appropriate percentage of holdings.
8 In addition to altering transaction economics, lenders (of whatever Tranche) may seek unanimous approval requirements for that Tranche to release collateral, adjust borrowing bases and other changes in respect of the collateral.
9 And trustees and technical consultants.

PART 10

INSURANCE

Introduction

There are two primary coverages provided by Aircraft insurances: *liability* insurances and *hull* insurances. Liability insurances protect the Aircraft Financiers (who are *not* operators of the equipment) from liability from passengers and third parties in connection with accidents and incidents involving the assets that they own or finance; by simply having an ownership (or Security Interest) in an asset, there are theories of liability (for example, Strict Liability) that may provide a basis to hold financiers liable for injury caused by such accidents and incidents[1] (see Part 7, 'Lessor Liability'). Hull insurances protect the Aircraft Financier from losses suffered as a result of the physical loss or damage to their financed assets. Similar to any hard asset, Aircraft Assets are subject to physical compromise due to any number of reasons, including a crash, a taxi clipping, an exploded Engine and so on. While the aircraft operator is typically required to pay for and/or replace the compromised asset, it may not have the financial wherewithal to do so. Accordingly, absent hull insurance, the Aircraft Asset's value as security for a financing would be replaced by the credit of the borrower/lessee. Aircraft Asset hull insurance, therefore, replaces the asset protection with the credit protection of highly rated insurers (as opposed to the bare credit of the operator).

Importantly, the basic hull and liability coverages described above exclude losses caused by 'War and Allied Perils' risks – those associated with war, terrorism, hijackings and so on (see 'AVN48B'). To provide coverage for these risks, operators procure, and Aircraft Financiers obtain, comparable endorsements as above for Hull and Liability War and Allied Perils risks.

Let us look at these basic forms of insurances in a little more detail.

Hull Insurances

Hull/Casualty Insurance may be for an agreed amount (Agreed Value) or for Replacement Value. Total Losses of Airframes (and Airframes and Engines) are typically insured on an Agreed Value basis. Engines, on the other hand, may be insured on either an Agreed Value basis or on an Insured Value or Replacement Value basis. If an item is insured on an Insured Value basis, the insurers will also have the right to replace that item. Partial Losses (as opposed to Total Losses) are insured in an amount necessary to repair the damaged asset, and the insurer will dole out funds to effect the repair as and when the repair is made.

As noted in the Introductory portion of this Part 10, Hull/Casualty Insurance is written as a basic coverage, from which War Risk and Allied Perils risks are excluded (under the AVN48B 'war' exclusions), and War Risk and Allied Perils insurance coverage, which is 'written back' over the AVN48B exclusions by, typically, the LSW 555D write-back. The LSW 555D write-back covers all of the AVN48B exclusions except:

- any emission of chemical, biological or biochemical materials;
- any radioactive contamination;
- any atomic/nuclear detonation;
- war between any of '5 great powers';
- confiscation for title or use by state of registration repossession;
- financial cause;
- delay, loss of use or consequential loss.

The chemical and radioactive exclusions are not total exclusions, meaning that some coverage is given. By contrast, LSW 555B does provide coverage for chemical, biological and biochemical events.

The amount of Hull/Casualty Insurance that a financier would normally require is at least 110 percent of the amount financed. The amounts over the 100 percent threshold are intended to cover interest accruals, breakage costs and other amounts owing to the financiers. The amount of Hull/Casualty Insurance that a lessor would normally require would, initially, include at least the amount of the Aircraft's purchase price (or fair market value – oftentimes 'richly' calculated), and that amount would decrease over time based on an agreed depreciation factor.

The Hull/Casualty Insurance program of an airline would be subject to aggregate fleet limits; that is, the insurers will cap the aggregate amount of exposure they will have to an airline for Hull/Casualty losses in a year. Such limits vary by airline and range from U.S.$400,000,000 to U.S.$1,000,000,000.

Hull all risk insurances may be subject to deductibles on recoveries although the deductibles would normally only apply for partial losses. Deductibles typically range between $500,000–$1,000,000 per occurrence, with the higher numbers applying to widebody aircraft. In addition to deductibles, hull all risk and war risk insurances may be subject to self-insurance (more on which later).

Another aspect of hull insurances that may concern an aircraft financier is the request by the airline-operator to have some amount of insurance proceeds paid to it directly in the event of a loss that is not a Total Loss. Airlines often seek to have this threshold amount paid directly to them rather than pass through the fingers of the financiers. This threshold amount is often tied to the value of an engine (but does not get paid directly to the airline if a default subsists).

Finally, hull all risk and hull war risk insurances usually are subject to a 50/50 sharing clause as per AVS103 (see, 50/50 clause, infra), which provides that for a hull loss as to which there is no agreement (within 21 days of the insurable event) as to whether the loss was due to a normal hull coverage event or a war risk hull coverage event, the hull insurers and hull war risk insurers will make a 50/50 payout to the insured, and settle-up among themselves once the cause of the event is ascertained.

Liability Insurances

Liability Insurances cover liability to third parties for accidental bodily injury or property damage involving an aircraft. Liability coverages apply to 'any one aircraft/ any one occurrence' and the amount typically ranges from $500,000,000 to $2,250,000,000, with the amount required a function of the seating capacity of the particular aircraft – the more seats (and, therefore, passengers at risk), the bigger the policy. As well, biases to higher levels of coverage would be appropriate in litigious jurisdictions (like the United States). It is worth noting that the levels of liability coverage required by a financier do not necessarily reflect the amounts of coverage procured by an airline-operator. These negotiated numbers are the floor of coverage; airlines typically obtain larger amounts of coverage (they are, after all, the party with primary liability). As in other instances, airlines try to keep the financier-imposed level of liability coverage to a minimum so as to maintain flexibility if liability rates skyrocket.

Like hull insurances, liability insurances have the 'standard' coverages and the 'war risk' coverages, with the standard liability coverages subject to the AVN48B exclusions. The liability insurance war risk coverages are those written back by AVN52; these write-backs do not cover:

- war between any of the '5 great powers';
- hostile atomic/nuclear detonation;
- aircraft requisition for title or use;

although coverage remains available until landing if an incident occurs in flight.

Lessors and financiers of Aircraft often seek to have the liability coverages extend for a 'tail period' after they no longer have an interest in the Aircraft. The purpose of this Tail Insurance is to provide coverage for liabilities that may arise after their interest is gone but nevertheless relates to a defect or issue that arose while their interest was extant.

Hull and Liability Insurance common issues

Hull and Liability Insurance coverages are subject to a number of common issues that are relevant for Aircraft Financiers:

- Self Insurance – See, Self Insurance, infra;
- Cancellation – Insurances are subject to cancellation by the insurers on 30 days' notice and, in the case of war risk insurances, seven days' notice;
- Government Coverages – Both Hull and Liability War Risk coverages may be provided by governmental authorities in the event such insurances are no longer available (for example, as happened in the immediate aftermath of the 9/11 terrorism episode) or are prohibitively expensive. See, for example, the FAA War Risk Policy, infra;
- Assignments of Insurance – In jurisdictions where the concept of 'third-party beneficiary' does not exist, it is common practice for aircraft financiers

to procure an Assignment of Insurances so as to be able to enforce the rights granted to them for coverage.

Endorsements for Coverage: AVN67B

These insurance coverages are procured by the borrower/lessee who is the operator of the Aircraft. In order to benefit from this coverage (which the operator is normally providing in any event for its own protection), Aircraft Financiers require that the related insurance policies inure to their benefit by the agreement of the insurers to specified policy endorsements. These endorsements have largely been standardized over the years (with the particular features more fully described in this Part 10). The most common endorsement is the AVN67B endorsement for insurances placed in the London Lloyds market; insurance placed in the U.S. market does not have a specified form of endorsement, but the U.S.-based airline/operator financier endorsements closely follow London's AVN67B endorsement (but with some, usually better (financier-friendly), language).[2]

As a matter of general practice, lessors and financiers pay particular attention to ensure that an AVN67B type endorsement exists and that the list of contracts and Contract Parties in the endorsement are complete and correct. Generally, an insurance certificate and a letter of undertaking from the airline's broker is the way in which insurance protection is shown. An important feature of AVN67B is the manner in which hull losses are paid. In contrast to U.S.-type loss payment provisions where a singular loss payee is identified in an AVN67B endorsement, 'settlement . . . shall be made to, or to the order of the Contract Parties'. That is, the insurers will 'cut the check' to all parties named as Contract Parties, and those parties are required to then pay to proceeds over to whomever shall be entitled to them under the financing agreements.

General matters

It is important to note that airlines procure insurance on a fleet-wide basis. Individual transactions cannot dictate requirements outside the airline's fleet-wide insurance arrangements. While most airlines may be just as risk-averse to the consequences of an accident as Aircraft Financiers, and, accordingly, maintain substantial levels of insurance, they often do seek to build in flexibility as relates to levels of contractually-mandated levels of liability coverages and self-insurance as well as deductible allowances in case rising insurance premium rates would dictate a need to change fleet-wide insurance program policy to maintain competitiveness and manage expenses.

The value of any insurance policy ultimately rests on the ability of the applicable insurers to satisfy claims. The responsibility of finding credit-worthy insurers rests with an operator's insurance broker. An operator will go on periodic road shows with its broker to engage a sufficient number of insurers to satisfy its insurance needs. These needs typically require a syndicate of insurers to be assembled, and not infrequently, multiple syndicates, to share the risks. Lloyds of London is one of the primary syndicate organizers, and other major U.S. and European insurance companies also participate. These insurers underwrite fixed percentages of the various insured risks, and their underwriting responsibility is several, not joint (see, for example, LSW 1001).

A number of jurisdictions worldwide require that insurance on locally registered Aircraft be procured in the local jurisdiction (as a means to buttress their local insurance markets). These local insurers are usually not of sufficient credit quality for the Financing Parties, so the insurance must be reinsured in the international markets, and the financiers would obtain the benefit of a Cut-Through endorsement.

Airlines frequently seek the ability to Self-Insure the risks for which insurance is mandated by the Financing Parties (over and above standard deductibles), not so much as those are risks they are willing to accept, but so as to be able to manage insurance costs when insurance premium costs skyrocket (as they sometimes do) as part of an industry cycle or event-driven occurrence. This Self-Insurance may also be effected by a Captive Insurance Company.

50/50 Clause

This is a policy clause that provides that, in the event of a hull loss where it is not readily determinable as to whether coverage would fall under the standard Hull Insurance or the War Risk hull coverage, the insurers covering the standard Hull Insurance and the insurers covering the War Risk hull coverage will each pay 50 percent of the Agreed Value, and then resolve the final apportionment among themselves at a later date if and when the cause of the loss is definitely ascertained. In this way, the beneficiary of the hull policies does not have to wait until there is a final determination that apportions responsibility as between the two policies. Historically, there have been a number of instances where it took a matter of weeks or months to determine whether an accident was the result of equipment malfunction/pilot error or an act of terrorism/sabotage. The 50/50 Clause in these instances assured a timely payout. A standard 50/50 Clause endorsement is AVS103.

Additional Insured

These are named persons, including lessors and financiers, who are provided coverage under an airline's/operator's Liability Insurances with respect to their interests.

One would expect all parties to an Aircraft Financing to be included as Additional Insureds. There is no additional cost to an airline/operator for including multiple parties as such (and their respective officers, agents, etc.), but it may be an administrative burden for an airline to have to monitor this list every insurance policy-year renewal period.

Agreed Value

This is the dollar amount the insurers agree to pay to the Loss Payee under a Hull Insurance policy in the event of a total loss of an Aircraft Asset.

> **PRACTICE NOTE**
>
> It is important to specify that hull losses are to be paid by the insurers on an 'agreed value' basis; otherwise, the insurers may be able to provide hull loss compensation in an amount equal to the Replacement Value of the lost asset (or actually replace the lost Aircraft Asset).

AVN48B

This is a Lloyd's Market Association and the International Underwriting Association clause that is used in airline insurance policies to exclude certain risks from standard coverage. Clause AVN48B lists those War Risks that are excluded from coverage under the airline's hull, spares, passenger, and third party liability policy. These include: (i) acts of war; (ii) nuclear detonation; (iii) riots; (iv) terrorist acts; (v) sabotage; (vi) governmental confiscation; and (vii) hijacking. See 'Appendix A-8' for the text of this endorsement.

AVN51

This is the Lloyd's endorsement that writes back War Risk hull coverage. See 'Appendix A-5' for the text of this endorsement.

AVN52E

This is the Lloyd's endorsement that writes back War Risk liability coverage. See 'Appendix A-6' for the text of this endorsement.

AVN67B

This is a policy endorsement that protects the interests of lessors and financiers under the hull and liability policies of an airline/operator. This endorsement provides lessors and financiers with protection against the risk of loss to their financed Aircraft Asset as well as protection from liability to passengers and third parties. The following features of AVN67B were innovative at the time of its introduction or are otherwise of interest.

- AVN67B is a 'stand-alone' endorsement that clearly lists the 'Contract Parties' and the 'Contracts', and overrides any conflicting provisions in the underlying insurance policy, at least as regards the express provisions of the endorsement. Thus financiers do not need to review the underlying policy.
- Conversely, AVN67B provides that: (i) except as expressly stated in the endorsement, the Contract Parties are subject to the terms, conditions, limitations, warranties, exclusions and cancellation provisions of the underlying policy; and (ii) the underlying policy is not varied by any

provisions in the Contract(s) which purport to endorse or amend the policy. Thus there is no need for insurers/brokers to review the lease/financing contracts.

- AVN67B provides Breach of Warranty and Severability of Interests provisions in favor of the financier.
- AVN67B requires the insurer to give the financier 30 days' notice of cancellation or material alteration of the policy (except in case of provision for cancellation or automatic termination stipulated in the policy). In the hull war-only version of AVN67B, this period is reduced to seven days (subject to any shorter periods in the policy).
- AVN67B provides that total loss payments are made to (or to the order of) the Contract Parties. Importantly, there is no single loss payee.
- Under AVN67B, the insurer's subrogation rights are only exercisable with consent.

See 'Appendix A-1' for the text of the AVN67B endorsement.

AVS103

This is a policy endorsement evidencing the 50/50 Clause. See 'Appendix A-2' for the text of this endorsement.

Breach of Warranty Clause

This is an endorsement contained in a financier's general policy endorsement that provides for coverage to the financier notwithstanding the breach by airline/operator who is the person maintaining the insurances of a warranty or other agreement with its insurers (and otherwise would invalidate the insurances or would be a defense of the insurers to make a payment). For example, most insurance policies require that the airline/operator comply with governmental regulations and operate in accordance with the applicable Aircraft's airworthiness certificate. Policies also contain specific notification requirements and policy declarations that need to be accurate and complied with. The following is the Breach of Warranty clause from the AVN67B endorsement.

> The cover afforded to each Contract Party by the Policy in accordance with this Endorsement shall not be invalidated by any act or omission (including misrepresentation and non-disclosure) of any other person or party which results in a breach of any term, condition or warranty of the Policy provided that the Contract Party so protected has not caused, contributed to or knowingly condoned the said act of omission.[3]

Broker's Letter

This is a letter from the insurance broker who procured the Liability and Hull Insurance on behalf of an airline/operator to a lessor and/or financier: (i) acknowledging the interests of the lessor/financier; (ii) agreeing to maintain the insurance through the current policy period and advise of any notice of cancellation from the underwriters; and (iii) agreeing to advise the lessor/financier of any adverse change in coverage. In addition, such brokers may be asked to opine as to the following:

- the brokers have reviewed the insurance provisions of the related financing documents and the insurance carried by the airline/operator is in conformity with such insurance provisions; and
- the insurance carried by airline/operator is of a type consistent with industry practice for other operators flying comparable equipment and routes.

Captive Insurance Company

For an airline, this is an insurance company owned by that airline established to cover certain insurable risks associated with such airline's operations. Captive Insurance Companies must be properly capitalized in compliance with local law requirements; they have historically re-insured a substantial portion of the risks they have underwritten. In addition to airline-owned captives, there may be ones that are industry-owned.[4]

PRACTICE NOTE

An Aircraft Financier will want to review the underwriting level of a Captive Insurance Company in an airline's insurance program in respect of the requisite insurances required in a financing. Insofar as a Captive Insurance Company may not constitute an 'internationally recognized' insurer, the level of a captive's underwriting may constitute self-insurance.

Certificate of Insurance

This is a certificate provided by the insurance broker that summarizes the insurance coverage for an Aircraft Asset (types of coverage, levels of coverage and liability limits) and includes agreed endorsements (such as AVN67B (or its equivalent)).

PRACTICE NOTE

Aircraft Financiers are careful to review the Certificates of Insurance produced for any Aircraft Financing. These certificates are reviewed by their in-house experts, outside insurance consultants or legal counsel. Mistakes that must be corrected are inevitably found.[1]

[1] The inevitability of errors has precipitated one person to comment: 'Those who can't do, teach. Those who can't teach, teach gym. Those who can't teach gym, prepare insurance certificates.'

Contingent Insurance

This is insurance coverage obtained by lessors/financiers that provides for them the same coverage as the airline/operator on comparable terms in the event the insurances obtained by the airline/operator fail to respond or the coverage proves to be inadequate. This type of coverage protects against the following types of risks: (i) the airline/operator fails to pay premiums on the primary insurances; (ii) the airline/operator flies outside of geographical limits; (iii) the broker of the airline/operator has caused an error or omission; and (iv) a failed Cut-Through clause. A Contingent Policy that is in place can easily have added to it a 'Repossessed Aircraft Endorsement' in the event the lessor/financier needs to repossess the Aircraft.

Contract

In the case of an AVN67B endorsement, Contracts are the lease and financing documents relating to a subject Aircraft Asset. These Contracts will be listed in the AVN67B endorsement.

Contract Party

In the case of an AVN67B endorsement, this is any of the persons who are parties to Contracts relating to the subject Aircraft Asset as those afforded coverage under the related airline/operator's insurance as Additional Insureds and Loss Payees.

Contribution Waiver

This is an insurance endorsement that waives an insurers right to require an insured to seek contribution from other insurance policies in respect of a claim. Aviation insurance policies often contain an endorsement by which insurers waive any right of contribution. Such waiver is sought because insurers are entitled to claim contribution from any other valid insurances available for the same loss. As such, insurance policies will often limit the liability of the insurers to the proportion of the loss or liability that the limit of coverage under such policy bears to the aggregate of the limit of coverage under all insurances available in respect of such loss or liability. Aircraft Financiers, therefore, often seek this endorsement so as to have a single source of recovery for losses without concern that the primary insurers will seek to limit their claims on this basis.

Cut-through

This is an endorsement to an insurance policy that is issued on a Reinsurance basis that provides that the reinsurer will pay insurance claims directly to the relevant finance parties (as opposed to the primary insurer subject to the reinsurance). Airline operators are required by Aircraft Financiers to reinsure into the international aviation insurance market if the primary insurers maintain (or are mandated to maintain) their Hull and Liability Insurance coverages in the local insurance markets with local insurers.[5] In the event of an insurable Total Loss of the Aircraft, the flow of funds would normally act in reverse, with the reinsurers paying the Agreed Value to the local insurance company who would in turn pay the Aircraft Financier Loss Payee. This flow of funds could result in delay of payment. The solution was the 'Cut-through' clause whereby reinsurers agree to pay the loss payee directly (see 'Appendix A-9' for a sample Cut-through Clause).

Deductible

This is the specified portion of an insured loss that is borne by the insured before the insurer becomes liable for payment. Deductibles on Hull Insurance losses typically apply for partial losses but will not apply to Total Losses. The amount of a Hull Deductible is largely dependent on the size of the related Aircraft. Narrow-bodies often have a deductible of U.S.$500,000 to U.S.$750,000 per occurrence, while wide-bodies may be twice as much.

PRACTICE NOTE

The amount of any Deductible converts, on a dollar-for-dollar basis, asset risk to the credit risk of the operator.

Deductible Buy-down

This is the payment of a supplemental premium to insurers that negates all or a specified portion of a Deductible.

FAA War Risk Policy

This is the hull and liability War Risk policy provided by the FAA in lieu of similar War Risk policies provided in the commercial market. After the terrorist events of September 11 in 2001, the FAA began issuing third party Liability and Hull War Risk Insurance to U.S. air carriers and required continued provision of this insurance (see 'Appendix A-7' for a copy of the FAA's policy).

Hull/Casualty Insurance

This is insurance that covers losses due to the physical loss, damage or destruction suffered by an Aircraft. This is colloquially referred to as 'metal' coverage. A person must have an insurable interest in an Aircraft in order to obtain and gain the benefit of hull insurance. An insurable interest would most likely include an operational, ownership or security interest in an Aircraft.

See the introductory portion to this Part 10 for a more comprehensive discussion of Hull/Casualty Insurance.

Liability Insurance

This is insurance that covers losses to passengers, baggage, cargo and third parties, including for bodily injury and damage to persons and property. Aircraft Financiers typically make liability insurance a key element of their negotiations involving Aircraft Finance documentation. Their overall concern is that due to their perceived 'deep pockets', they will be a likely target in any litigation resulting from the operation of their financed Aircraft.

There are three main sources of liability risk for Aircraft Financiers.

1   Aircraft third party/passenger legal liability: the possibility that the financier may be named in litigation relating to an operational loss of an Aircraft.
2   General legal liability: liability arising out of claims not directly caused by an accident involving an Aircraft.
3   Aircraft products liability: liability arising out of losses alleged to have been caused by a defect in an Aircraft Asset supplied by an Original Equipment Manufacturer (OEM) or operating lessor.

For insurance purposes, these risks are normally divided into the following categories:

*   passengers liability (including baggage);
*   cargo liability;
*   mail liability; and
*   aircraft third party liability.

The last category provides coverage in respect of damage caused by the aircraft to third party persons and property, whereas the first three refer to persons or goods carried on board the aircraft itself. The remaining liabilities are often referred to as 'airline general third party'. These are the liabilities arising from the ground operations of the airline. Airlines will often develop substantial operations for maintenance and other services which generate additional revenues for them.

There are two primary purposes for requiring an airline to procure liability insurance. One is to provide protection for the Aircraft Financier for the risks stated below. The second is to ensure the airline is properly, from a credit perspective, protecting itself against what could be catastrophic losses associated with an accident.

See the introductory portion of this Part 10 for a more comprehensive discussion of Liability Insurance.

PRACTICE NOTE

Liability insurance coverage typically varies by Aircraft size – the larger the Aircraft, the higher the per-occurrence coverage (in the light of the greater number of passengers at risk). Aircraft Financiers typically require coverage of at least U.S.$2 million per seat. Airlines usually have liability insurance

---

coverages well in excess of that coverage level, but seek to retain the flexibility to go lower if there are developments in the insurance market (such as a spike in premium charges) that necessitate lowering the liability insurance coverage level.

Liability Limits

In the case of any airline/operator, this is the maximum Hull Insurance liability and Liability Insurance liability that the underwriters will accept in any policy year.

Loss Payee

This is the person named in a Hull Insurance policy (as may be modified by any endorsements) as the party to whom casualty loss proceeds are to be paid. In many cases, airlines and financiers agree that hull losses below certain specified limits get paid directly to the airline while losses above that limit get paid to the financier.

PRACTICE NOTE

Importantly, there is no individual Loss Payee concept under the AVN67B endorsement. Rather, hull losses are payable to all of the named Contract Parties, collectively. As among the Contract Parties, their agreement (and security arrangements) would govern the disposition of loss proceeds.

Letter of Undertaking (LOU)

The Letter of Undertaking (LOU) covers any ongoing undertakings contained in the Broker's Letter.

LSW 555D

This is the insurance endorsement that provides coverage for hull War Risk coverage (see 'Appendix A-3' for the text of this endorsement).

Notice of Cancellation

This is an endorsement in an insurance policy that requires the insurers to provide advance notice to additional insureds and loss payees if such policy is going to be cancelled, terminated or materially changed. Due to the importance of insurances in addressing an Aircraft Financier's property and liability risks, Aircraft Financiers require a period following provision of a notice that the insurers will be cancelling, terminating or otherwise materially changing a policy in which the Aircraft Financiers have an interest in order to obtain their own coverage. Aircraft Financiers accordingly typically obtain an endorsement from the insurers that provides to them prior notice of at least 30 days (seven days or such lesser period as may be customarily available in the case of war risk and allied perils coverage) of cancellation or material change before such cancellation or change of the policy is effective as against them. The time limitations typically commence when the notices are *given* by the insurers, not when they are *received* by the specific insured party. It is quite common to require the airline/operator's insurance broker to agree to provide such a notice too. However, such notice is typically not required to be given by an insurer (or the insurance broker) in the event that a policy terminates (without renewal) on its scheduled expiry date.

Political Risk Insurance

Political Risk Insurance (or Repossession Insurance) is insurance obtained by a lessor or mortgagee for an Aircraft registered in a particular jurisdiction to cover the risk that, notwithstanding the laws in place in that jurisdiction, such lessor or mortgagee is unable to repossess such Aircraft within an agreed time frame from that jurisdiction (see Annex A-4 for AVN147 which is a standard form for Political Risk Insurance).

PRACTICE NOTE

In the case of the standard AVN147 policy form, the political risk insurers do not make it easy to collect on a policy, requiring the insured to perform numerous tasks for repossession prior to any possible collection. These myriad requirements will affect the way in which the insured party will handle defaults in its transactions.

Reinsurance

This is the insurance procured by a primary insurer with another insurance company (a reinsurer) to lay off risk. In the context of Aircraft Finance, Reinsurance is typically required when the government of registry requires the airline/operator to maintain the airline's Hull and Liability Insurances with local insurers (whose credit may not be acceptable). That insurance is what would be reinsured (to the maximum permitted amount) in the international aviation insurance markets. In order to obtain a direct benefit from the reinsurance, Finance Parties typically require a Cut-through endorsement to have payments made to them directly by the reinsurer.

PRACTICE NOTE

Some jurisdictions do not allow more than a specified percentage of the local insurance to be reinsured.

Replacement Value

In respect of an Aircraft or Engine, this is the value insured under a Hull Insurance policy that is the cost to replace the lost equipment. Aircraft are usually insured on an Agreed Value basis. Engines and Spare Parts are typically insured for Replacement Value.

Residual Value Insurance

This is an insurance policy that protects the insured against a decline in market value of tangible assets. Residual Value Insurance is used in the Aircraft finance market in respect of Aircraft Assets. It is a common feature in Japanese Operating Lease (JOL) financings.

Self-insurance

This is the retention by an airline/operator of a portion of the risk of loss in respect of Hull Insurance and/or Liability Insurance. Self-insurance in the airline context is often evidenced by the management of Deductibles and first-loss coverage. The idea of Self-insurance is that by retaining risks, and paying the resulting self-insured portion of claims or losses from captive or on-balance sheet financial provisions, the overall cost to the airline may be cheaper than buying commercial insurance from a commercial insurance company in respect of the entire set of risks.

PRACTICE NOTE

Aircraft Financiers abhor Self-insurance for the obvious reason that the amount self-insured converts third party (highly creditworthy) insurance company risk (that itself is a conversion from asset risk) to the credit risk of the airline/operator. Accordingly, Aircraft Financiers seek to limit the ability of the airline/operator to Self-insure. There are typically two baskets of Self-insurance

1   *Standard Deductibles* (which are applicable on a per-incident basis) (see 'Deductibles'). It is fairly customary that standard Deductibles apply in case of partial losses (although they could apply for Total Losses too). These Deductibles may range from U.S.$750,000 to U.S.$1 million for narrowbody Aircraft and U.S.$1.5 million to U.S.$2 million for widebody Aircraft. Airline/operators may seek to allow for greater Deductibles than these in case of insurance market changes.

2   *First loss coverage.* This type of Self-insurance would provide that the airline/operator is responsible for first losses in any calendar year up to the specified level. While allowances for this type of Self-insurance is not typical in the Aircraft Finance marketplace, a number of U.S. major airlines have been able to insist on broad allowances, allowing for the lesser of: (i) 1.5 percent of the airline's insurable fleet value;

---

and (ii) 50 percent of the insurable value of the largest Aircraft in the airline's fleet, in any one year policy period. Given fleet sizes, clause (ii) would normally control and could, therefore, be in excess of U.S.$1 billion (given, for example, Boeing 777 or 787 values); thus an airline's first loss may be entirely self-insured.[1]

Self-insurance may also be achieved by an airline's utilization of its own Captive Insurance Company as an insurer.

---

[1] We understand that the U.S. majors do not actually, as a matter of course, take advantage of this Self-insurance allowance, but want the ability to do so in case market pricing would make this a compelling alternative.

Sending

Sending is the transit or shipment of Aircraft Assets. Insurers view the risk of loss during the movement of Engines or Spare Parts during a shipping process as much higher than a static situation, such as when in storage in a warehouse. As such they will limit their maximum exposure accordingly.

Set-off Rights Waiver

This is an endorsement that waives an insurers' right to set-off from claims it may have against an insured from amounts it must payout on a policy claim. An insurer would typically seek to have the right to set-off unpaid premiums as against any claims made under the applicable insurances. This could obviously be a significant amount, particularly if the airline/operator insures its entire fleet under a standard fleet policy that covers hundreds of Aircraft. In order to protect against any diminution in the amount of any insurance claim an Aircraft Financier may have, it is customary for an Aircraft Financier to obtain an endorsement that the insurer waives any right to set off or counterclaim in respect of unpaid premiums for all Aircraft covered under such policy (although in certain cases, the insurers do require set-off rights in respect of the premiums payable in respect of the particular Aircraft that suffered or was the subject of the loss).

Severability of Interests Clause

A Severability of Interests Clause is an endorsement that expressly provides that the Liability Insurance will operate in all respects, except in respect of the limit of liability, as if each party insured was the subject of a separate policy. The following is the Severability of Interests Clause from the AVN67B endorsement.

> Subject to the provisions of this Endorsement, the Insurance shall operate in all respects as if a separate Policy had been issued covering each party insured hereunder, but this provision shall not operate to include any claim howsoever arising in respect of loss or damage to the Equipment insured under the Hull or Spares Insurance of the Insured. Notwithstanding the foregoing the total liability of Insurers in respect of any and all Insureds shall not exceed the limits of liability stated in the Policy.

Spare Parts Coverage

This is Hull Insurance coverage for off-wing Engines, and Spare Parts and equipment while uninstalled. Spare Parts Coverage is usually maintained on a Replacement Value basis, although Engines may be insured on an Agreed Value basis (in the case of an Engine financing, for example).

Stipulated Loss Value (SLV)

In Lease financings, this term is often used to stipulate the required Agreed Value of the Aircraft Asset subject to the lease for Hull Insurance purposes. 'Agreed Value' may be a term similarly used for such purpose.

Storage Insurance

This is the Hull and Liability Insurances maintained on an Aircraft Asset while the Aircraft Asset is in Storage. Insofar as Aircraft Assets that are in Storage are at much lessened risk of loss due to their immobility, the scope of coverages and levels of liability coverage can be reduced.

Tail Coverage

Tail Coverage Liability Insurance coverage is often required by lessors after they have sold their interests in an Aircraft to cover liability claims for a (usually) two-year period ('tail') following such sale. The requirement to maintain any such tail coverage is typically limited to a period expiring on the earlier of a set period of time (such as two years) and the occurrence of the first heavy maintenance visit. The primary motivation of such requirement is the receipt by the financier of the benefit of the products liability component of the broader liability coverage during such period, as this covers the financier for anything that the operator may have done to the Aircraft while it was in the operator's care, custody, and control and that is later found to be the reason for an accident.

PRACTICE NOTE

Secured parties/lenders sometimes seek Tail Coverage from airlines for whom they are providing financing. Such coverage is simply not necessary or appropriate because there is no legal theory of liability for lenders in a post-return situation (unless they had previously operated the Aircraft).

Top-up Insurance

This is Hull and/or Liability Insurance coverage obtained or required by lessors or financiers to provide insurance coverage in addition to the primary coverage ordinarily procured by the relevant airline/operator.

Total Loss Only

Total Loss Only (TLO) is an insurance endorsement that provides that particular Hull Insurance coverage is payable only on a Total Loss of the Aircraft.

Waiver of Subrogation

This is an agreement between two parties in which one party agrees to waive subrogation rights against another in the event of a loss. The intent of the waiver is to prevent one party's insurer from pursuing Subrogation rights against the party that may be the credit on an Aircraft Finance transaction. The following is a Waiver of Subrogation clause from a typical U.S. major insurance broker.

> The Insurers waive their rights of subrogation against the Additional Insureds but only to the extent the Named Insured has waived its rights of recovery under the Contract(s).

In contrast, here is the AVN67B endorsement for Waiver of Subrogation.

> Upon payment of any loss or claim to or on behalf of any Contract Party(ies), Insurers shall to the extent and in respect of such payment be thereupon subrogated to all legal and equitable rights of the Contract Party(ies) indemnified hereby (but not against any Contract Party). Insurers shall not exercise such rights without the consent of those indemnified, such consent not to be unreasonably withheld. At the expense of Insurers such Contract Party(ies) shall do all things reasonably necessary to assist the Insurers to exercise said rights.

War Risk Insurance (Hull)

This is insurance that covers the same loss-risks as Hull Insurance but covers War Risks.

War Risk Insurance (Liability)

This is insurance that covers the same loss-risks as Liability Insurance but covers War Risks.

War Risks

The following are 'war risk and allied perils' risks.

- War (whether declared or not, including war between Great Powers),[6] invasion, acts of foreign enemies, warlike hostilities, civil war, rebellion, revolution, insurrection, martial law, exercise of military or usurped power, or any attempt at usurpation of power.
- Any hostile detonation of any weapon of war, including any employing atomic or nuclear fission and/or fusion or other similar reaction of radioactive force or matter.
- Strikes, riots, civil commotions, or labor disturbances.

- Any act of one or more persons, whether or not agents of a sovereign power, for political or terrorist purposes and whether the loss or damage resulting therefrom is accidental or intentional, except for ransom or extortion demands. Payments in response to ransom or extortion demands.
- Any malicious act or act of sabotage, vandalism or other act intended to cause loss or damage.
- Confiscation, nationalization, seizure, restraint, detention, appropriation, requisition for title or use by or under the order of any foreign government (whether civil or military or de facto) or foreign public or local authority. This policy will not cover any lawful government seizures of aircraft or spare parts that are the result of outstanding legal debts, taxes, fines, or unlawful acts committed with the knowledge of airline officials or the unlawful operation of such aircraft by the named insured.
- Hijacking or any unlawful seizure or wrongful exercise of control of the aircraft or crew (including any attempt at such seizure or control) made by any person or persons onboard the Aircraft or otherwise, acting without the consent of the Insured.
- The discharge or detonation of any weapon or explosive device while on an Aircraft covered by the Policy of Insurance.

PRACTICE NOTE

Aircraft, unfortunately, are a favorite target of terrorists. Terrorist attacks and attempted attacks involving Aircraft are numerous: Lockerbie, the Underwear Bomber, the Shoe Bomber, 9/11, to name but a few. Terrorism counter-measures are very much outside the control of the airline operators; responsibility for security falls on government agencies. Accordingly, while Aircraft insurers are generally able to work with airlines to promote safety (which would lower risk of loss), there is precious little that insurers can do to promote War Risk event prevention. During periods of heightened international tensions when terrorist activity may increase, War Risk insurers naturally need to increase premiums to protect against the increased risk of loss. Since these premium increases may make the cost of such insurance prohibitive, governments often step in to provide War Risk coverage at cheaper rates. (See, for example, 'FAA War Risk Policy'.)

Notes

1   To be sure, liability risks are far higher for lessor/owners, than secured parties. However, secured parties are potential owners of the finances asset should Foreclosure become necessary.
2   The AWG has been pushing for the industry to adopt 'AVN67C', which includes some refinements to AVN67B. Adoption of AVN67C is moving very slowly.
3   In contrast, the following is typical of a U.S. endorsement:

Such insurance as is afforded the Additional Insureds shall not be invalidated and shall protect their interests regardless of any action or inaction of Lessee or any other Insured (other than any action of an Insured as the case may be, and the only as against such particular Insured) whether or not such action or inaction is a breach or violation of any warranties, declarations or conditions of the policies, but in no event shall this clause apply in the event of exhaustion of policy limits, nor to losses, claims, expenses and so on, excluded from coverage under the policies.

4   The Air Transport Association formed Equitime, which was intended to provide War Risk Insurance for member airlines, with FAA backdrop support.
5   Normally a small retention is kept by the operator's local insurance company and the rest is reinsured.
6   U.S., Great Britain, France, Germany, Russia and China.

PART 11

VALUATION

Introduction

The valuation of Aircraft Assets is a critical consideration for Aircraft Financiers when analyzing whether an Aircraft Asset is to be financed, and if so, in what amount. The advance rate – the percentage of an asset's value the financer is willing to advance – is of course a fundamental matter in the course of an Aircraft Asset financing. The 'value' determinant is the somewhat subjective element of this equation. This determination may be based on the actual purchase price of the asset (as is done in Export Credit Agency (ECA) Financings)[1] but more frequently is entrusted to ISTAT-certified appraisers, who would be advised of the type of appraised value sought; whether Base Value, Market Value, Maintenance-adjusted and so on, each of which is described below. As an alternative for appraised value (or purchase price) determination, a financer may use as a benchmark or an online appraisal value database (which data in turn may be further refined based on that financer's internally-assessed adjustments).

Valuations may also be used in the course of a transaction, for example:

- to assess Loan to Value (LTV) in, say, a Warehouse, where a failure to satisfy an LTV test may require: (i) the transaction to Turbo; (ii) the inclusion of additional collateral, or (iii) lower prospective advance rates;
- to assess LTVs in, say, an Enhanced Equipment Trust Certificate (EETC) where the most senior classes will receive all distributions on account of principal following certain default events (triggering events) until prescribed LTV levels are satisfied (which is a form of Turbo);
- to assess LTVs in a Back-leveraged Operating Lease with a Balloon so as to ascertain whether there is sufficient collateral value to pay off that Balloon;
- to assist Lessors in understanding values of Aircraft Assets coming off Operating Leases;
- to determine an Early Buy-out Option (EBO) price; and
- to assess Lease damages on a Marked-to-Market basis, whether comparing contract Stipulated Loss Values (SLVs) with Fair Market Values (FMVs) or contract rental rates to current market rental rates.

A few important points to make here:

- valuations of a particular asset will vary by very wide margins depending on the type of appraisal required;

- even using the same type of appraisal methodology, different appraisers may have significantly different valuation assessments of the identical equipment;[2]
- there is much to be said for performing valuations based on actual maintenance status and condition; recognizing that for an in-operation aircraft asset, such condition is fleeting (but could provide valuable guidance as to decay curves from that point); and
- Maintenance Reserves on deposit with a lessor should necessarily be taken into consideration when assessing an Aircraft Asset's value on a maintenance status basis.

Appraisers will likely look at the same variables for making their value assessments, such as:

- Aircraft market trading values;
- if actual maintenance condition is to be considered:
  - overall Aircraft maintenance condition;
  - overhaul status of Aircraft Assets, Landing Gear (LG) and other major components;
  - status of maintenance documentation;
  - status of Airworthiness Directive (AD) and Service Bulletin (SB) compliance;
  - Aircraft utilization (as compared with industry averages); and
  - history of accidents or incidents;
- Aircraft specifications;
- cabin configuration;
- Aircraft modifications (as compared with industry standards);
- recent sales prices and lease rates;
- in-production versus out-of-production;
- market penetration;
- engine variance;
- future order book;
- life-cycle stage;
- freighter convertibility; and
- profile of operator base.[3]

However, the provision of appraisal values for Aircraft Assets by appraisers has a large 'art' (as opposed to 'science') component. There will surely be differences in how they weigh the previously noted variables. In the light of this inherent uncertainty, financers are wise to be wary of appraised value numbers and take those values with a grain of salt; financers accordingly often provide their own haircuts to presented valuations.[4]

The primary definitions for valuations in Part 11 were culled from *Aircraft Appraisers' Program and Appraisers' International Board of Governors* by ISTAT Appraisers' International Board of Governors, revised 30 January 2013 (seventh revision) (Appraisal Guide).

Exhibit 11.1 Criteria associated with principal types of appraised values

	Base value	Market value	Distress value
The Aircraft			
Time status (if not new)	Usually mid-time, mid-life	Usually as found	Usually as found
Physical condition (if not new)	Usually average, considering type and age	Usually as found	Usually as found
Highest best use valuation	Yes	Yes	Maybe
The Parties			
Willing, able, prudent and knowledgeable	Yes	Yes	Yes
Lack of pressure	Yes	Yes	No; seller is usually 'motivated'
The Market			
Open, unrestricted	Yes	Yes	Yes
State of market balance	Reasonably stable, reasonably balanced	As perceived at the time	As perceived at the time
The Transaction			
Arm's-length	Yes	Yes	Yes
Sale for cash or equivalent	Yes	Yes	Yes
Adequate time for effective exposure on market	Yes	Yes	Yes
Single-unit sale	Yes	Usually	Maybe

Source: ISTAT Appraisal Guide

The Appraisal Guide includes the chart shown in Exhibit 11.1, which is helpful for comparison purposes.

Adjusted Market Value

If Market Value for an Aircraft is ascertained based on mid-time, mid-life, then Adjusted Market Value adjusts that value to take into account the actual technical status and maintenance condition of the Aircraft.

Base Value

Base Value is an appraiser's opinion of the underlying economic value of an Aircraft Asset in an open, unrestricted, stable market environment with a reasonable balance of supply and demand, and an assumption that full consideration of the asset's 'highest and best use' is given. An Aircraft Asset's Base Value is founded in the historical trend of values and in the projection of value trends and presumes an arm's-length, cash transaction between willing, able and knowledgeable parties, acting prudently, with an absence of duress and with a reasonable period of time available for marketing.

In most cases, the Base Value of an Aircraft Asset assumes its physical condition is average for an asset of its type and age, and its maintenance time status is at mid-life, mid-time (or benefiting from an above-average maintenance status if it is new or nearly new, as the case maybe). Because it is related to long-term market trends, the Base Value definition is commonly applied to analyses of historical values and projections of residual values.

COMMENT

Since Base Value pertains to a somewhat idealized Aircraft Asset and market combination it may not necessarily reflect the actual value of the Aircraft Asset in question, but is a nominal starting value to which adjustments may be applied to determine an actual value.

Book Value

Book Value is the value at which an asset is carried on a balance sheet. Book Value of an asset is determined by taking the cost of such asset and subtracting the accumulated depreciation taken on such asset. For Aircraft operating lessors, their Aircraft Book Value may vary greatly relative to a Fair Market Value assessment, since these assets are not Marked-to-Market.

Comprehensive Appraisal

This is an appraisal that includes a detailed inspection of the Aircraft and its Records. Sufficient detail is required, for example, to insure that the Records are in sufficiently good order to allow for the re-registration of the Aircraft in a different country.

Desktop Appraisal

This is an appraisal which does not include any inspection of the Aircraft Asset or review of its maintenance records. It is based upon assumed Aircraft Asset condition and maintenance status or information provided to the appraiser or from the appraiser's own database. A desktop appraisal would normally provide a value for a mid-time, mid-life Aircraft Asset. A Desktop Appraisal may also assume that a particular Aircraft is a typical example of its type, model and age, is in generally good condition with no damage history, and that the Aircraft is in compliance with all ADs and significant Service Bulletins. Other assumptions may include:

- the airframe and engine manufacturers will continue to support the particular models;

226

- the Aircraft Asset is utilized in a typical fashion (taking into consideration a 'normal' Hour/Cycle ratio and sector length); and
- the Aircraft Asset has not been customized in such a way as to require unusually high transition costs.

Distress Value

This is an appraiser's opinion of the price at which an Aircraft Asset could be sold under abnormal conditions, such as an artificially limited marketing time period, the perception of the seller being under duress to sell, an auction, a liquidation, commercial restrictions, legal complications, or other such factors that significantly reduce the bargaining leverage of the seller and give the buyer a significant advantage that can translate into heavily discounted actual trading prices. Apart from the fact that the seller is uncommonly motivated, the parties to the transaction are otherwise assumed to be willing, able, prudent and knowledgeable, negotiating at arm's-length, normally under the market conditions that are perceived to exist at the time, not an idealized balanced market.

COMMENT

While the Distress Value normally implies that the seller is under some duress, there are occasions when buyers, not sellers, are under duress or time pressure and, therefore, willing to pay a premium value.

Extended Desktop Appraisal

This is a Desktop Appraisal that includes consideration of maintenance status information that is provided to the appraiser from the client, aircraft operator, or in the case of a second opinion, possibly from another appraiser's report. An Extended Desktop Appraisal would normally provide a value that includes adjustments from the mid-time, mid-life baseline to account for the actual maintenance status of the Aircraft Asset.

Financial Appraisal

This is an appraisal that determines the value of an Aircraft to an investor based upon the income earning potential from its leases and residual value. A financial appraisal may be done in conjunction with either Desktop or Full Appraisals.

Fair Market Value

Fair Market Value (FMV) is a term synonymous with Market Value, and similarly Current Fair Market Value is synonymous with Current Market Value because the criteria typically used in those documents that use the term 'Fair' reflect the same criteria set out in the definition of Market Value.

Full Appraisal

Full Appraisal includes an inspection of the Aircraft Asset and its maintenance records. This inspection is aimed solely at determining the overall condition of the Aircraft Asset and records to support the value opinions of the appraiser, and would not, for example, include opening of inspection panels on the Aircraft Asset or a detailed review of record archives. A Full Appraisal would normally provide a value that includes adjustments from the mid-time, mid-life baseline to account for the actual maintenance status of the Aircraft Asset, and possibly other adjustments to reflect the findings of the inspection of the Aircraft Asset and its records.

Future Base Value

Future Base Value is an appraiser's forecast of Base Value of an Aircraft as at a date in the future.

Loan to Value (LTV)

Loan to Value (LTV) is the ratio of an Aircraft Asset loan to the value of that asset. This is a metric for assessing the amount of a loan in relation to the assessed value of the Aircraft Asset securing that loan. If the ratio is in excess of 100 percent, that loan is perceived as 'underwater', since the value of the collateral is insufficient to pay off the loan if that collateral were to be liquidated. The 'value' side of the equation is typically an appraised value using one of the formulations supplied in Part 11.

PRACTICE NOTE

In addition to setting LTV tests at the beginning of a financing to determine so-called advance rates, a number of transactions (mostly in the nature of Non-recourse transactions) require periodic LTV testing of the financed Aircraft portfolio with consequences for failure to satisfy the text including a Turbo feature until compliance is restored.

Market Value

Market Value or Current Market Value, if the value pertains to the time of the analysis, is an appraiser's opinion of the most likely trading price that may be generated for an Aircraft Asset under the market circumstances that are perceived to exist at the time in question. Market Value assumes that the Aircraft Asset is valued for its highest, best use, that the parties to the hypothetical sale transaction are willing, able, prudent and knowledgeable, and under no unusual pressure for a prompt sale, and that the transaction would be negotiated in an open and unrestricted market on an arm's-length basis, for cash or equivalent consideration, and given an adequate amount of time for effective exposure to prospective buyers.

COMMENT

The Market Value of a specific Aircraft Asset will tend to be somewhat consistent with its Base Value in a stable market environment, but where a reasonable equilibrium between supply and demand does not exist, trading prices, and therefore Market Values, are likely to be at variance with the Base Value of that Aircraft Asset. Market Value may be based upon either the actual (or specified) physical condition and maintenance time status of the Aircraft Asset, or alternatively upon an assumed average physical condition and mid-life, mid-time maintenance time status, depending on the nature of the appraisal assignment. The actual basis for the Aircraft's technical status used in determining the value should be set out in the appraiser's report.

Residual Value

This is the value of an Aircraft Asset at a future date, usually in connection with the conclusion of a Lease term. Not the same as Salvage or Scrap Value (see p. 230).

Salvage Value

This is the actual or estimated selling price of an Aircraft, Engine or major assembly based on the value of marketable parts and components that could be salvaged for re-use on other Aircraft or Engines. The value should be determined and stated in such a way to make clear whether it includes adjustment for removal costs. Salvage Value is not the same as Scrap Value.

COMMENT

Salvage Value (Parting-out Value) becomes applicable when disassembly for parts would most probably result in the highest cash yield for the asset 'as-is' as compared with the Market Value of the asset as a whole. For high-value items such as Engines and landing gears, the salvage value might be estimated on the basis of the remaining 'good time' before the item would require a major inspection or overhaul. While such disassembly for parts may result in the highest cash yield that can be generated in the marketplace, an owner may elect to reinvest in the asset to restore it as a working Aircraft, Engine or major assembly because the asset has a 'value-in-use' to him that exceeds the Salvage Value.

PRACTICE NOTE

In addition to its meaning as an appraisal term above, Salvage Value is also an accounting term for the value of an asset when it has been fully depreciated over its book depreciation period. In that context, Salvage Value is not synonymous with Market Value.

Scrap Value

This is the actual or estimated market value of an Aircraft Engine or major assembly based solely on its metal or other recyclable material content with no saleable reusable parts or components remaining. The scrap value is usually expressed as net of removal and disposal costs. In some cases, scrap value could be zero if the dismantling and disposal costs are high, as for example hazardous materials or composite assemblies that might be impossible to recycle.

Securitized Value

This is an appraiser's opinion of the value of an Aircraft Asset, under Lease, given a specified lease payment stream (rents and term), and estimated future Residual Value at lease termination, and an appropriate discount rate. Also called 'Lease-Encumbered Value'.

COMMENT

The Securitized Value may be more or less than the appraiser's opinion of Current Market Value. Moreover the appraiser may not be fully aware of the credit risks associated with the parties involved, nor all relevant factors such as the time-value of money to those parties, provisions of the lease that may pertain to items such as security deposits, purchase options at various dates, term extensions, sublease rights, repossession rights, reserve payments and return conditions.

Single-unit Sale

This is a comment in an Appraisal opinion that such opinion either does or does not assume a single-unit transaction wherein the Aircraft would be sold by itself, not as part of a wholesale lot or a large portfolio of Aircraft that would be sold en masse in a transaction where some 'volume discount' might typically apply.

Notes

1 Other than in ECA Financings, purchase prices for Aircraft are highly confidential.
2 In many Aircraft Financing transactions where LTVs are important determinants from the outset, such as an EETC, multiple appraisers will be used and their results averaged.
3 The appraiser may take into consideration the stability of the airlines operating the related Aircraft make and model. The stable event of Southwest Airlines, the largest operator of the Boeing 737NG class model, no doubt helps support Boeing 737NG prices given the relative unlikelihood of Southwest failing and glutting the market with this model.
4 It should be noted that certain appraisers are known to be more asset optimistic than others, and their valuations would therefore tend to be at higher levels than their peers. The optimistic appraisers would, accordingly, be more popular with the borrowers and lessors than the others.

CONTRACTS AND INSTRUMENTS

Introduction

Aircraft Finance is replete with contracts, agreements and instruments necessary to evidence the commercial arrangements among the parties. The complexity of transactional structures reviewed in Part 2 results in complex, and lengthy, contractual documentation. Part 12 reviews the principal contracts drawn-up for these purposes. Importantly, there is no agreed 'form' of any of these Contracts. Each is separately negotiated. While there may be a general market practice as to certain terms, the wording in each deal is necessarily honed to the particular parties and transaction.

Account Control Agreement

This is a tripartite agreement among an Aircraft Financier, the bank holding the Bank Account in which the financier wants an interest and the debtor pursuant to which the financier is granted 'control' over that account so as to comply with Perfection requirements of the UCC, so that the financier can perfect its Security Interest in that bank account.

Assignment of Insurances

Assignment of Insurance is an agreement by which the airline which procures the insurance with respect to an Aircraft assigns that insurance to the finance parties so that the finance parties will be in privity with the insurers, and therefore be in a position to enforce the insurances. This assignment is necessary in those jurisdictions where being named as a third-party beneficiary on insurances is not sufficient for enforcement purposes.

Common Terms Agreement (CTA)

Common Terms Agreement (CTA) is an umbrella-type agreement often used by leasing companies to cover the leasing of multiple Aircraft Assets to a single lessee, with all of the general provisions (that is, matters that would cover the Aircraft Asset across the board) included in the CTA, with Aircraft Asset-specific matters, such as asset identification and economic terms, dealt with in a supplement.

Deficiency Guarantee

This is a Guarantee that provides an Aircraft Financier with protection should the sale proceeds of the Aircraft Asset be insufficient to pay all (or part of) the balance of a loan following liquidation of that Aircraft Asset at a specified time (such as at Lease expiry).

PRACTICE NOTE

Deficiency Guarantors often try to create as many hurdles as possible for the beneficiary to actually be able to make a claim thereunder. A failure to provide a requisite notice to the Deficiency Guarantor may prove to be fatal for the Aircraft Financier. Close attention, therefore, needs to be paid to embedded timelines for the provision of notices and the taking of particular actions.

Depository Agreement

This is an agreement used in pre-funded Enhanced Equipment Trust Certificate (EETC) transactions pursuant to which the Depository arrangements (discussed above) are evidenced.

Guarantee

This is a contractual agreement issued by a person ('Guarantor') in favor of a creditor to support the payment and/or performance obligations of a debtor (lessee or borrower).

PRACTICE NOTE

Guarantees, similarly, come in many stripes and flavors. They may provide for automatic payment/performance undertakings of the Guarantor. They may require demand. Or, they may require the passage of time or the taking of certain action in order to compel Guarantor performance. In addition, Guarantees may be limited to only particular obligations of a debtor, may be capped as to amount and may be qualified as to time and place for drawing.

General Terms Agreement (GTA)

General Terms Agreement (GTA) is an umbrella-type agreement often used in Engine purchase agreements to cover warranties and other general terms applicable to an Engine sale (see also 'Common Terms Agreement').

Head Lease Agreement

In a Lease/Head Lease Structure, the Head Lease Agreement is between the lessor/owner and the intermediate lessee (not the operator).

Irrevocable De-registration and Export Request Authorization (IDERA)

An Irrevocable De-registration and Export Request Authorization (IDERA) is an authorization which gives a designated party the authority to seek de-registration and export of an Aircraft; for Cape Town purposes this instrument facilitates the exercise of the remedies of de-registration and export provided by Article IX(d) of the Protocol. The debtor must complete this form and submit the completed form to the applicable Aviation Authority for recordation. By virtue of the debtor's submission of the authorization, the person in whose favor the authorization has been issued (that is, the creditor), or its certified designee, becomes entitled (and is the only person entitled) to procure the de-registration and export of an Aircraft in accordance with the terms of the authorization. Appendix I sets out the prescribed Cape Town form of an IDERA.

PRACTICE NOTE

IDERAs are not commonly used in the U.S. for U.S.-registered Aircraft. The FAA accepts a 'Certificate of Repossession' from a creditor issued at the time de-registration is sought.

Intercreditor Agreement

This is a multi-party agreement among creditors that details their respective Intercreditor rights and obligations (see Part 9, 'Intercreditor', and related terms).

Lease Agreement/Lease

This is an agreement between a lessor/owner and lessee evidencing a Lease Financing.

Lessee Consent or Lease Assignment Acknowledgment

This is an instrument used in Back-leveraged Operating Lease financings whereby the lessee in that transaction, among other things, acknowledges and consents to the collateral assignment of the related Lease Agreement and provides certain estoppel assurances, all for the benefit of the financing parties.

PRACTICE NOTE

Here are some items typically covered in a Lessee Consent:

- Lessee agrees to send rent directly to the financing parties;

- Lessee agrees to name financing parties as Contract Parties/Additional Insureds under the Insurances, and to name them as indemnified parties in the indemnity provisions of the Lease;
- Lessee agrees to listen to instructions of financing parties, to the exclusion of the lessor, if it is sent an enforcement notice;
- Lessee confirms amount of rent, lease terms and so on, as an estoppel; and
- Financing parties provide to Lessee Quiet Enjoyment rights.

The English law equivalent is commonly known as a 'lease assignment and acknowledgment'.

Liquidity Facility Agreement

This is an agreement or instrument taking the form of a revolving credit agreement or a Standby Letter of Credit (SBLOC) that reflects a Liquidity Facility.

Loan/Credit Agreement

This is an agreement that provides for one or more financiers to make one or more loans to a borrower, the economic terms thereof, the conditions for making the loan(s), the making of representations and warranties by the borrower/obligors, covenants binding on the borrower/obligors, events of default that would allow the financiers to call the loan(s) and certain other provisions. In the context of Aircraft Finance, a Loan/Credit Agreement would typically be used in a Mortgage Financing.

Letter of Moral Intent (LOMI)

A Letter of Moral Intent (LOMI) is an undertaking by a credit support party that falls short of a guarantee. A LOMI is offered when that credit support party is not willing, able to or empowered to provide a guarantee. It, as the term indicates, provides a modicum of moral impetus for the provider of the LOMI to ensure that the related debtor to which the LOMI relates does not default. LOMIs can be quite varied. They may:

- contain an agreement to ensure that the debtor always has a positive net worth;
- contain an agreement to at all times own 100 percent of the debtor; and
- contain a statement simply recognizing that the financing parties have made a loan to the debtor.

As to be expected, LOMIs tend to be rather relationship-oriented.

Management/Servicing Agreement

This is an agreement by a servicer to provide, for a fee, certain services to a financier in relation to the management of Aircraft Assets, typically in the context of those Aircraft Assets being subject to Lease(s). Those services may include those specified in Appendix H.

PRACTICE NOTE

In addition to agreeing to the fees and other compensation to be provided to the Servicer and the services to be provided, Servicing Agreements must address a few important characterizations of the Servicer/Servicee relationship.

- What is the Servicer's standard of liability?
- What is the Servicer's standard of care for its performance?
- What is the Servicer's conflict of interest policy?
- On what matters must the Servicer take directions from the Servicee?

Mortgage

This is an agreement that evidences the granting of a Security Interest in Aircraft Assets and other collateral and is used in a Mortgage Financing. This term is used interchangeably with Security Agreement or Trust Indenture.

Omnibus Agreement

This is a single agreement that seeks to amend multiple agreements in one fell swoop.

Participation Agreement

The same name (unfortunately) for agreements that may be either: (i) a multi-party agreement in a structured financing such as Leveraged Lease which dictates the terms and conditions, among other things, for each party's participation in the financing and provides a road-map for their participation; or (ii) an agreement to evidence a Participation.

Pledge Agreement

This is an agreement pursuant to which a debtor (or guarantor thereof) grants a Security Interest in Ownership Interests (or other instruments).

Purchase Agreement

This is the primary agreement that evidences Purchase Documentation. Purchase Agreements with Original Equipment Manufacturers (OEMs) are usually hundreds of pages long because they contain the detailed specifications for the Aircraft Assets, as well as the Warranties.

Purchase Option Agreement

This is an agreement that provides to a party the right (but not the obligation) to purchase an Aircraft Asset (or an Ownership Interest relating thereto) on certain defined terms, including price, place and time.

Reimbursement Agreement

This is an agreement by the account party of an SBLOC to repay the bank issuing the Letter of Credit if there is a drawing under it.

Remarketing Agreement

This is an agreement to provide remarketing services in respect of an Aircraft Asset if that asset has been repossessed or returned following the expiration of a Lease term. (See the items specified in Part 2 of Appendix H.)

Reservation of Rights Agreement

This is an agreement evidencing the Reservation of Rights in respect of a particular financing transaction.

Residual Value Agreement/Guarantee (RVG)

A Residual Value Agreement/Guarantee (RVG) is an agreement by a person in favor of an Aircraft Asset owner or financier to guarantee that such Aircraft Asset will have at least a particular residual value when that Aircraft Asset comes off Lease or is sold.

PRACTICE NOTE

See Practice Note for 'Deficiency Guarantee'.

Standby Letter of Credit (SBLOC)

A Standby Letter of Credit (SBLOC) is a letter of credit issued by a bank that provides for that bank to make payments to the beneficiary of that instrument upon the presentation of specified documentation which, often, contain assertions as to the occurrence of specified events.

PRACTICE NOTE

SBLOCs are often used to replace cash used to fund Security Deposits and/or Maintenance Reserves in an Operating Lease. These are factors to be considered if an SBLOC is to replace cash.

- Expiry Date of SBLOC – if the SBLOC is scheduled to expire before the end of the lease term, then the SBLOC should be drawable prior to its expiry with the proceeds deposited in a Bank Account of the lessor or financier.
- Place of Drawing – the jurisdiction should be physically convenient to the lessor/financier.
- Issuing Bank – the issuing bank should have an acceptable credit rating or the SBLOC should be confirmed by a bank with an acceptable credit rating.
- Assignability – the SBLOC should be assignable to: (i) a financer; and (ii) a new lessor.

Security Agreement

This is an agreement that evidences the granting of a Security Interest in Collateral and is used in a Mortgage Financing. This term is used interchangeably with Mortgage or Trust Indenture.

Standstill Agreement

This is an agreement among creditors (only) or creditors and debtors by which (certain of) the creditors agree to refrain from exercising remedies in respect of a financing on stated terms and conditions, and for a limited time frame.

Sublease Agreement

This is an agreement evidencing a lease transaction between a person that holds a leasehold interest in an Aircraft Asset (that is, that is not the owner thereof) and a sublessee. Sublease Agreements are used in primarily two contexts: (i) an operating lessee/airline has obtained an Aircraft Asset pursuant to a Lease Agreement but that Aircraft Asset represents excess capacity so such airline finds another airline to use that asset; or (ii) in the context of a Lease/Head Lease Structure. Sublease Agreements may be Subject and Subordinate to the related Head Lease Agreement.

Tax Indemnity Agreement (TIA)

A Tax Indemnity Agreement (TIA), largely used in connection with U.S. Leveraged Leases, is an agreement by the related lessee to indemnify the related equity investor (the Owner Participant) for the failure by the Owner Participant to realize the tax benefits associated with the ownership of the related Aircraft Asset on particular terms and condition.

Term Sheet

The Term Sheet for a financing is a document that highlights the basic business understanding for that financing. A Term Sheet would likely cover the following topics:

- General Description/Summary
- Parties
- Collateral
- Basic Economic Terms
  - In the case of a Lease
    - Rent
    - Maturity Date
    - Termination/Stipulated Loss/Insured Value
    - Early Termination Options
    - Purchase/Renewal Options
    - Maturity Date
  - In the case of a Loan
    - Principal Amount of Loan
    - Amortization
    - Interest Rate
    - Maturity Date
    - Fees (Commitment/Up-front)
    - Prepayment
- Delivery Conditions (Lease)
- Return Conditions (Lease)
- Insurance Requirements
- Special Covenants (as applicable)
  - LTV Coverage (Loan)
  - Financial covenants
  - Turbo Events (Back-Leveraged Loan)
- General Covenants
  - Leasing/Subleasing
  - Reregistration
  - Inspection
  - Maintenance/ADs (and AD-Cost Sharing, if a Lease)
- Conditions Precedent (which may include internal approvals by the parties)
- Events of Default
- Transaction Expenses
- Governing Law
- Confidentiality

In addition to these basic items, an important purpose of a Term Sheet is to ferret out any headline issues prior to the commencement of documentation and to have a meeting of the minds on those issues prior to the commencement of documentation. It is best practice, for example, to flesh out Intercreditor terms (see Part 9) if that is pertinent to the transaction.

A Term Sheet may be issued as a binding contract, as a LOMI or somewhere in-between – where only certain provisions are binding (such as transaction expenses and confidentiality provisions).

Termination Agreement

This is an agreement that evidences the termination of a Lease Agreement (or Mortgage Financing).

Trust Agreement

This is an agreement evidencing an Owner Trust structure.

PRACTICE NOTE

Trust Agreements used for U.S.-registered Aircraft should be submitted in advance to the FAA's aeronautical counsel for review and approval, together with any other agreement (such as an operating agreement) legally affecting a relationship under a trust (14 CFR 47.7(c)(2)(i)).

Trust Indenture

This is an agreement that evidences the granting of a Security Interest in Aircraft Assets and other collateral and is used in a Mortgage Financing. This term is used interchangeably with Mortgage or Security Agreement.

Warranty Agreement

This is an agreement providing for the grant of Warranties.

Warranty Assignment

This is an agreement for the assignment to financing parties of a Warranty Agreement. The related OEM would typically need to consent to such assignment.

SPECIAL CONTRACTUAL PROVISIONS

Introduction

Aircraft documentation as reviewed in Part 12 contain the contractual provisions addressed in the respective terms, generally speaking. Aircraft documentation reviewed in Part 12 may also contain the contractual provisions defined below. There are a number of contractual terms that are in use in Aircraft Finance transactions that are industry jargon terms that may need explanation. Part 13 reviews these terms.

Concentration Limits

In a Securitization or Warehouse, Concentration Limits are restrictions imposed on the borrower/issuer to preserve the benefits of diversification. These restrictions are geared to lower transaction risk by requiring the issuer to limit 'placing all its eggs in one basket'. The restrictions limit concentration of the following.

- Aircraft Assets by type:
  - cargo versus commercial;
  - narrow body versus wide body;
  - regional jets versus non-regional jets;
  - Airbus versus Boeing; and
  - old versus new.
- Lessees:
  - geographical:
    - by continent – Asia versus North America versus Europe and so on; and
    - by country;
  - by credit quality; and
  - by size (for example, revenue).
- Leases by type:
  - short-term versus long-term;
  - fixed versus floating: and
  - Finance Lease versus Operating Lease.

PRACTICE NOTE

Due regard will need to be given to facility ramp-up periods when Concentration Limits will inevitably be out-of-whack.

Cross-collateralization

This is a contractual provision providing that multiple Aircraft Assets serve as collateral security for a single liability. Cross-collateralization may arise in a single transaction that involves the financing of multiple Aircraft Assets, and may arise across multiple transactions when the Aircraft Assets serving as collateral under each of them will support the obligations of the other unrelated transactions.

PRACTICE NOTE

To properly reflect Cross-collateralization, the 'other' collateral must be covered both in the Granting Clause and in the Waterfall. As to the Waterfall, there will need to be agreement on just where the proceeds of the 'other' collateral are placed (for example, who gets final dibs). Another item to address is whether the Cross-collateralization is 'one way' or 'two-way'. Looking at two separate transactions, for example, does the collateral on deal No. 1 secure the obligation on deal No. 2 only (one-way) or does the collateral on deal No. 1 secure the obligations on deal No. 2 *and* the collateral on deal No. 2 secure the obligations on deal No. 1 (two-way). Finally, transactions with the benefit of Cross-collateralization need to address the prospect of individual Aircraft covered thereby falling out of a transaction (by reason of, for example, sale or Event of Loss), since there may be concerns about cherry-picking and loss of the stronger collateral.[1]

> [1] For example, a transaction may require a prepayment of 125 percent of the loan associated with an Aircraft sale, with the excess 25 percent applied to the loans associated with the other financed Aircraft on a pro rata basis.

Cross-default

This is a contractual provision in a particular Lease Agreement or Mortgage providing that such Lease Agreement or Mortgage will be in default if the lessee/borrower defaults on its obligations under certain other specified contracts or obligations.

Cross-defaults can be an especially powerful tool in the context of a U.S. airline bankruptcy when Section 1110 is applicable. In a multi-Aircraft financing, absent Cross-default, if an airline debtor is subject to Chapter 11 bankruptcy proceedings, the airline can pick and choose which Aircraft to keep (under Section 1110(a)), reject/surrender (Section 1110(c)) or renegotiate (Section 1110(b)), which leaves the Aircraft Financier at risk with being stuck with less than desirable Aircraft. However, if the Aircraft are Cross-defaulted, the airline is faced with an all-or-nothing choice since the rejection/return of any one 'underwater' Aircraft would default all of the other Aircraft, thereby requiring the airline to return all of the Aircraft, even those that it wanted to keep. Aircraft Cross-defaulted to a core-Aircraft, the debtor will not be able to return/reject the underwater Aircraft if it wants to keep the core Aircraft, and would need to be careful to perform on that underwater Aircraft's Financing so as to have continued possession of its core Aircraft. This Cross-defaulting can be effected not only in a single facility but also across multiple facilities.

PRACTICE NOTE

Cross-defaults can be tricky insofar as there are a large number of variables that can be implicated when assessing exactly what the clause is to look like. Here are variables to consider.[1]

- Cross-default to *what*?
  - Type – debt for borrowed money? GAAP debt? Leases? Commodity Hedges? Interest Rate Swaps? Limit the debt obligation held by the particular lender/lessor? Other?
  - Dollar thresholds – how much? A simple dollar amount, or an amount determined based on another variable, say, percentage of Net Worth? However determined, is the amount:
    - original principal amount?
    - outstanding principal amount?
    - amount not paid when due?
    - measured individually or cumulatively?
    - limited to debt/leases held by specified parties?
    - how to calculate values for Operating Leases, Hedges and Swaps?
- What *type* of default?
  - Failure to pay when due? Any amount or just principal and interest?
  - Failure to pay after grace periods run?
  - Any incipient default?
  - Any full-blown 'event of default' (that is, grace periods have run, and lenders/lessors are entitled to exercise remedies)?
  - Any of the preceding matters having occurred and X number of days have elapsed from such occurrence?
  - In case of debt, acceleration required?
  - In case of a lease, exercise of remedies – which ones?
- Should the financier bootstrap to another particular transaction, say, a debtor's primary credit facility?
  - Who are the holders of that debt? Do they have aligning interests?
  - What happens if modified? Waived?
  - What happens if terminated?

Another item to address is whether the Cross-default is 'one way' or 'two-way'. Looking at two separate transactions, for example, does default on deal No. 1 default the financing of deal No. 2 only (one-way) or does default on deal No. 1 default the financing of deal No. 2 *and* default on deal No. 2 default the financing of deal No. 1 (two-way).

---

[1] Kroft, SR, 'Cross-default provisions in financing and derivatives transactions', *The Banking Law Journal*, March 1996.

Defaulting Lender

In credit facilities with multiple lenders, the borrower runs the risk that a lender will not fund its portion of a committed loan at the time of intended drawdown. This risk surfaced during the financial meltdown of 2007/2008 when the world's financial institutions suffered credit and liquidity issues, and as a result, a number of these institutions failed to fund their commitments. Banks based in Iceland were especially susceptible to defaulting on their loan commitments at this time. In light of this experience, credit facilities with multiple lenders having forward (i.e., future) lending commitments often have detailed provisions on how to deal with lenders who have defaulted in their funding obligations (a 'Defaulting Lender'). Provisions relating to Defaulting Lenders address matters such as:

(i) limiting voting rights of a Defaulting Lender;
(ii) blocking payments to a Defaulting Lender; and
(iii) interest charges assessed on a Defaulting Lender.

PRACTICE NOTE

The concept of a Defaulting Lender is quite a recent phenomenon, largely driven by the bank liquidity problems stemming from the market meltdown of 2008. A number of banks (for example, Icelandic banks) during that period were unable to meet their funding obligations.

Debt Services Coverage Ratio (DSCR)

In a closed-system structured financing (such as a Warehouse or Securitization), the Debt Services Coverage Ratio (DSCR) is the measure of whether required Debt Service has sufficient coverage from Lease rental cash flows. Failure to satisfy a DSCR test may force a transaction to go Turbo.

PRACTICE NOTE

A DSCR test will have two components for any applicable period:

1   the Debt Service component; and
2   the Lease rentals component.

The Debt Service component will, of course, cover the Debt Service for the relevant period, as adjusted by any Hedge inflows (which would be a deduction) or Hedge outflows (which would be an addition). In addition, to ensure properly whether the rental income is sufficient for a relevant period, all costs payable ahead of, or on par with, Debt Service pursuant to the relevant Waterfall should be included as well; this would include servicing fees, legal costs, trustee costs, reserve payments and so on.

The Lease rentals component should be fairly straightforward.

The servicing period on a retrospective back basis should be long enough to smooth out any aberrations. Accommodations will need to be made for any facility ramp-up period.[1]

---

[1] Some financiers may view a look-forward calculation as a better gauge of portfolio health since that can take into consideration recently defaulted leases.

Early Buy-out Option (EBO)

Early Buy-out Option (EBO) is a right of the lessee in a Lease Financing to purchase the leased asset prior to scheduled lease termination at a fixed or current-market Fair market value (FMV) price.

Earnings Before Interest, Taxes, Depreciation and Amortization (EBITDA)

Earnings Before Interest, Taxes, Depreciation and Amortization (EBITDA) is a financial condition measuring metric that is intended to test the operating cash flow of an operating company. EBITDA is often measured against another metric such as Debt Service.

Event of Loss/Total Loss

This is a condition that parties to an Aircraft Asset financing agree results in the loss of such Aircraft Asset that either requires re-payment of a loan under a related Mortgage Financing or the payment of Stipulated Loss Value under a Lease Financing or, if an applicable option, requires the mortgagor/lessee to replace the lost Aircraft Asset with a comparable asset in satisfaction with documentation requirements. Events of Loss for any Aircraft Asset may include any of the events with respect to such property specified in Appendix J.

The most typical (and obvious) type of event covered by this term is when an Aircraft is so damaged as to be beyond economical repair, having regard to its value; and this is the type of loss Hull Insurance policies (such as AVN67B and its U.S. equivalent) will cover. This term also picks up the situation where the insurers of an Aircraft enter into a settlement agreement with the insureds as to the determination that such an event has occurred. The matter of whether a Total Loss has occurred is not necessarily cut and dried. While a catastrophic crash is obviously a Total Loss, other events may not be so obvious. For example, if an Aircraft runs off a runway and suffers structural damage, or has had a Hard Landing, the insurers may seek to find that no Total Loss has occurred, electing to provide funds instead to repair the Aircraft. Indeed, whether or not an Aircraft suffering an 'event' has damaged it beyond economical repair is a matter where maintenance experts – and litigators – can get involved.

A closer look at the different events that may constitute an Event of Loss as per Appendix J will show that some of the events are not insurable losses; that is, the

Exhibit 13.1 Event of loss
Source: Getty Images: #171543012

standard insurance policies do not cover the specific event – in which case the financier would be looking to the credit of the relevant airline. As well, certain of these events would not be insured under a standard Hull Insurance policy, but rather a Hull War Risk Insurance policy (e.g., confiscation by a government).

PRACTICE NOTE

Many of the standard Events of Loss events are *not* events that are insurable losses.

Excluded Countries

These are jurisdictions specified in financing documents in which financed Aircraft may not be operated, leased/subleased into or re-registered in. Financiers seek to prevent their financed Aircraft from being located in specified jurisdictions for a number of reasons.

- Certain 'bad' countries are actually named on OFAC, UN or EU-sanction or other verboten lists, making it illegal to conduct business in those jurisdictions.
- Certain countries have inadequate laws or enforcement to protect the interests of Aircraft lessors and mortgagees (see Part 7, 'Country Risk').

- Certain countries, while having adequate laws to protect Aircraft Financiers, have poor records in the enforcement of laws generally or in the enforcement of Aircraft repossession laws particularly[1] (see Part 7, 'Country Risk').
- Financiers may want to limit the concentration of risk in a particular jurisdiction (taking into account other Aircraft in their portfolio, and not necessarily the particular lessor they are financing).

PRACTICE NOTE

In the case of the financing of Aircraft for operating lessors, these lessors seek to have as few restrictions as possible on the permissible jurisdictions in which they can deploy or redeploy an Aircraft insofar as they want to preserve their options to take advantage of as many jurisdictional opportunities as is possible worldwide. Lessors, therefore, prefer including a list of Excluded Countries, as compared with 'Permitted Countries' (which is typically more narrow) (see 'Permitted Countries').

Material Adverse Change (MAC)

Material Adverse Change (MAC) clauses are representations by the debtor and/or conditions precedent as to the lack of the occurrence of a MAC. They would typically measure whether there has been a Material Adverse Change in a debtor's financial condition (business, operations or prospects) as from a benchmark date (which is usually the date of the debtor's most recently published audited financial statements). The theory here is that financiers do not want to be lending or leasing into a problem. In a number of non-U.S. jurisdictions, a MAC clause may be included as an event of default.[2] While a 'financial condition' MAC clause would entail an examination of the debtor's latest financial condition against the one at the benchmark date (with a MAC, therefore, needing to show a significant erosion of net worth and/or cash balance position),[3] a MAC that included 'prospects' can take into account recent events, such as SARS – or 9/11 – type events that can have an impact on future financial viability.

In addition to the credit-related MAC described above, it is not uncommon to see 'Market MACs'. These MACs provide funding and/or underwriting outs for financiers if there is a MAC in the relevant LIBOR-funding markets (in the case of banks) (see Part 4, 'Funding MAC') or loan syndication markets (in the case of underwriters).

PRACTICE NOTE

MAC clauses may also be oriented as 'material adverse effect' provisions. The usage of 'effect' rather than 'change' may have a more expansive application beneficial to a financier. 'Calling' a MAC requires very careful consideration by its beneficiary. If the MAC is not tied to hard and objective financial

---

covenants, the beneficiary must be very comfortable that a MAC has really occurred. The risk to a beneficiary of calling a MAC, of course, is that such action may have catastrophic results for the MAC's target; it may trigger cross-defaults (in the case of MAC that is an event of default) or trigger a default on, say, an Original Equipment Manufacturer (OEM) purchase agreement (in the case of MAC condition precedent). That target may dispute the occurrence of the MAC, and if it wins that dispute (most likely in court), the beneficiary may face large damage claims against it. While 'I know it when I see it'[1] may be an approach, the beneficiary of a MAC will need to express a clear case for calling it. In the world of airline finance, cash balances may be a good proxy for a deteriorating credit, with, say, a 50 percent drop in such balances as a clear case of the occurrence of a MAC (unless a 50 percent drop is normal due to seasonal fluctuations).

---

[1] Justice Potter Stewart used this phrase when the U.S. Supreme Court was considering a definition for pornography/obscenity. See his concurring opinion in *Jacobellis v Ohio*, 378 U.S. 184 (1964), regarding possible obscenity in *The Lovers*.

Mandatory Document Terms

These are specified documentary terms required to be included in lease or mortgage financing documents in respect of Aircraft Assets that have been financed or are committed for financing and prior to the actual negotiation, drafting, execution and delivery of such documents. A typical context for this would be in the case of Aircraft financed in a Warehouse or Collateralized Lease Obligation (CLO), where Aircraft may come off-lease and need to be redeployed under new leases. To protect the Aircraft Financier, Mandatory Document Terms would typically be included covering certain baseline requirements for Aircraft-related (and other) provisions pertaining to:

- insurance (including minimum liability and hull coverages);
- lien-lifting;
- maintenance;
- possession; and
- events of default and remedies.

These parameters seek to ensure that the Aircraft collateral securing the investments held by the Aircraft Financiers is properly preserved.

Mandatory Economic Terms

These are specified economic terms required to be included in lease or mortgage financing documents in respect of Aircraft Assets that have been financed or are committed for financing and prior to the actual negotiation, drafting, execution and delivery of such documents. A typical context for this would be in the case of Aircraft financed in a Warehouse or CLO, where Aircraft may come off-lease and need to be redeployed under new leases. To protect the Aircraft Financier, Mandatory Economic Terms would typically be included covering certain baseline economic requirements addressing:

- minimum rent;
- minimum agreed value; and
- minimum/maximum term.

These parameters seek to ensure, among other things, that the Aircraft Financiers maintain sufficient cash flow to service their investments.

Most Favored Nation (MFN)

A Most Favored Nation (MFN) provision is one where a debtor agrees to provide a creditor with terms and conditions no less favorable than the best deal it has offered to any of its other creditors.

Net Lease

A Net Lease clause provides that all costs in connection with the use of an Aircraft Asset are paid by the lessee and are not obligations of the lessor/owner. Most Operating Leases and Finance Leases are net leases. These leases are often also called 'triple net' leases (referring to taxes, insurance and maintenance). In contrast, there are 'full service' leases, where these costs are for the account of the lessor, but these full service leases are seldom found in commercial Aircraft Finance (other than for cargo Aircraft as in Aircraft, Crew, Maintenance and Insurance (ACMI)).

Permitted Countries

These are jurisdictions specified in financing documents in which financed Aircraft may be operated, leased/subleased in or re-registered. Financiers seek to limit the universe of countries their financed Aircraft are located for the reasons specified in 'Excluded Countries'. In contrast to Excluded Countries, which specifically address these jurisdictions in which the financed Aircraft may *not* be located, a Permitted Country list specifically addresses the more limiting list of countries in which the financed Aircraft *may* be located.

PRACTICE NOTE

A Permitted Country list may be the more likely approach when dealing with a financing directly with an airline which is not likely to be too focused on alternative Aircraft deployment options insofar as that airline is expecting to need the contracted-for lift for the entire term of the financing. This is in contrast to an operating lessor which is always concerned about redeployment options in case it must take back an Aircraft prior to the end of its scheduled term.

Purchase Option

This is the option to buy leased property at the end of the Lease term (or, at stated dates during the lease term). The Purchase Option price may be at a fixed price or at FMV. In the U.S., if the tax characteristics of a true lease are to be protected, the purchase option may not be at a price less than the asset's fair market value at the time the right is exercised. (See also Part 7, 'Bargain Purchase Option'.)

Renewal Option

This is a right whereby the lessee has the option to renew a lease for an additional period (or periods) after the original termination date at an agreed rate and for a specified term.

Trigger Event

This is an event which, when occurring, has consequences to a financing. Trigger Events come in two categories.

1   *Credit* – if there is some credit support for a deal and that credit party violates any cash, EBITDA, Threshold Rating or Net Worth threshold, then that party may have to provide additional credit support. In the context of Back-leveraged Operating Leases, this may mean that the borrower needs to turn over to the lender Maintenance Reserves and/or Security Deposits.
2   *Collateral* – in a secured financing, the inability to satisfy a Loan to Value (LTV) test or a DSCR test may force a Turbo Event.

Turbo

Typically caused by a Trigger Event, a Turbo is the transition of a Securitization or Warehouse to an accelerated repayment profile. This would shift payments that would otherwise go to an equity investor or a junior Tranche holder to the senior debt Tranche until the Trigger Event is subsequently 'cured'.

Notes

1   Several banks and operating lessors have had bad experiences in India recently in the retrieval of Aircraft from a failed Indian carrier. Even though India is a Contracting State to the Cape Town Convention, local authorities would not allow banks and lessors to de-register their Aircraft or, in those cases where de-registration was permitted, the airport authorities demanded millions of dollars in unpaid dues from the banks and lessors as a condition to allowing the Aircraft to fly (even within India). Needless to say, India now appears more regularly on the list of Excluded Countries.
2   For U.S. purposes, it is usually viewed as too vague for calling an event of default, thereby subjecting the creditor to potential 'lender liability' risks. Similarly, lenders are rather cautious about invoking this clause as a loan-blocker. For a lender to safely invoke this clause, the material adverse change must be extremely clear and apparent.
3   The latter being especially relevant for airlines.

PART 14

OTHER

Introduction

There is much industry jargon that is not characterizable in any of the preceding Parts of this Handbook, hence the lumping of a number of them in this 'Other' category. These jargon terms cover, for example, industry organizations and generally used acronyms that are ubiquitous in the Aircraft Finance world's dialogue.

Aviation Working Group (AWG)

The Aviation Working Group (AWG) began work in 1994, at the request of the International Institute for the Unification of Private Law (UNIDROIT), as an ad hoc industry group to contribute to the development of the Cape Town Convention on International Interests in Mobile Equipment and the Protocol to the Convention on International Matters Specific to Aircraft Equipment, which was signed in 2001. In 2002, the AWG became a not for profit legal entity and its scope of activity has significantly expanded. It now addresses a wide range of topics affecting international aviation financing, including issues relating to insurance. For example, AWG was heavily involved in the discussions to update AVN67B to AVN67C in 2007. AWG is also a forum for considering new issues as they emerge and develop, as well as continuing to monitor issues of importance to the international aviation financing and leasing community. AWG is co-chaired by Airbus and Boeing, and its members and affiliates comprise the major aviation manufacturers and financial institutions, including most of the world's largest leasing companies. AWG's members and their affiliates manufacture substantially all modern commercial aircraft and engines, and lease and finance a substantial majority of such new equipment. The AWG website (AWG.aero) is a useful resource.

Civil Reserve Air Fleet (CRAF)

The Civil Reserve Air Fleet (CRAF) is part of the U.S. mobility reserves. Selected aircraft from U.S. airlines, contractually committed to the Civil Reserve Air Fleet, support U.S. Department of Defense airlift requirements in emergencies when the need for airlift exceeds the capability of military aircraft. To join the Civil Reserve Air Fleet, carriers must maintain a minimum commitment to CRAF of 30 percent of its Civil Reserve Air Fleet capable passenger fleet and 15 percent of its Civil Reserve Air Fleet capable cargo fleet. Aircraft committed must be U.S. registered and carriers must also commit and maintain at least four complete crews for each Aircraft.

Three stages of incremental activation allow for tailoring an airlift force suitable for the contingency at hand. Stage I is for minor regional crises, Stage II would be used for major theatre war and Stage III for periods of national mobilization.

CRAF has been activated once as part of Operation Desert Shield and once as part of Operation Iraqi Freedom.

During any period an Aircraft is activated under CRAF and is in the possession of the government of the United States of America or an instrumentality or agency of the government, a secured party or lessor may not, on account of any event of default, be entitled to exercise repossessory-type remedies in such manner as to limit the operator's control of such Aircraft, unless at least 60 days' (or such lesser period as may then be applicable under the Military Airlift Command program of the government of the United States of America) prior written notice of default shall have been given by the lessor or secured party by registered or certified mail to the operator with a copy addressed to the Contracting Office Representative for the Military Airlift Command of the U.S. Air Force under the contract with the operator relating to such Aircraft.

CUSIP

A nine character U.S.-based security identifier assigned by Standard & Poor's CUSIP Service Bureau. CUSIP is an acronym for Committee on Uniform Securities Identification Procedures. In case of a security acquired by an insurance company, the security designation would be a PPN rather than a CUSIP.

Department of Transportation (DOT)

The Department of Transportation (DOT) is a Department in the Executive Branch of the U.S. government.

Double Tax Treaty

A Double Tax Treaty is a tax treaty between two countries that addresses the problem of double taxation by two jurisdictions on the same declared income (in the case of income taxes), asset (in the case of capital taxes), or financial transaction (in the case of sales taxes). It is not unusual for a business or individual who is resident in one country to make a taxable gain (profits) in another. This person may find that he is obliged by domestic laws to pay tax on that gain locally and pay again in the country in which the gain was made. Since this is inequitable, many nations make bilateral double taxation agreements with each other. In some cases, this requires that tax be paid in the country of residence and be exempt in the country in which it arises. In the remaining cases, the country where the gain arises deducts taxation at source ('withholding tax') and the taxpayer receives a compensating foreign tax credit in the country of residence to reflect the fact that tax has already been paid. To do this, the taxpayer must declare himself (in the foreign country) to be non-resident there.

European Aeronautic Defense and Space Company (EADS)

The European Aeronautic Defense and Space Company NV is an international conglomerate based in Europe[1] that is a major participant in aerospace, defense and related services. EADS owns Airbus SAS, the manufacturer of Airbus commercial Aircraft.

Emission Trading Scheme (ETS)

The Emission Trading Scheme (ETS) is the EU's directive for controlling greenhouse gas emissions. The ETS was effected pursuant to Directive 2003/87/EC (13 October 2003) (establishing a scheme for greenhouse gas (GHG) emission allowance trading within the EU and amending Council Directive 96/61/EC) (the 2003 Directive) and Directive 2008/101/EC (19 November 2008) (amending the 2003 Directive to include aviation activities in the scheme for GHG emission allowance trading within the Community) (the 2008 Directive). The ETS purports to have an extraterritorial reach, with covered operators required to monitor and surrender (and pay for, if necessary) allowances corresponding to CO_2 emissions from every point of their flights to, from or within the EU – whether on the ground in, or over, non-EU territory or international water. There is much resistance by U.S., and other non-European, carriers to this extraterritorial 'grab'.

Eurocontrol

Eurocontrol, the European Organization for the Safety of Air Navigation, is the organization that co-ordinates and plans air traffic control for all Europe. This involves working with national authorities, air navigation service providers, civil and military airspace users, airports, and other organizations. Its activities involve all gate-to-gate air navigation service operations: strategic and tactical flow management, controller training, regional control of airspace, safety-proofed technologies and procedures, and collection of air navigation charges (see Part 7, 'Eurocontrol/Airport Charges').

Foreign Account Tax Compliance Act (FATCA)

The Foreign Account Tax Compliance Act (FATCA) is U.S. legislation that is a portion of the 2010 Hiring Incentives to Restore Employment (HIRE) Act that requires individuals to report their financial accounts held overseas and foreign financial institutions to report to the IRS about their American clients. FATCA was designed primarily to combat off-shore tax evasion and recoup federal tax revenues.

Funding Gap

See introduction to Part 3.

Generally Accepted Accounting Principles (GAAP)

Generally Accepted Accounting Principles (GAAP) are the standards, conventions and rules accountants follow in the U.S. in recording and summarizing transactions and in the preparation of financial statements.

International Air Transport Association (IATA)

The International Air Transport Association (IATA) is the trade association for the world's airlines, representing some 240 airlines or 84 percent of total air traffic. IATA supports many areas of aviation activity and helps formulate industry policy on aviation issues.

International Civil Aviation Organization (ICAO)

International Civil Aviation Organization (ICAO) is a specialized agency of the UN, established to promote safe and orderly development of international civil aviation throughout the world. ICAO sets standards and regulations necessary for aviation safety, security, efficiency and regularity, as well as for aviation environmental protection. It serves as the forum for co-operation in all fields of civil aviation among its 191 member states.

International Financial Reporting Standards (IFRS)

International Financial Reporting Standards (IFRS) are the standards, conventions and rules accountants follow worldwide (other than the U.S.) in recording and summarizing transactions and in the preparation of financial statements.

Incident Report

This is the notification to the National Transportation Safety Board (NTSB) by the operator of an Aircraft of any 'accident' and certain 'incidents'. See 'National Transportation Safety Board'.

Investment Grade

This is a rating that indicates that a security (or issuer) has a relatively low risk of default. Bond rating firms, such as Standard & Poor's, use different designations consisting of upper and lower-case letters 'A' and 'B' to identify a bond's credit quality rating. 'AAA' and 'AA' (high credit quality) and 'A' and 'BBB' (medium credit quality) are considered investment grade. Credit ratings for bonds below these designations ('BB', 'B', 'CCC' and so on) are considered low credit quality, and are commonly referred to as 'junk bonds'. If an agency downgrades a company's bonds from 'BBB' to 'BB', such downgrade will cause a reclassification of such company's debt from investment grade to 'junk' status with just a one-step drop in quality. The repercussions of such an event can be highly problematic for the issuer and can also adversely affect bond prices for investors. Certain investors are permitted to invest in and/or hold only certain rated securities.

International Society of Transport Aircraft Traders (ISTAT)

International Society of Transport Aircraft Traders (ISTAT) is an industry body for people involved in the trading of commercial aircraft. ISTAT also certifies aircraft appraisers. The society runs multiple conferences each year. See their website, www.ISTAT.org, for more details.

Know Your Customer (KYC)

Know Your Customer (KYC) covers the procedures required by banks and other Aircraft Financiers to do due diligence on their counterparties to make sure that the debtors are not on any OFAC-type list or are otherwise 'not legit'. KYC requirements have taken on elevated scrutiny in recent years with the advent of strict anti-money laundering vetting rules imposed by government authorities. These rules require that Aircraft Financiers obtain significant amounts of information on their counterparties, authorized signatories and owners, and mandate the continued monitoring of their counterparties in financing transactions.

Loan Market Association (LMA)

The Loan Market Association (LMA) is a London-based association of financial and other institutions (primarily European and London-based) whose objective is to improve the liquidity, efficiency and transparency of the primary and secondary syndicated loan markets in Europe, the Middle East and Africa. The LMA publishes documents forms frequently used by industry participants in the English law market.

Letter of Intent (LOI)

Letters of Intent (LOIs) are often used as an indication of interest in the purchase and sale of Aircraft Assets. They may be legally binding, or they may not be. LOIs will usually allow for an inspection of the Aircraft Assets and the payment of a good faith deposit. They may also require the potential seller to pull the asset off the market for a specified period.

Loan Syndication and Trading Association (LSTA)

The Loan Syndication and Trading Association (LSTA) is a New York-based association of financial and other institutions (primarily U.S.-based) whose objective is to promote a fair, orderly, efficient and growing corporate loan market that provides leadership in advancing and balancing interests of all market participants. The LSTA publishes New York law governed forms frequently used by industry participants, especially in the trading of 'aircraft loan paper'.

Mark-to-Market

Mark-to-Market is the act of recording the price or value of a security, portfolio or account to reflect its current market value rather than its book value. Problems can arise when the market-based measurement does not accurately reflect the underlying asset's true value. This can occur when a company is forced to calculate the selling price of these assets or liabilities during unfavorable or volatile times, such as a financial crisis. For example, if liquidity is low or investors are fearful, the current selling price of a bank's assets or collateral could be much lower than the true value.[2] Aircraft Assets serving as collateral may be Marked-to-Market as part of a Loan to Value (LTV) analysis.

National Association of Insurance Commissioners (NAIC)

The National Association of Insurance Commissioners (NAIC) is the U.S. standards setting and regulatory support organization created and governed by the chief insurance regulators from the 50 states, the District of Columbia and five U.S. territories. Through the NAIC, state insurance regulators establish standards and best practices, conduct peer reviews and co-ordinate their regulatory oversight. NAIC staff supports these efforts and represents the collective views of state regulators domestically and internationally. NAIC members, together with the central resources of the NAIC, form the national system of state-based insurance regulation in the U.S.

National Association of Insurance Commissioners' Rating

These are ratings of securities provided by the NAIC's Securities Valuations Office (SVO). The SVO is responsible for the day-to-day credit quality assessment and valuation of securities owned by state regulated insurance companies. Insurance companies report ownership of securities to the Capital Markets and Investment Analysis Office when such securities are eligible for filing on Schedule D or DA of the NAIC Financial Statement Blank. The Capital Markets and Investment Analysis Office conducts credit analysis on these securities for the purpose of assigning an NAIC designation and/or unit price. These designations and unit prices are produced solely for the benefit of NAIC members who may utilize them as part of the member's monitoring of the financial condition of its domiciliary insurers. Unlike the ratings of nationally recognized statistical rating organizations, NAIC designations are not produced to aid the investment decision-making process and therefore are not suitable for use by anyone other than NAIC members. See 'Appendix D' for NAIC Ratings as compared with Rating Agency Ratings.

National Transportation Safety Board (NTSB)

The National Transportation Safety Board (NTSB) is a division of the DOT, the NTSB is charged with investigating Aircraft-related accidents and incidents. Although the NTSB delegates some accident investigation to the FAA, the notification required by Part 830 must be made to the NTSB. Under the FAR Part 830.2 (49 CFR 830), an 'Accident' is an occurrence associated with the operation of an aircraft which takes place between the time any person boards the aircraft with the intention of flight and all such persons have disembarked, and in which any person suffers death or serious injury, or in which the aircraft receives substantial damage.

'Substantial damage means damage or failure which adversely affects the structural strength, performance, or flight characteristics of the aircraft, and which would normally require major repair or replacement of the affected component.' Substantial damage does not include: engine failure or damage limited to an engine if only one engine fails or is damaged, bent fairings or cowling, dented skin, small punctured holes in the skin or fabric, ground damage to rotor or propeller blades, and damage to landing gear, wheels, tires, flaps, engine accessories, brakes or wingtips.

An 'incident' is defined as 'an occurrence other than an accident, associated with the operation of an aircraft, which affects or could affect the safety of operations'. A Hard Landing would qualify as an incident.

Incidents involving large, multiengine aircraft (more than 12,500 pounds maximum certificated take-off weight) must be reported if they involve: (i) in-flight failure of electrical systems which requires the sustained use of an emergency bus powered by a back-up source such as a battery, auxiliary power unit, or air-drive generator to retain flight control or essential instruments; (ii) in-flight failure of hydraulic systems that results in sustained reliance on the sole remaining hydraulic or mechanical system for movement of flight control surfaces; (iii) sustained loss of the power or thrust produced by two or more engines; and (iv) an evacuation of an Aircraft in which an emergency egress system is utilized.

PRACTICE NOTE

It is not uncommon for Return Conditions to stipulate that an Aircraft Asset be returned without any Incident Report of record.

Over the Top (OTT)

Over the Top (OTT); the acronym is used when someone is suggesting something unusual or off-market (or off-color).

Part-out

This is the dismantling of an Aircraft Asset, and the sale of the parts and metal. This would be done if an Aircraft Asset is not saleable or leasable.

Pool Balance

Pool Balance is a term used in EETC financings to reflect, fundamentally, how much allocated principal (in relation to the Equipment Notes in a Pass Through Trust) remains outstanding.

PPN

A PPN is a private placement number assigned by Standard & Poor's CUSIP Service Bureau in cooperation with the NAIC. A PPN may be assigned to privately placed securities purchased by insurance companies.

Private Export Funding Corporation (PEFCo)

The Private Export Funding Corporation (PEFCo) is a private sector lender established to assist in the financing of U.S. exports by supplementing the financing available from commercial banks and other lenders. Shareholders include commercial banks, industrial companies and financial services companies.

Rating Agency

A company that assigns credit ratings on particular debt securities (or preferred stock) – rating the debtor's ability to pay back the debt by making timely interest payments and of the likelihood of default. A Rating Agency may rate the creditworthiness of issuers of debt obligations, or the debt instruments themselves.

In Aircraft Finance, the Rating Agencies are most apt to be rating Enhanced Equipment Trust Certificates (EETCs), Collateralized Debt Obligations (CDOs) and Collateralized Lease Obligations (CLOs) (and related products).

Some investors (like insurance companies) are unable to invest in unrated securities, so a credit rating for them is imperative. A credit rating also permits – or makes much more easy – the trading of securities in the secondary market. The level of credit rating on a particular security will have a bearing on the interest rate borne by that security, with a higher rating leading to a lower interest rate.

The Rating Agencies are usually rather transparent about the methodologies utilized in their analysis of securities issued in the Aircraft Finance sector. These methodologies may be on their websites. See, for example, Standard & Poor's The Rating Process for Aircraft Financing – Standard & Poor's Structured Finance – Aircraft Securitization Criteria available at www.standardandpoors.com/ratings/criteria/en/us.

Credit rating is a highly concentrated industry with the three largest Rating Agencies – Moody's Investors Service, Standard & Poor's and Fitch Ratings – controlling most of the ratings business. Kroll Bond Rating Agency has made some headway in the rating of certain Aircraft Finance Securities in recent years.

All of the major Rating Agencies qualify as a Nationally Recognized Statistical Rating Organization (NRSRO), a designation made by the U.S. Securities and Exchange Commission. The credit ratings obtained by insurance companies on the securities they acquire must be provided by NSROs in order to allow their regulators to qualify such securities for reserve relief in respect of the insurance companies' investment portfolio.

Rating Agency Credit Scale

This is the scale of ratings offered by Rating Agencies on the credit of an issuer or asset protection afforded by the collateral securing a financing (see 'Appendix D').

Rats and Mice

These are obligations owing to a lender in a Mortgage Financing that do not include principal and interest. Such amounts include amounts payable on account of Yield Protection, LIBOR Breakage, Liquidity Breakage, Make-Whole, Prepayment Fee and Hedge Breakage Loss.

PRACTICE NOTE

Insofar as Rats and Mice are not predictable, they may be hard to factor in an analysis of cash flow coverage from, say, an Operating Lease. Accordingly, lenders may ask otherwise Non-recourse lessors to be recourse on these items.

Remarketing Agent

This is the person undertaking the responsibilities as such under a Remarketing Agreement. A Remarketing Agent is typically a party with experience in remarketing a particular asset. It may be an Original Equipment Manufacturer (OEM).

Servicer

This is the person taking on the responsibilities as such under a Servicing Agreement. A Servicer is typically an operating lessor with the requisite experiences, structure or staff to service an Aircraft Asset portfolio. They are typically paid fees that are a percentage of collected revenue.

Sharia'a

Sharia'a is Islamic law, which is pertinent for transactions structured under an Islamic Finance model (see, for example, Part 3 '*Sukuk*').

Sister Companies

Sister Companies are legal entities that share a common parent. In Aircraft Finance Securitizations, each Aircraft Asset is often placed in its own legal entity, such as a limited liability company, all of which are commonly owned by the issuer. See, for example, Special Purpose Entities (SPEs) owned by the issuer in Exhibit 3.5.

Threshold Rating

This is a rating level on the Rating Agency Credit Scale that, once crossed, has consequences to a financing. In an Enhanced Equipment Trust Certificate (EETC), the Liquidity Provider must maintain its credit rating above the assigned Threshold Rating; if it does not do so, that Liquidity Provider must place with the transaction trustee an amount equal to its committed liability under the related Liquidity Facility.

U.S. Export–Import Bank

The U.S. Export–Import Bank (U.S. Ex-Im) is the official U.S. export-credit agency, and its obligations are backed by the full faith and credit of the U.S. Government. The air transportation sector accounts for 46.3 percent of U.S. Ex-Im's total exposure, with the highest geographic concentration in Asia. During the fiscal year 2012, U.S. Ex-Im authorized more than U.S.$11.6 billion to support the export of 154 commercial Aircraft manufactured in the U.S., to 22 airlines and seven Aircraft leasing companies located in 21 different countries.

U.S. Major

This is a U.S. airline with revenue in any fiscal year in excess of U.S.$1 billion.

Value Added Tax (VAT)

Value Added Tax (VAT) is a form of consumption tax that is prevalent in Europe and many other jurisdictions (but not, currently, in the U.S.). VAT is similar to a sales tax in that ultimately only the end consumer is taxed. It differs from the sales tax in that, with the sales tax, the tax is collected and remitted to the government only once, at the point of purchase by the end consumer. VAT is relevant at the time an Aircraft Asset is sold in a VAT jurisdiction.

Withholding Taxes

These are taxes on a payee that are withheld by the payor (or its agent) on account of interest, rent or other earnings at prescribed rates. Withholding Taxes typically occur in cross-border transactions where there is no Double Tax Treaty or other available exemptions that would allow an avoidance of avoid tax withholding. Aircraft Finance transactions are often structured so as to eliminate any requirement for withholding. (See Part 3, 'Lease/Head Lease Structure'.) Aircraft Financiers would typically require a debtor to gross-up payments so that the received funds of interest or rent are paid in full.

Notes

1 Headquartered in both France and Germany.
2 But what, then, is true value?

PART 15

RISK

Before Aircraft Financiers put money out the door, their senior management necessarily asks: what are the risks for a particular transaction? Exhibit 15.1 schematically displays many of the risks inherent in an aircraft financing transaction.[1]

These risks can be broken up into three broad categories – Credit Risk, Asset Risk and Insurance Risk, with a fourth category – Systemic Risk – impacting both Asset Risk and Credit Risk. While Exhibit 15.1 largely speaks for itself, there are a few line items that may need further clarification.

In the case of Credit Risk –

60 Minutes Risk is the risk that some media outlet – like the widely-watched '60 Minutes' television program aired in the United States – would do some exposé on an airline which would impact ridership.

Valujet Risk is the risk that some untoward operational event occurs at an airline adversely affecting ridership. This risk is named after a U.S. airline by the name of Valujet that had the operational misfortune in 1996 of putting into the cargohold of a passenger aircraft oxygen canisters that were not fully depleted and secured, and such canisters caught fire bringing down the aircraft in the Florida Everglades, with the loss of 110 passengers and crew. The temporary suspension of that airline's operating certificate and attendant bad press doomed the airline (with its eventual merger into another U.S. airline, AirTran). Less egregious events like JetBlue's Valentine's Day Crisis in 2007, where an ice storm paralyzed the airline and left passengers trapped on aircraft for hours, can likewise have credit effects on an airline. In the aftermath of that crisis, JetBlue's founder and CEO, David Neeleman, was sacked.

In the case of Asset Risk –

Appraisal Bias is the risk that the appraiser(s) designated to perform appraisal(s) on the aircraft to be financed are too value optimistic. To be sure, different appraisers have different views on aircraft values. In order to smooth out such biases, some transactions require multiple appraisals to be obtained.

Documentation Risk is the risk that, heaven forbid, the lawyers did not properly draft the legal documents. The misdrafting can affect economic matters if the transaction's economic-specific terms are not properly reflected or legal matters (such as proper perfection language and processes) if counsel has not properly reflected local law requirements.

Aircraft Finance Risk

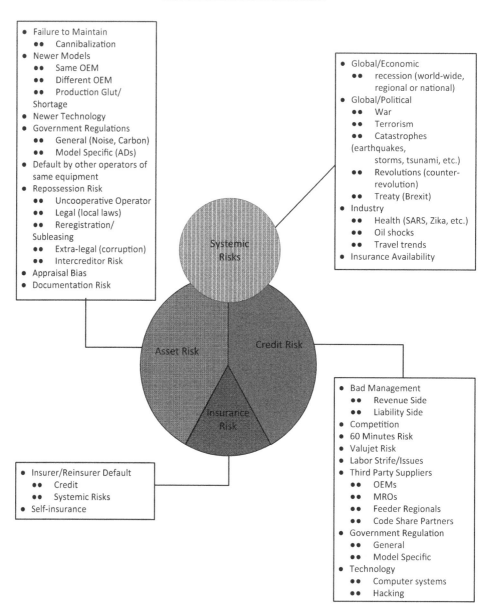

- Failure to Maintain
  - Cannibalization
- Newer Models
  - Same OEM
  - Different OEM
  - Production Glut/ Shortage
- Newer Technology
- Government Regulations
  - General (Noise, Carbon)
  - Model Specific (ADs)
- Default by other operators of same equipment
- Repossession Risk
  - Uncooperative Operator
  - Legal (local laws)
  - Reregistration/ Subleasing
  - Extra-legal (corruption)
  - Intercreditor Risk
- Appraisal Bias
- Documentation Risk

- Global/Economic
  - recession (world-wide, regional or national)
- Global/Political
  - War
  - Terrorism
  - Catastrophes (earthquakes, storms, tsunami, etc.)
  - Revolutions (counter-revolution)
  - Treaty (Brexit)
- Industry
  - Health (SARS, Zika, etc.)
  - Oil shocks
  - Travel trends
- Insurance Availability

Systemic Risks

Asset Risk   **Credit Risk**

Insurance Risk

- Insurer/Reinsurer Default
  - Credit
  - Systemic Risks
- Self-insurance

- Bad Management
  - Revenue Side
  - Liability Side
- Competition
- 60 Minutes Risk
- Valujet Risk
- Labor Strife/Issues
- Third Party Suppliers
  - OEMs
  - MROs
  - Feeder Regionals
  - Code Share Partners
- Government Regulation
  - General
  - Model Specific
- Technology
  - Computer systems
  - Hacking

Exhibit 15.1 Aircraft finance risk

Financier Risk

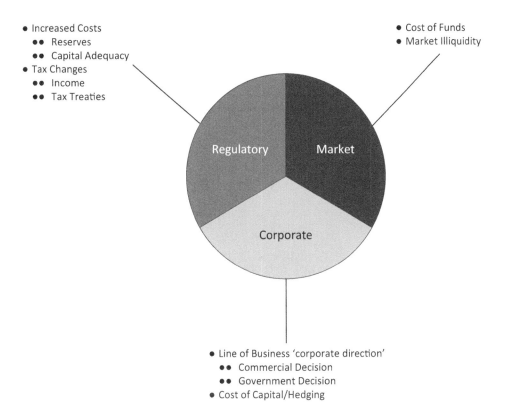

- Increased Costs
  - •• Reserves
  - •• Capital Adequacy
- Tax Changes
  - •• Income
  - •• Tax Treaties

- Cost of Funds
- Market Illiquidity

Regulatory   **Market**

Corporate

- Line of Business 'corporate direction'
  - •• Commercial Decision
  - •• Government Decision
- Cost of Capital/Hedging

Exhibit 15.2 Financier risk

In addition to the risks attendant to a transaction, financiers themselves face endemic risks to their doing business. Exhibit 15.2 is a schematic showing these risks.

The Regulatory Risks, of course, have greater impact on financiers that are regulated entities, especially banks, but can also impact insurance companies and other regulated lenders.[2]

Let me say a few words about the Corporate Risk items. Line of Business 'corporate direction' comes in two parts. First, are those that are part of a commercial decision. In this case, a financial institution may decide to exit, for example, the aircraft finance sector altogether, as a number of English banks did in 2008. Second, are those that are part of a government decision. In this second case, governments may direct a bank to exit a line of business, as a number of German states did in the aftermath of the 2007/2008 financial crisis when German lending institutions known as Landesbanks (which are German-state owned) were directed to exit the business of (among other) financing aircraft.

The risk identified as Cost of Capital/Hedging is the risk associated with a particular financial institution appropriately hedging, for example, their interest rate positions with the proper mix of long-term and short-term debt.

Notes

1 These identified risks are necessarily generic. Particular transactions will have their own particular risks based on transaction structures and other deal-specific factors.
2 Under the Dodd-Frank legislation in the United States, if an institution is designated as 'too big to fail' it is subject to a host of regulations that can impair its business.

APPENDICES

INSURANCE PROVISIONS AVN67B

AIRLINE FINANCE/LEASE CONTRACT ENDORSEMENT

It is noted that the **Contract Party(ies)** have an interest in respect of the **Equipment** under the Contract(s). Accordingly, with respect to losses occurring during the period from the Effective Date until the expiry of the Insurance or until the expiry or agreed termination of the **Contract(s)** or until the obligations under the **Contract(s)** are terminated by any action of the Insured or the **Contract Party(ies)**, whichever shall first occur, in respect of the said interest of the **Contract Party(ies)** and in consideration of the **Additional Premium** it is confirmed that the Insurance afforded by the Policy is in full force and effect and it is further agreed that the following provisions are specifically endorsed to the Policy:-

1   **Under the Hull and Aircraft Spares Insurances**

1.1   In respect of any claim on **Equipment** that becomes payable on the basis of a Total Loss, settlement (net of any relevant **Policy Deductible**) shall be made to, or to the order of the **Contract Party(ies)**. In respect of any other claim, settlement (net of any relevant **Policy Deductible**) shall be made with such party(ies) as may be necessary to repair the **Equipment** unless otherwise agreed after consultation between the Insurers and the Insured and, where necessary under the terms of the **Contract(s)**, the **Contract Party(ies)**.

Such payments shall only be made provided they are in compliance with all applicable laws and regulations.

1.2   Insurers shall be entitled to the benefit of salvage in respect of any property for which a claims settlement has been made.

2   **Under the Legal Liability Insurance**

2.1   Subject to the provisions of this Endorsement, the Insurance shall operate in all respects as if a separate Policy had been issued covering each party insured hereunder, but this provision shall not operate to include any claim howsoever arising in respect of loss or damage to the **Equipment** insured under the Hull or Spares Insurance of the Insured. Notwithstanding the foregoing the total liability of Insurers in respect of any and all Insureds shall not exceed the limits of liability stated in the Policy.

2.2 The Insurance provided hereunder shall be primary and without right of contribution from any other insurance which may be available to the **Contract Party(ies)**.

2.3 This Endorsement does not provide coverage for the **Contract Party(ies)** with respect to claims arising out of their legal liability as manufacturer, repairer, or servicing agent of the **Equipment**.

3   Under ALL Insurances

3.1 The **Contract Party(ies)** are included as Additional Insured(s).

3.2 The cover afforded to each Contract Party by the Policy in accordance with this Endorsement shall not be invalidated by any act or omission (including misrepresentation and non-disclosure) of any other person or party which results in a breach of any term, condition or warranty of the Policy PROVIDED THAT the **Contract Party** so protected has not caused, contributed to or knowingly condoned the said act or omission.

3.3 The provisions of this Endorsement apply to the **Contract Party(ies)** solely in their capacity as financier(s)/lessor(s) in the identified **Contract(s)** and not in any other capacity. Knowledge that any **Contract Party** may have or acquire or actions that it may take or fail to take in that other capacity (pursuant to any other contract or otherwise) shall not be considered as invalidating the cover afforded by this Endorsement.

3.4 The **Contract Party(ies)** shall have no responsibility for premium and Insurers shall waive any right of set-off or counterclaim against the **Contract Party(ies)** except in respect of outstanding premium in respect of the **Equipment**.

3.5 Upon payment of any loss or claim to or on behalf of any **Contract Party(ies)**, Insurers shall to the extent and in respect of such payment be thereupon subrogated to all legal and equitable rights of the **Contract Party(ies)** indemnified hereby (but not against any **Contract Party**). Insurers shall not exercise such rights without the consent of those indemnified, such consent not to be unreasonably withheld. At the expense of Insurers such **Contract Party(ies)** shall do all things reasonably necessary to assist the Insurers to exercise said rights.

3.6 Except in respect of any provision for Cancellation or Automatic Termination specified in the Policy or any endorsement thereof, cover provided by this Endorsement may only be cancelled or materially altered in a manner adverse to the **Contract Party(ies)** by the giving of not less than Thirty (30) days' notice in writing to the **Appointed Broker**. Notice shall be deemed to commence from the date such notice is given by the Insurers. Such notice will NOT, however, be given at normal expiry date of the Policy or any endorsement.

EXCEPT AS SPECIFICALLY VARIED OR PROVIDED BY THE TERMS OF THIS ENDORSEMENT:

1   THE CONTRACT PARTY(IES) ARE COVERED BY THE POLICY SUBJECT TO ALL TERMS, CONDITIONS, LIMITATIONS, WARRANTIES, EXCLUSIONS AND CANCELLATION PROVISIONS THEREOF.

2   THE POLICY SHALL NOT BE VARIED BY ANY PROVISIONS CONTAINED IN THE CONTRACT(S) WHICH PURPORT TO SERVE AS AN ENDORSEMENT OR AMENDMENT TO THE POLICY.

SCHEDULE IDENTIFYING TERMS USED IN THIS ENDORSEMENT

1   **Equipment** (Specify details of any aircraft, engines or spares to be covered):

2   **Policy Deductible** applicable to physical damage to the **Equipment** (insert all applicable Policy deductibles):

3   (a) Contract Party(ies): **AND (b)**, in addition, in respect of Legal Liability Insurances:

4   Contract(s):

5   **Effective Date** (being the date that the Equipment attaches to the Policy or a specific date thereafter):

6   Additional Premium:

7   Appointed Broker:

INSURANCE PROVISIONS AVS103

50/50 PROVISIONAL CLAIMS SETTLEMENT CLAUSE

WHEREAS the Insured has in full force and effect

A) a 'Hull All Risks' policy which inter alia contains the War Hijacking and Other Perils Exclusion Clause (AVN 48B) / the Common North American Airline War Exclusion Clause, and

B) a 'Hull War Risks' policy which inter alia covers certain of the risks excluded by AVN 48B / the Common North American Airline War Exclusion Clause in A) above

NOW IT IS HEREBY UNDERSTOOD AND AGREED THAT

in the event of loss of or damage to an aircraft identified on the schedule of aircraft forming part of this policy and where agreement is reached between the 'Hull All Risks' Insurers and the 'Hull War Risks' Insurers that the Insured has a valid claim under one or other policy where nevertheless it cannot be resolved within 21 days from the date of occurrence as to which policy is liable, each of the aforementioned groups of insurers agree, WITHOUT PREJUDICE to their liability, to advance to the Insured 50% of such amount as may be mutually agreed between them until such time as final settlement of the claim is agreed

PROVIDED ALWAYS THAT

(i) the 'Hull All Risks' and 'Hull War Risks' placing slips are identically endorsed with this provisional claims settlement clause

(ii) within 12 months of the advance being made all Insurers specified in (i) above agree to refer the matter to arbitration in London in accordance with the Statutory provision for arbitration for the time being in force

(iii) once the arbitration decision has been conveyed to the parties concerned, the 'Hull All Risks' Insurers or the 'Hull War Risks' Insurers as the case may be shall repay the amount advanced by the other group of Insurers together with interest for the period concerned which is to be calculated using the London Clearing Banks' Base Rate

(iv) if the 'Hull All Risks' and 'Hull War Risks' policies contain differing amounts payable, the advance will not exceed the lesser of the amounts involved. In the event of Co-insurance or risks involving uninsured proportion(s), the appropriate adjustment will be made.

INSURANCE PROVISIONS LSW555D

AVIATION HULL 'WAR AND ALLIED PERILS' POLICY

SECTION ONE: LOSS OF OR DAMAGE TO AIRCRAFT

Subject to the terms, conditions and limitations set out below, this Policy covers loss of or damage to the Aircraft stated in the Schedule against claims excluded from the Assured's Hull 'All Risks' Policy as caused by:

(a) War, invasion, acts of foreign enemies, hostilities (whether war be declared or not), civil war, rebellion, revolution, insurrection, martial law, military or usurped power or attempts at usurpation of power.

(b) Strikes, riots, civil commotions or labour disturbances.

(c) Any act of one or more persons, whether or not agents of a sovereign power, for political or terrorist purposes and whether the loss or damage resulting therefrom is accidental or intentional.

(d) Any malicious act or act of sabotage.

(e) Confiscation, nationalisation, seizure, restraint, detention, appropriation, requisition for title or use by or under the order of any government (whether civil, military or de facto) or public or local authority.

(f) Hi-jacking or any unlawful seizure or wrongful exercise of control of the Aircraft or crew in flight (including any attempt at such seizure or control) made by any person or persons on board the Aircraft acting without the consent of the Assured. For the purpose of this paragraph (f) only, an aircraft is considered to be in flight at any time from the moment when all its external doors are closed following embarkation until the moment when any such door is opened for disembarkation or when the aircraft is in motion. A rotor-wing aircraft shall be deemed to be in flight when the rotors are in motion as a result of engine power, the momentum generated therefrom, or autorotation.

Furthermore this Policy covers claims excluded from the Hull 'All Risks' Policy from occurrences whilst the Aircraft is outside the control of the Assured by reason of any of the above perils. The Aircraft shall be deemed to have been restored to the control of the Assured on the safe return of the Aircraft to the Assured at an airfield not excluded by the geographical limits of this Policy, and entirely suitable for the operation of the Aircraft (such safe return shall require that the Aircraft be parked with engines shut down and under no duress).

SECTION TWO: EXTORTION AND HI-JACK EXPENSES

1   This Policy will also indemnify the Assured subject to the terms, conditions, exclusions and limitations set out below, and up to the limit stated in the Schedule, for 90% of any payment properly made in respect of:

    (a)   threats against any Aircraft stated in the Schedule or its passengers or crew made during the currency of this Policy.

    (b)   extra expenses necessarily incurred following confiscation, etcetera (as Section One clause (e)) or hi-jacking, etcetera (as Section One clause (f)) of any Aircraft stated in the Schedule.

2   No cover will be provided under this Section of the Policy in any territory where such insurance is not lawful, and the Assured is at all times responsible for ensuring that no arrangements of any kind are made which are not permitted by the proper authorities.

SECTION THREE: GENERAL EXCLUSIONS

This Policy excludes loss, damage or expense caused by one or any combinations of any of the following:

    (a)   war (whether there be a declaration of war or not) between any of the following States: the United Kingdom, the United States of America, France, the Russian Federation, the People's Republic of China; nevertheless, if any Aircraft is in the air when an outbreak of such war occurs, this exclusion shall not apply in respect of such Aircraft until the said Aircraft has completed its first landing thereafter;

    (b)   confiscation, nationalisation, seizure, restraint, detention, appropriation, requisition for title or use by or under the authority of the Government(s) stated in the Schedule, or any public or local authority under its jurisdiction;

    (c)   the emission, discharge, release or escape of any chemical, biological or biochemical materials or the threat of same but this exclusion shall not apply;

        (i)   if such materials are used or threatened to be used solely and directly in:

            1   the Hi-jacking, unlawful seizure or wrongful exercise of control of an Aircraft in flight and then only in respect of loss of or damage to such Aircraft the subject of a valid claim under clause (i) Section One above; or

            2   any threat against an Aircraft stated in the Schedule or its passengers or crew and then only in respect of payments as are insured under Section Two above;

        (ii)   other than as provided for in sub-paragraph 1 above, to loss of or damage to an Aircraft if the use of such materials is hostile and originates solely and directly;

            1   on board such Aircraft, whether it is on the ground or in the air.

            or

    2   external to such Aircraft and causes physical damage to the Aircraft whilst the Aircraft's wheels are not in contact with the ground.

Any emission, discharge, release or escape originating external to the Aircraft that causes damage to the Aircraft as a result of contamination without other physical damage to the Aircraft exterior is not covered by this Policy.

(d)   any debt, failure to provide bond or security or any other financial cause under court order or otherwise;

(e)   the repossession or attempted repossession of the Aircraft either by any title holder, or arising out of any contractual agreement to which any Assured protected under this Policy may be party;

(f)   delay, loss of use, or except as specifically provided in Section Two any other consequential loss; whether following upon loss of or damage to the Aircraft or otherwise;

(g)   any use, hostile or otherwise, of radioactive contamination or matter but this exclusion shall not apply to loss of or damage to an Aircraft if such use is hostile and originates solely and directly;

    (i)   on board such Aircraft, whether it is on the ground or in the air, or

    (ii)   external to such Aircraft and causes physical damage to the Aircraft whilst the Aircraft's wheels are no longer in contact with the ground. Any such use originating external to the Aircraft that causes damage to the Aircraft as a result of contamination without other physical damage to the Aircraft exterior is not covered by this Policy.

(h)   any use, hostile or otherwise, of an electromagnetic pulse but this exclusion shall not apply to loss of or damage to an Aircraft if such use originates solely and directly on board such Aircraft, whether it is on the ground or in the air;

(i)   any detonation, hostile or otherwise, of any device employing atomic or nuclear fission and/or fusion or other like reaction, and notwithstanding (g) and (h) above, any radioactive contamination and electromagnetic pulse resulting directly from such detonation is also excluded by this Policy.

SECTION FOUR: GENERAL CONDITIONS

1   This Policy is subject to the same warranties, terms and conditions (except as regards the premium, the obligations to investigate and defend, the renewal agreement (if any), the amount of deductible or self insurance provision where applicable AND EXCEPT AS OTHERWISE PROVIDED HEREIN) as are contained in or may be added to the Assured's Hull 'All Risks' Policy.

2   Should there be any Material Change in the nature or area of the Assured's operations, the Assured shall give immediate notice of such Change to the Underwriters; no claim arising subsequent to a Material Change over which the Assured had control shall be recoverable hereunder unless such change has been accepted by the Underwriters.

'Material Change' shall be understood to mean any change in the operation of the Assured which might reasonably be regarded by the Underwriters as increasing their risk in degree or frequency, or reducing possibilities of recovery or subrogation.

3   The due observance and fulfilment of the terms, provisions, conditions and endorsements of this Policy shall be conditions precedent to any liability of the Underwriters to make any payment under this Policy: in particular the Assured should use all reasonable efforts to ensure that he complies and continues to comply with the laws (local or otherwise) of any country within whose jurisdiction the Aircraft may be, and to obtain all permits necessary for the lawful operation of the Aircraft.

4   Subject always to the provisions of Section Five, and the Schedule, Underwriters hereon agree to follow the Hull 'All Risks' Policy in respect of Breach of Warranty Cover, Hold Harmless Agreements and Waivers of Subrogation.

SECTION FIVE: CANCELLATION REVISION AND AUTOMATIC TERMINATION

Amendment of Terms or Cancellation:

1   (a)   Underwriters may give notice, effective on the expiry of 7 days from midnight G.M.T. on the day on which notice is issued, to review the rate of premium and/or the geographical limits. In the event of the review of the rate of premium and/or geographical limits not being accepted by the Assured then at the expiry of the said 7 days, this Policy shall become cancelled at that date.

Automatic Review of Terms or Cancellation

(b)   Notwithstanding 1(a) above, this Policy is subject to automatic review by Underwriters of the rate of premium and/or conditions and/or geographical limits effective on the expiry of 7 days from the time of any hostile detonation of any weapon of war employing atomic or nuclear fission and/or fusion or other like reaction or radioactive force or matter wheresoever or whensoever such detonation may occur and whether or not the insured Aircraft may be directly affected. In the event of the review of the rate of premium and/or conditions and/or geographical limits not being accepted by the Assured then at the expiry of the said 7 days, this Policy shall become cancelled at that date.

Cancellation by Notice

(c)   This Policy may be cancelled by the Assured or Underwriters giving notice not less than 7 days prior to the end of each period of 3 months from inception.

Automatic Termination	2	Whether or not such notice of cancellation has been given this Insurance shall **TERMINATE AUTOMATICALLY** Upon the outbreak of war (whether there be a declaration of war or not) between any of the following States, namely, the United Kingdom, the United States of America, France, the Russian Federation, the People's Republic of China.

PROVIDED THAT if the Aircraft is in the air when such outbreak of war occurs then this insurance, subject to its terms and conditions and provided not otherwise cancelled, terminated or suspended, will be continued in respect of such Aircraft until the said Aircraft has completed its first landing thereafter.

THE SCHEDULE Policy Number:

Assured:

Address of Assured:

Additional Assured(s):

Approved Lienholder(s) for Breach of Warranty protection:

Aircraft hereby insured:

Manufacturer	**Model**	**Registration**	**Agreed Value**

Geographical Limits:

Excluding Confiscation, etcetera by Government(s) of:

Period of Policy:

> **From:**

> **To:**

> Both days inclusive

Extortion and Hi-jack Expenses:

Limit of Policy:

90% of any one loss and in all (WARRANTED REMAINING 10% UNINSURED)

Premium:

Immediate notice of changes in risk or of circumstances likely to give rise to a loss hereunder to be communicated to:

Dated in London:

04/06

LSW 555D

INSURANCE PROVISIONS LSW147

LSW 147-REPOSSESSION OF LEASED EQUIPMENT INSURING CONDITIONS

INSURING CLAUSE

Whereas the Assured has entered into a Lease Agreement dated _____, 20___ to supply the Lessee with the Insured Equipment.

Now this Policy is to pay the Assured the Agreed Value in the event of the Insured Equipment being lost as a consequence of the occurrence during the period of this Policy of an Insured Peril, subject to the Definitions, Exclusions, Conditions and Warranties below.

DEFINITIONS

1   INSURED PERIL

Means an action taken by, or refusal or failure by, the Foreign Government, being:

a) Confiscation, seizure, appropriation, expropriation, nationalisation, restraint, detention or requisition for title or use of the Insured Equipment by the Foreign Government.

or b) Refusal or failure of the Foreign Government to allow the Assured to exercise its rights to repossess the Insured Equipment in accordance with the terms and conditions of the Lease Agreement.

or c) Refusal or failure of the Foreign Government to allow the Assured to remove the Insured Equipment from the Foreign Country following the Assured's exercise of its rights to repossess the Insured Equipment in accordance with the terms and conditions of the Lease Agreement (or its mortgage over the Insured Equipment).

or d) Refusal or failure of the Foreign Government to allow the Insured Equipment to be de-registered from the aviation register of the Foreign Country following the Assured's exercise of its rights to repossess the Insured Equipment in accordance with the terms and conditions of the Lease Agreement.

or e) Refusal or failure of the Foreign Government, following a compulsory sale or other compulsory disposal or divestiture of the Equipment in the Foreign Country, to allow the Assured to obtain the proceeds of sale, disposal or divestiture in United States dollars or another currency which is freely convertible into U.S. dollars in the international exchange markets, following Perils as per 1. a), b), and c) above.

provided always that Underwriters hereon shall not be liable for any loss or damage to the Insured Equipment unless the action by the Foreign Government is effective for a period not shorter than the applicable Waiting Period specified in the Schedule hereto.

2 INSURED EQUIPMENT

Means the Aircraft specified in the Schedule hereto.

This policy also covers loss of all or any of the Engines specified in the Schedule hereto from any of the Insured Perils (notwithstanding that the airframe does not suffer any such loss).

This policy also covers loss of the technical records for the aircraft or engines from any of the Insured Perils (notwithstanding that the airframe and engines do not suffer any such loss). In the event of such loss, Underwriters will pay the Assured an amount equal to the costs incurred by the Assured in reconstituting such technical records and carrying out any maintenance, checks or repairs necessary to effect the same, but only following a loss as per 1. a), b), c) and d) above, subject to separate schedule limits endorsed hereto.

3 FOREIGN COUNTRY

Means . . .

4 LESSEE

Means the entity to which the Insured Equipment has been leased in the Foreign Country.

5 FOREIGN GOVERNMENT

Means the present or any succeeding government or governmental authority (whether civil, military or de facto) of the Foreign Country, or any definable region thereof, provided that such governing authority exercises effective executive control therein, and without having regard to the manner of its accession.

6 AGREED VALUE

Means the amount specified in the Schedule hereto.

Underwriters' maximum total liability shall not in any event exceed the Agreed Value specified for the Aircraft in the Schedule.

7 LEASE AGREEMENT

Means the contract specified in the Schedule hereto and all supporting documentation, under which the Assured agrees to lease the Insured Equipment to the Lessee on terms and conditions seen by Underwriters hereon and under which the Assured retains title to the Insured Equipment.

8 WAITING PERIOD

Means the period specified in the Schedule hereto, and prior to the expiry of which Underwriters hereon shall not be liable for any loss or damage to the Insured Equipment; provided that if the Assured satisfies Underwriters that there is no reasonable prospect of the action by the Foreign Government being reversed or cancelled during the Waiting Period, Underwriters will pay the Agreed Value to the Assured upon being so satisfied, notwithstanding that the Waiting Period has not expired.

The Waiting Period shall commence from the date of advice to Underwriters of an event likely to give rise to a claim under this Policy.

EXCLUSIONS

1 MATERIAL DEFAULT BY THE ASSURED

Excluding any loss arising from material default by the Assured (or any agent, sub- or cocontractor of the Assured) in the performance of their obligations under the Lease Agreement.

2 LOSS OR DAMAGE DURING WAITING PERIOD

Excluding loss or damage to the Insured Equipment arising from any cause whatsoever prior to the expiry of the Waiting Period.

3 WAR

Excluding any loss arising from destruction or physical damage directly or indirectly occasioned by, happening through or in consequence of war, invasion, acts of foreign enemies, hostilities (whether war be declared or not), civil war, rebellion, revolution, insurrection, military or usurped power.

4 NON-COMPLIANCE WITH LAWS

Excluding any loss arising from any failure of the Assured to comply with the laws of the Foreign Country in force at inception of this policy (other than any laws whose purpose is to empower the Foreign Government to confiscate, seize, appropriate, expropriate, nationalize, restrain, detain or requisition assets), or of the Assured's own country.

5 NECESSARY PERMITS

Excluding any loss arising from any failure of the Assured to obtain all permits and authorisations necessary at inception of this policy and/or make every endeavor reasonably practicable to keep such permits and authorisations in force during currency and/or make every endeavor reasonably practicable to obtain such new permits and authorisations as may be stipulated by the Foreign Government during currency.

CONDITIONS

1 NON-ASSIGNMENT

The Assured shall not (without Underwriters' prior written consent) assign or transfer this Policy or the benefits or obligations thereof to any other person, provided always that the Assured can, with the Underwriters' prior agreement, require any claims payments hereunder to be made to a named loss payee, all the Assured's obligations under this Policy remaining unaffected.

2 DUE DILIGENCE

The Assured shall use due diligence and do and concur in doing all things reasonably practicable to avoid or diminish any loss herein insured. The Assured shall be deemed to have complied with this condition including but not limited to if the Assured shall have acted in accordance with the express written directions of the Underwriters during the Waiting Period.

3 NOTICE OF LOSS

Upon the discovery of any event likely to give rise to a claim under this Policy, the Assured shall within a reasonable period thereafter give notice thereof to the Underwriters hereon.

4 ONUS OF PROOF

In any claim, and/or any action, suit or proceeding to enforce a claim for loss hereunder, the burden of proving that the loss is recoverable under this Policy, that no Warranty has been breached and that no Exclusion applies shall fall upon the Assured.

5 COINSURANCE

The Assured shall bear any coinsurance requirement of all losses hereunder at their own risk and uninsured.

6 EXAMINATION OF THE ASSURED

The Assured shall submit to examination under oath, and shall produce for examination, at such reasonable place as is designated by the Underwriters or their representative, all documents in their possession or control which relate to the specific matters in question, and shall permit extracts and copies thereof to be made.

The foregoing requirement shall not oblige the Assured to produce or copy any document, if to do so would constitute a breach by the Assured of any obligation of confidentiality or a breach of copyright or similar right, until the Assured has obtained requisite consents from third parties to avoid such a breach being committed.

7 OTHER INSURANCES

This insurance does not cover any loss or damage which at the time of the happening of such loss or damage is insured by or would, but for the existence of this policy, be insured by any other existing policy or policies except in respect of any excess beyond the amount which would have been payable under such other policy or policies had this insurance not been effected. The insurance of other exposures under the Lease Agreement in excess of or in addition to the coverage provided hereunder shall not be affected by the Assured without the prior approval of Underwriters hereon.

8 SUBROGATION

The Assured under this Policy shall at the request and at the expense of the Underwriters do and concur in doing and permit to be done all such acts and things as may be necessary or reasonably required by the Underwriters at any time after the Waiting Period has commenced for the purpose of enforcing any rights and remedies, or of obtaining relief or indemnity from other parties to which the Underwriters shall or would become entitled or subrogated upon their paying for any loss under this Policy.

9 RECOVERIES

After payment of a claim hereunder, all subsequent salvage and recoveries made by the Assured shall inure to the benefit of Underwriters. The Assured hereby acknowledges that it will receive such sums recovered in trust for Underwriters and that it will Pay all sums upon receipt by it and/or by any person on its behalf to Underwriters.

10 NO CANCELLATION AND PREMIUM RETURN

No cancellation and no return of premium for short interest, unless specifically agreed at inception.

11 HEADINGS

Headings in this Policy are included herein for convenience of reference only and shall not affect the interpretation of this Policy.

12 GOVERNING LAW AND ARBITRATION

The construction, validity and performance of this Policy and all matters arising therefrom shall be governed by English law and all disputes which may arise under, out of, in connection with or in relation to this Policy or to the determination of the amount of loss hereunder shall be submitted to arbitration at the London Court of International Arbitration in accordance with its rules at the date of such submission. The award rendered by the arbitrator(s) shall be final and binding upon all parties and judgment thereon may be entered in any court having jurisdiction.

WARRANTIES

1 LEASE AGREEMENT

Warranted that the Lease Agreement is signed by or on behalf of all necessary parties and is in full force and effect at inception hereof. This warranty shall be deemed satisfied (as to the validity and enforceability of the Lease Agreement against the Lessee) by the terms of the independent legal opinion referred to under warranty 4. below. In addition, it shall not constitute a breach of this warranty if the Assured agrees to waive or defer fulfilment of any conditions precedent or subsequent described in the Lease, but this qualification shall not apply to those conditions enumerated under the Material Change warranty of this Policy.

2   DISCLOSURE

Warranted that at inception hereof the Assured have no knowledge or information of any matter, fact or circumstance which is likely to give rise to a loss hereunder.

3   HULL AND LIABILITIES INSURANCE

Warranted that the Lease Agreement requires hull and war and allied perils insurances and liabilities insurances to be maintained by the lessee throughout the term of the leasing thereunder, on the terms specified therein.

4   LEGAL OPINION

Warranted that a written independent legal opinion of a qualified lawyer in the Foreign Country shall be supplied by the Assured to Underwriters confirming that the laws of the Foreign Country in effect at the date it is submitted do not prevent or hinder the Assured's exercise of its rights under the Lease Agreement. Underwriters confirm that the opinion of _____ dated __, 20__ satisfies the warranty set forth herein and the Underwriters agree to endorse a copy of such opinion and attach it to the Policy upon receipt thereof by Underwriters from the Assured.

5   MATERIAL CHANGE

Warranted that immediate notice of any material change in the Lease Agreement shall be given to Underwriters. No claim arising subsequent to such change shall be recoverable hereunder unless such change has been accepted by Underwriters.

1) Terms of payments
2) Events of default
3) Legal status of lease contract in host country
4) Uses of insured aircraft/routes
5) Maintenance provisions
6) Period of Agreement
7) Cross-collateralization provisions, if any
8) Identity of Lessee/sub-Lessee
9) Interest change in nature of Lease
10) Arbitration/Law
11) Change in insurance requirement
12) Any change which could reasonably be foreseen by lessor to ultimately affect collateral value of the aircraft

INSURANCE PROVISIONS AVN51

EXTENDED COVERAGE ENDORSEMENT (AIRCRAFT HULLS)

Notwithstanding the contents of the War, Hi-jacking and Other Perils Exclusion Clause forming part of this Policy, IT IS HEREBY UNDERSTOOD AND AGREED that this Policy is extended to cover claims caused by the following risks:

(i)   Strikes, riots, civil commotions or labour disturbances;
(ii) Any malicious act or act of sabotage; and
(iii) Hi-jacking or any unlawful seizure or wrongful exercise of control of the Aircraft or crew in Flight (including any attempt at such seizure or control) made by any person or persons on board the Aircraft acting without the consent of the Insured

PROVIDED ALWAYS THAT

1   The above extension shall only apply to the extent that the loss or damage is not otherwise excluded by (a), (b), (d) and (f) of the War, Hi-jacking and Other Perils Exclusion Clause;

2   the limits of Insurers' liability in respect of any or all of the risks covered under this endorsement shall not exceed the sum of . . . . (in the aggregate during the Policy period);

3   the Insured has paid or has agreed to pay the additional premium of . . . . required by the Insurers in respect of this extension; and

4   the insurance provided by this endorsement may be cancelled by the Insurers giving notice effective on the expiry of seven days from midnight GMT on the day on which notice is issued.

AVN 51 1.10.96

INSURANCE PROVISIONS ANV52E

EXTENDED COVERAGE ENDORSEMENT (AVIATION LIABILITIES)

1 WHEREAS the Policy of which this Endorsement forms part includes the War, HiJacking and Other Perils Exclusion Clause (Clause AVN 48B), IN CONSIDERATION of an Additional Premium of . . . ., it is hereby understood and agreed that with effect from . . . ., all sub-paragraphs other than . . . . of Clause AVN 48B forming part of this Policy are deleted SUBJECT TO all terms and conditions of this Endorsement.

2 EXCLUSION applicable only to any cover extended in respect of the deletion of subparagraph (a) of Clause AVN 48B.

Cover shall not include liability for damage to any form of property on the ground situated outside Canada and the United States of America unless caused by or arising out of the use of aircraft.

3 LIMITATION OF LIABILITY

The limit of Insurers' liability in respect of the coverage provided by this Endorsement shall be . . . . or the applicable Policy limit whichever the lesser any one Occurrence and in the annual aggregate (the "sub-limit"). This sub-limit shall apply within the full Policy limit and not in addition thereto.

To the extent coverage is afforded to an Insured under the Policy, this sub-limit shall not apply to such Insured's liability:

(a) to the passengers (and for their baggage and personal effects) of any aircraft operator to whom the Policy affords cover for liability to its passengers arising out of its operation of aircraft;

(b) for cargo and mail while it is on board the aircraft of any aircraft operator to whom the Policy affords cover for liability for such cargo and mail arising out of its operation of aircraft.

4 AUTOMATIC TERMINATION

To the extent provided below, cover extended by this Endorsement shall TERMINATE AUTOMATICALLY in the following circumstances:

(i) **All cover**
- upon the outbreak of war (whether there be a declaration of war or not) between any two or more of the following States, namely, France, the People's Republic of China, the Russian Federation, the United Kingdom, the United States of America

(ii) **Any cover extended in respect of the deletion of sub-paragraph (a) of Clause AVN 48B**
- upon the hostile detonation of any weapon of war employing atomic or nuclear fission and/or fusion or other like reaction or radioactive force or matter wheresoever or whensoever such detonation may occur and whether or not the Insured Aircraft may be involved

(iii) **All cover in respect of any of the Insured Aircraft requisitioned for either title or use**
- upon such requisition

PROVIDED THAT if an Insured Aircraft is in the air when (i), (ii) or (iii) occurs, then the cover provided by this Endorsement (unless otherwise cancelled, terminated or suspended) shall continue in respect of such an Aircraft until completion of its first landing thereafter and any passengers have disembarked.

5 REVIEW AND CANCELLATION

(a) **Review of Premium and/or Geographical Limits (7 days)**
Insurers may give notice to review premium and/or geographical limits such notice to become effective on the expiry of seven days from 23.59 hours GMT on the day on which notice is given.

(b) **Limited Cancellation (48 hours)**
Following a hostile detonation as specified in 4 (ii) above, Insurers may give notice of cancellation of one or more parts of the cover provided by paragraph 1 of this Endorsement by reference to sub-paragraphs (c), (d), (e), (f) and/or (g) of Clause AVN 48B such notice to become effective on the expiry of forty-eight hours from 23.59 hours GMT on the day on which notice is given.

(c) **Cancellation (7 days)**
The cover provided by this Endorsement may be cancelled by either Insurers or the Insured giving notice to become effective on the expiry of seven days from 23.59 hours GMT on the day on which such notice is given.

(d) Notices
All notices referred to herein shall be in writing.

AVN 52E 12.12.01

INSURANCE PROVISIONS FAA
WAR RISK POLICY

**UNITED STATES OF AMERICA DEPARTMENT OF
TRANSPORTATION FEDERAL AVIATION ADMINISTRATION**
Policy No. _____

**WAR RISK INSURANCE FOR AIRCRAFT HULL, PASSENGER AND
CREW LIABILITY, AND THIRD PARTY LIABILITY ISSUED
PURSUANT TO CHAPTER 443 OF TITLE 49 OF THE UNITED STATES
CODE**

**DESCRIPTION OF COVERAGE UNDER THIS CONTRACT OF
INSURANCE AND THOSE TERMS, CONDITIONS AND DEFINITIONS
APPLICABLE TO ALL PARTS OF THIS CONTRACT OF INSURANCE**

I   COVERAGE

The United States of America (hereinafter, the Insurer), represented by the
Administrator of the Federal Aviation Administration, acting for the Secretary of
Transportation (hereinafter the Administrator or FAA), hereby provides coverage
of this War Risk Insurance Policy to _____ (hereinafter the
Insured), in accordance with applicable provisions of law and subject to all limitations
thereof, and upon the payment of premiums, pursuant to the provisions of chapter
443 of title 49 of the United States Code (49 U.S.C. Sections 44301 through 44310
et seq.) (hereinafter, chapter 443) consisting of the following three parts:

A   Part I: Hull Insurance, covering physical damage to Equipment (including
aircraft, aircraft spare parts and engines);

B   Part II: Comprehensive Liability Insurance, covering Premises, Products/
Completed Operations, Hangarkeepers, Cargo and Mail Liabilities and
Passenger and Crewmember Personal Injury, Bodily Injury or Death
(excluding Third Party War Risk Liability); and

C   Part III: Third Party War Risk Liability.

II   TERMS AND CONDITIONS

A   The sections in this part shall apply to Part I (Hull Insurance), Part II
(Comprehensive Insurance) and Part III (Third Party Insurance) of this
Contract of Insurance.

B   Parts I, II, and III, set forth below, shall be considered to be separate elements of this War Risk Insurance Policy, and the deletion, cancellation, or any other means through which any one Part might be terminated or suspended shall have no effect upon the validity of any remaining Part or Parts.

C   The terms and conditions set forth in each Part shall be independent of any other Part, and shall not be construed to affect any other Part, provided, however, that the aggregate amount of insurance provided under all three Parts of this War Risk Insurance Policy shall not exceed U.S.$_____ per occurrence/per aircraft incurred by the Insured for the higher of third party losses or passenger liability limits set forth in Parts II or III of this Policy of Insurance.

D   The premiums for each Part of this War Risk Insurance Policy shall be separately determined, provided, however, that for air carriers who did hold a policy of insurance with the FAA on June 19, 2002, the total premium paid for this War Risk Insurance Policy shall not exceed twice the premium (annual and pro rata) paid by the Insured for Third Party War Risk Liability Insurance issued by the Administrator as of June 19, 2002, and for air carriers who did not hold a policy with the FAA on June 19, 2002, the total premium paid by the Insured for all three Parts of this War Risk Insurance Policy shall not exceed twice the premium for Third Party War Risk Liability Insurance as stipulated in Part III, Article IX, of this contract of insurance.

    The Insured shall provide any and all records and information requested by the Administrator for the purpose of calculating the premium for each Part of this War Risk Insurance policy. Failure to provide such records and information in a reasonable time as directed by the Administrator, shall void this War Risk Insurance Policy, and be cause for cancellation or non-renewal of this Policy of Insurance at the sole discretion of the Administrator.

E   This Policy of Insurance may be amended by the Administrator or terminated by either Party, and shall be either amended or terminated through the occurrence of any condition under chapter 443 that would affect the operation of this agreement.

F   Coverage of any aircraft listed as insured under this Policy of Insurance shall be held in abeyance if that aircraft is either (1) under charter by an agency of the United States or (2) is under the control of the Department of Defense during activation of the Civil Reserve Air Fleet (CRAF); and in either case, nonpremium insurance is available pursuant to 49 U.S.C. Section 44305 of chapter 443. Coverage in full under this Policy of Insurance for such aircraft shall be restored when the aircraft completes said charter operations or is released from CRAF activation and nonpremium insurance becomes unavailable under 49 U.S.C. Section 44305.

G   Coverage under this Policy does not apply to the following aircraft operations: (1) those that are intentionally conducted into or within geographic areas prohibited by a FAA Special Federal Aviation Regulations (SFAR); (2) operations that are not authorized by the air carriers' operation specifications issued in compliance with Part 119 of Title 14 of the Code of Federal Regulations; or (3) any operations that are excluded by separate endorsement to this policy. If a waiver or exclusion is granted to an SFAR, this Policy of Insurance shall not be effective while the Insured is operating aircraft pursuant to the waiver or exclusion.

H The Insured shall notify the Insurer, in writing, of a pending Material Change which will affect the operational control of aircraft. Notification of the actual Material Change shall occur no later than 48 hours after implementation.

Notification by electronic mail will satisfy the reporting requirement so long as the Insured supplies the Insurer with a paper copy of this notice as soon as practicable thereafter.

III RECONCILIATION OF ACTUAL AND ESTIMATED PREMIUMS

Within 30 days of the expiration or termination of this Policy of Insurance, the insured shall calculate and submit data to reconcile the actual premium owed with the deposit premium estimated.

1 If the premium owed is greater than the deposit premium paid by the Insured, the Insured shall pay the premium difference to the Insurer.

2 If the premium owed is less than the deposit premium paid by the Insured, the Insurer will refund the premium difference.

IV ACTIVATION, AMENDMENT AND TERMINATION

A Insurance coverage shall commence upon activation by the FAA, and this Policy of Insurance shall remain in effect until amended by the Administrator or terminated by either party.

B This Policy of Insurance shall automatically terminate:
1 Upon effective expiration of the authority of the Secretary of Transportation, subject to retroactive reauthorization, to provide insurance pursuant to chapter 443 of Title 49 of the United States Code; or
2 Seven (7) days after notice of termination has been directly communicated to the Insured, in writing, by the Administrator. A confirmation copy of this notice may be sent by facsimile or other additional means of communication or delivery to the Insured.

C Termination of the entirety of this of Policy of Insurance shall terminate each and every Part of this War Risk Insurance Policy.

D The Insurer has the right at its sole discretion to terminate this Contract of Insurance as of the date of a Material Change in the status of the Insured, if such Material Change to the Insured's status results in a condition which exceeds the statutory authority of the Insurer to provide insurance.

E The Administrator may terminate this Policy of Insurance for default if the Insured fails to pay the premium or comply with the terms of the policy. In the event of default, the Insurer will forward a Notice of Default to the Insured. The Insured agrees to:

        1   Calculate the actual premium for all operations conducted by the Insured between acceptance of the Insured's application for insurance and the date and time of termination for default, and submit the calculation for the actual premium to the Insurer within thirty (30) days of the date on the Notice of Default; and

        2   Pay the actual premium for all operations conducted by the Insured between acceptance of the Insured's application and the date and time of termination for default.

        3   Produce a list of all Other Insured Parties in accordance with paragraph VIII. A. of this part of the Policy of Insurance.

F   The Insured shall provide any and all records and information requested by the Administrator, including (but not limited to) copies of its commercial aviation liability and hull insurance policies applicable to the period of insurance and all schedules and attachments thereto, for the purpose of verifying the terms and scope of applicable commercial insurance coverage. The Insurer will maintain such Insured-provided commercial aviation liability and hull insurance information as confidential. Failure to provide such records and information in a reasonable time when requested shall be grounds for termination of this Policy of Insurance.

V DEFINITIONS

A   'Air Transportation Business' means:

        1   the ownership, maintenance, sale or use of aircraft by an air operator, or

        2   operations necessary or related to the providing of air transportation by the Insured, or

        3   operations, including maintenance and supply of goods or services provided for, or by others which are necessary or related to the provision of air transportation by the Insured, or others approved by the Insured, or

        4   all non-revenue operations involving the operation of aircraft by the Insured, including company sponsored activities, events, promotions, award programs and other events or happenings designed to further the Air Transportation Business, image and good will of the Insured.

B   'Air Operator' means the party exercising operational control (authority) over initiating, conducting or terminating a flight operation.

C   'Bodily Injury' means Bodily Injury sustained by any person caused by an Occurrence during the policy period, including sickness, disease, mental anguish, trauma, fright, disability or death at any time resulting therefrom or resulting from the apprehension thereof.

D   'Personal Injury' means:

        1   false arrest or imprisonment, delay, detention, malicious prosecution, discrimination, wrongful entry to or eviction from any Premises or Aircraft or vehicle or other invasion of the right of private occupancy; and

2   incidental medical malpractice, error or mistake by any physician, surgeon, nurse, medical technician or other person performing medical services on behalf of the Named Insured in the provision of immediate medical relief occasioned by a War Risk Occurrence.

E   'Policy Territory/Geographical Limits' means anywhere in the world.

F   'Property Damage' means (1) injury to or destruction of tangible property and (2) under Part III Third Party coverage, loss of use of tangible property, which has been physically injured or destroyed, provided such loss of use is caused by a War Risk Occurrence.

G   'War Risk Occurrences' (also 'Occurrences' or 'Occurrence') are defined as any loss or damage directly or indirectly arising from, or occasioned by, or happening through or in consequence of:

1   War (whether declared or not, including war between Great Powers), invasion, acts of foreign enemies, warlike hostilities, civil war, rebellion, revolution, insurrection, martial law, exercise of military or usurped power, or any attempt at usurpation of power.

2   Any hostile detonation of any weapon of war, including any employing atomic or nuclear fission and/or fusion or other like reaction of radioactive force or matter.

3   Strikes, riots, civil commotions, or labor disturbances.

4   Any act of one or more persons, whether or not agents of a sovereign power, for political or terrorist purposes and whether the loss or damage resulting therefrom is accidental or intentional, except for ransom or extortion demands. Payments in response to ransom or extortion demands are hereby specifically denied under this policy.

5   Any malicious act or act of sabotage, vandalism or other act intended to cause loss or damage.

6   Confiscation, nationalization, seizure, restraint, detention, appropriation, requisition for title or use by or under the order of any foreign government (whether civil or military or de facto) or foreign public or local authority. This policy will not cover any lawful government seizures of aircraft or spare parts that are the result of outstanding legal debts, taxes, fines, or unlawful acts committed with the knowledge of airline officials or the unlawful operation of such aircraft by the named insured.

7   Hijacking or any unlawful seizure or wrongful exercise of control of the aircraft or crew (including any attempt at such seizure or control) made by any person or persons onboard the aircraft or otherwise, acting without the consent of the Insured.

8   The discharge or detonation of any weapon or explosive device while on an aircraft covered by this Policy of Insurance.

H   'Passenger', when mentioned under any liability provision of this Policy of Insurance, means a person who enters into a contract of transportation or other agreement by which the person is to be transported by the Insured, and who has acted upon that contract or other agreement by checking in for transportation and receiving a boarding pass or other means of identification for that transportation, whose subsequent movements are made in direct response to the places, times and means of transportation that are directly involved with, made as a consequence of, and thus governed by, the air operations conducted by the Insured. A person shall cease to be a passenger when that person's movements are no longer governed by the air operations conducted by the Insured pursuant to the contract of transportation or other agreement with the Insured. A person who is identified, at any time and in any way, as a knowing participant in the commission of a War Risk Occurrence shall not be considered as a passenger for the purposes of this Contract of Insurance.

I   'Additional Insureds' are persons with whom the Insured has entered into contracts of indemnity for coverage pursuant to Paragraphs I. B, Parts II and III of this Contract of Insurance.

J   'Material Change' is any substantial change in corporate ownership, financial or operational structure which affects the normal operation of insured aircraft, the operational control of insured aircraft by the Insured, or a change in the type of flight operation of aircraft operated by the Insured.

K   'Aircraft Accident' means Property Damage and/or Bodily Injury resulting from an Occurrence which has not been determined to be the consequence of items 1 through 8 as defined in 'G' under the Definition section of this Policy.

VI   COVERAGE FOR AIRCRAFT OUTSIDE THE CONTROL OF THE INSURED

This Policy, subject to the exclusions contained herein, covers claims arising while an aircraft under the operational control of the Insured, is outside the control of the Insured by reason of any of the above perils. The aircraft shall be deemed to have been restored to the control of the Insured on the safe return of the aircraft to the Insured at an airfield which is entirely suitable for the operation of the aircraft. Such safe return shall require that the aircraft be parked with engines shut down with the crew under no duress. The Insurer waives no rights of subrogation or indemnification by virtue of this clause.

VII   NOTIFICATION TO THE FAA

In all instances when notification to the FAA by the Insured is required by Article IV of Parts I, II, and III of this policy of insurance, it shall be made to the following:

Manager, Management Staff
Office of Policy, International Affairs, and Environment
600 Independence Ave., SW
FOB 10B 6E1500
Washington, DC 20591
Phone: (202) 385–8869
FAX: (202) 267–7179

Insurance Staff, Aviation Insurance Program Office APL-20
Federal Aviation Administration
600 Independence Ave., SW FOB 10B 6E1500
Washington, DC 20591
Phone: (202) 385–8875; (202) 385–8851; (202) 385–8839; (202) 285–8960
FAX: (202) 267–7179

At any time that notification cannot be made to the above numbers, the Insured shall contact the FAA Washington Operations Center at (202) 267–3333. The caller shall give the name of the Insured, a contact name, title, and telephone number.

VIII   OTHER INSURED PARTIES

A   Definition and Identification of Other Insured Parties. For purposes of Parts I, II and III of this Policy of Insurance, the term 'Insured' shall include Other Insured Parties, which are legal or private persons (1) that are aircraft lessors, lendors, lienholders, or other persons to the extent of their ownership or interest in an aircraft operated by _____ (hereinafter, Other Insured Parties) and (2) that are listed in the commercial insurance policy of _____ provided, however, that _____ shall certify in writing that its agreements with the Other Insured Parties require the inclusion of the Other Insured Parties in the war risk insurance coverage obtained by _____. Upon request by the Insurer, _____ shall, within ten (10) working days, identify in writing all such Other Insured Parties to the Insurer.

B   Warranties Applicable to Other Insured Parties. The following warranties shall apply to Other Insured Parties identified under part A of this Article VIII:

1   Respecting equipment loss coverage under Part I to this Policy of Insurance, the Other Insured Parties shall be named as loss payees as their respective interests may appear;

2   Provisions of this Policy of Insurance, including this Article VIII, shall apply worldwide and have no territorial restrictions or limitations;

3   Respecting the interests of the Other Insured Parties in this Policy of Insurance, the insurance shall not be invalidated or impaired by any act or omission (including misrepresentation and nondisclosure) by the Insured or any other person (including use for illegal purposes of any Insured Equipment), and shall insure the Other Insured Parties regardless of any breach or violation of any representation, warranty, declaration, term, or condition contained in such policies by the Insured or any other Person;

4   If the Insurer terminates this Policy of Insurance for any reason whatsoever, or if it is allowed to lapse for nonpayment of premium, or if any material change is made in the Policy of Insurance which adversely affects the interest of any of the Other Insured Parties, such cancellation, lapse, or change shall not be effective as to the Other

291

Insured Parties for seven (7) days after receipt by the Other Insured Parties of written notice from the Insured of such cancellation, lapse or change, or publication of notice of cancellation by the Insurer in the Federal Register;

5   The Insurer waives any rights of subrogation or any right of setoff (including for unpaid premiums), recoupment, counterclaim, or other deduction, whether by attachment or otherwise, against each Other Insured Party;

6   Insurance proceeds from this Policy of Insurance shall be primary without right of contribution from any other insurance that may be available to any Other Insured Parties;

7   All of the liability insurance provisions of the Policy of Insurance, except the limits of liability, shall operate in all respects as if a separate policy had been issued covering each Other Insured Party;

8   None of the Other Insured Parties shall be liable for any insurance premium; and

9   The Policy of Insurance authorizes a 50/50 Clause per Lloyd's Aviation Underwriters' Association Standard Policy Form AVS103, or its equivalent.

IX SUPPLEMENTAL COVERAGES

This Policy of Insurance shall pay, within the Limits of Liability stated elsewhere herein, the following expenses incurred by the Insured, provided, however, with respect to paragraphs A, B, C, D and E below, the Insurer's Limit of Liability for A, B, C, D and E combined shall not exceed the greater of $25,000 per passenger or $5,000,000 for any one War Risk Occurrence unless approved in advance by the Insurer.

A   Reasonable expenses incurred as a result of a War Risk Occurrence in respect of:

1   The repatriation of Passengers;

2   the repatriation of the body of any Passenger for burial and/or cremation;

3   the funeral of any Passenger;

4   any necessary first aid, hospital, dental, mental health services, nursing treatment and medical services;

5   the search for and/or recovery and/or identification of bodies of Passengers;

6   where possible, the reasonable transportation of a relative or friend of a Passenger killed or injured in a War Risk Occurrence to and from a place near the scene of the War Risk Occurrence or to and from a place where the Passenger has been removed or taken including necessary living expenses incurred by the relative or friend including but not limited to food, lodging, local transportation, and incidentals while they are away from home for a period not exceeding one year from the date of the War Risk Occurrence; and

7   other acts of humanity reasonably incurred.

B   Reasonable search and rescue operations for an aircraft insured under this Policy of Insurance which is determined to be missing and unreported after the computed maximum endurance of the flight has been exceeded.

C   Any attempted or actual removal of wreckage of an Aircraft insured under this Policy of Insurance, including the transportation, security costs, and storage or said wreckage and fees incurred by the Insured or its contractors to accomplish these tasks.

D   The foaming of a runway to prevent or mitigate possible loss or damage.

E   Any public inquiry or inquiry by the Federal Aviation Administration, National Transportation Safety Board or similar domestic or foreign governmental agency having investigative authority into an Occurrence involving an Aircraft insured under this Policy of Insurance that the Insured is called upon to pay.

F   For each War Risk Occurrence, all reasonable expenses, incurred by the Insured and/or the Insured's contractors, not to exceed the greater of $25,000 per passenger or $5,000,000, unless approved by the Insurer in advance, for each Aircraft/War Risk Occurrence incurred by the Insured not otherwise payable under this Policy, which arise out of and are a result of efforts expended by the Insured and/or assessed against the Insured solely by virtue of, 1) the requirements of the Aviation Disaster Family Assistance Act of 1996 (as may be amended from time to time) and any regulations imposed by the appropriate federal governmental agency as a result thereof, and/or 2) the 'Principles of Understanding between ATA Carriers and the NTSB Regarding Certain Aviation Expenditures Related to the Recovery and Identification of Aviation Accident Victims'.

G   For each War Risk Occurrence, all reasonable expenses not to exceed $5,000,000 unless approved by the Insurer in advance, incurred by the Insured and/or the Insured's contractors for the Occurrence site investigation (not otherwise covered by paragraph IX.E), security, and remediation.

X   ALLOCATION OF COSTS BETWEEN THE INSURER AND OTHER INSURERS WITH WHOM THE INSURED HAS A POLICY OF INSURANCE

A   If in the event of an Aircraft Accident, the information at the time of the response does not clearly identify the event as being a War Risk Occurrence, the Insured shall have the right to use airline service providers including the commercial aviation insurance company that would respond on behalf of the Insured to an Aircraft Accident to provide services, activities, or actions that are usual and customary following an Aircraft Accident. If the Aircraft Accident is subsequently determined to be a War Risk Occurrence, the FAA will reimburse the Insured for expenses the Insured incurred for items that are identified in Supplemental Coverages, Article IX, of this policy, subject to adherence to the limits and preapproval conditions specified therein.

B If, in addition to this Policy of Insurance which, *inter alia*, covers the risks excluded by AVN48B/the Common North American Airline War Exclusion Clause, the Insured has another policy of insurance in full force and effect that has a 'Hull All Risks' and/or a 'Liability All Risks' policy which, *inter alia*, contains the War Hi-Jacking and Other Perils Exclusion Clause (AVN 48B) and/or the Common North American Airline War Exclusion Clause, the insurer agrees that in the event of an Occurrence involving loss or damage to an aircraft on the schedule of aircraft forming part of this Policy of Insurance and where agreement is reached between the 'Hull All Risks' and/or the 'Liability All Risks' insurers and the Administrator that the Insured has a valid claim under one or the other policy where nevertheless it cannot be resolved within twenty-one (21) days from the date of the Occurrence as to which policy is liable, each of the aforementioned groups of insurers agree, without prejudice to their liability, to advance to the insured fifty percent (50%) of such amount as may be mutually agreed between them until such time as final settlement of the claim is agreed, provided that:

1 the 'Hull All Risks' and/or the 'Liability All Risks' policies, and this Policy of Insurance are identically endorsed with this provisional claims settlement clause;

2 within twelve (12) months of the advance being made all insurers specified above in this paragraph agree to refer the matter to arbitration in the United States in accordance with the Statutory provision for arbitration for the time being in force;

3 once the arbitration decision has been conveyed to the parties concerned, the 'All Risks' insurers or the Insurer, as the case may be, shall repay the amount advanced by the other group of insurers together with interest for the period concerned, which is to be calculated using the London Clearing Banks' Base Rate;

4 if the 'Hull All Risks' and/or the 'Liability All Risks' policies and this Policy of Insurance contain differing amounts payable, the advance will not exceed the lesser of the amounts involved. In the event of Co-insurance or risks involving uninsured portion(s), the appropriate adjustment will be made.

XI WARRANTY BY INSURED

The Insured warrants this Policy of Insurance to be free from any claim for loss, damage, or expense covered under any commercial policy in effect for the benefit of the Insured (except any insurance which may be mandated by the Terrorism Risk Insurance Act of 2002), and to be free from any claim for loss, damage, or expense not covered by any policy of insurance whatsoever.

XII FINAL GENERAL PROVISIONS

A Notwithstanding any other provision of this Policy of Insurance, no errors or omissions in furnishing notification or reports required by this Policy of

Insurance shall prejudice the protection afforded by this Policy of Insurance, but shall be corrected by the Insured when such errors or omissions in furnishing notification or reports are discovered.

B   The Administrator authorizes the Insured to enter into indemnity agreements (and this Policy of Insurance will insure) for war risk liability awards under Parts II and III of this Policy of Insurance with vendors, agents and subcontractors whose goods and services are necessary to the operation of aircraft by the Insured. Upon request, the Insured will provide copies of any or all such indemnity agreements.

XIII   EFFECTIVE DATE AND AUTHORIZED SIGNATURES

The entirety of this War Risk Insurance Policy (including Parts I, II and III below), becomes effective as of 00:00 GMT of the _____, subject to the submission of a Schedule of Aircraft as required by Part I, Article IIA and unless one or more Parts are terminated, shall remain in effect until 23:59 GMT on the _____, or until amended or terminated in accordance with the terms of this Policy of Insurance. In accordance with the Memorandum of Agreement of _____ and amendments thereof between the Insured and the United States of America regarding the power to bind a contract of insurance:

For the UNITED STATES OF AMERICA
By: Date:
Title: Manager, Management Staff
Aviation Policy, International Affairs, and Environment
Federal Aviation Administration

For the INSURED,
By an Individual Empowered to Bind the Insured By:
Date:
Title:

UNITED STATES OF AMERICA

DEPARTMENT OF TRANSPORTATION

FEDERAL AVIATION ADMINISTRATION

Policy No. _____

PART I: HULL INSURANCE

I   COVERAGE

The Insurer, represented by the Administrator acting for the Secretary of Transportation, shall provide by this Part I of the Policy of Insurance, in accordance with applicable provisions of law and subject to all limitations thereof, and upon the payment of a premium, pursuant to the provisions of chapter 443, physical damage insurance (hereinafter, Hull Insurance) for aircraft (including aircraft spare parts, engines and appliances, hereinafter collectively referred to as 'Equipment') in the amounts shown herein for War Risk Occurrences arising from the Air Transportation Business of _____ (hereinafter, the Insured), for the aircraft described in the attached 'Schedule of Equipment' (hereinafter the Schedule) for the account of the Insured for risks associated with the operation of such Equipment by the Insured (or operated by others but for which the Insured is responsible to another party or parties, by reason of contract or agreement, for insurance against physical loss of or damage to said Equipment.)

II SUM INSURED TO BE DETERMINED BY THE ADMINISTRATOR

A   The Equipment values insured by this Policy of Insurance shall be the same as, for air carriers who held an FAA policy as of November 25, 2002, those set forth in the Schedule from the Insured's commercial hull and/or equipment policies as of November 25, 2002, and operated by the Insured as of the date of issuance of this Policy of Insurance or as amended thereafter, and for those who did not hold an FAA policy on November 25, 2002, those set forth in the Schedule from the Insured's current commercial hull and/or equipment policies for those hulls and/or equipments operated by the Insured as of the date of issuance of this Policy of Insurance or as amended thereafter. The value of each item of Equipment set forth in the Schedule shall represent the amount of war risk physical damage insurance in effect for such Equipment and shall be deemed to be the Sum Insured; provided, however, the Sum Insured for each item of Equipment shall be determined by the Administrator to be the Agreed Value of the item of Equipment in the exercise of her discretion pursuant to 49 U.S.C. Sections 44302 and 44306. For purposes hereof, 'Agreed Value' means: (i) the amount in effect on the commercial policy immediately prior to the effective date of this Policy of Insurance as set forth in the Schedule and/or FAA pre-approved self-insurance, or (ii) the stipulated value required by a bona fide agreement reached at arm's length between the Insured and a third party lender or lessor, or (iii) for items of equipment that are not subject to encumbrance, Agreed Value means, at the discretion of the Administrator, the greater either of, (a) the amount that represents the fair and reasonable value of the aircraft, or (b) the net book value of the aircraft at the time of loss. The Insured shall provide the Administrator with a Schedule of Aircraft (the Schedule) from the Insured's current commercial hull policy, or other reasonable document certified as true by the Insured that reflects the aircraft in operation (or the future date of operation of) during the period of this Contract of Insurance within ten (10) calendar days of issuance of this War Risk Insurance Policy.

B   The Insured agrees that, if the Sum Insured of Equipment insurance carried against loss or damage from risks other than war risks is voluntarily reduced or increased by the Insured on its commercial policy to an amount other than the Sum Insured under this Policy, the Sum Insured under this Policy shall be considered to have been automatically amended to the new amounts at the time of such change on the commercial policy unless a specific value has been predetermined by the Insurer prior to a loss incident resulting from a War Risk Occurrence. Such variations shall conform to changes incurred on the insured's commercial policy schedule of Equipment valuations.

C   Insurance coverage of this policy is limited to operations of aircraft listed in the Schedule. It is the insured's responsibility to verify the accuracy of the Schedule and the Schedule must be revised by the Insured to add or delete Equipment, or to change aircraft registration numbers.

1   Any Equipment acquired by the Insured as Owner (or for which they have agreed to provide coverage) shall be attached hereto as of the precise time delivery occurs or title is vested in the Insured, or risk of loss transfers to the Insured, whichever occurs first. The Sum Insured of such Equipment shall be the purchase price to the Insured (unless another value shall be specifically declared to and agreed by the Insurer subject to the determination of the Administrator, in the exercise of her discretion pursuant to 49 U.S.C. Sections 44302 and 44306).

2   Any Equipment operated by the Insured under lease or other agreement under the terms of which the Insured assumes responsibility therefore or is to provide War Risks physical damage insurance therefore shall be attached hereunder as of the precise time such assumption of responsibility or such Insurance is to attach in accordance with the terms of such lease or other agreement. If no value has been specifically declared in the lease or other agreement and reported to the Insurer, then the Sum Insured shall be, the equitable and reasonable value agreed to between the Insured and the Administrator in the exercise of her discretion pursuant to 49 U.S.C. Sections 44302 and 44306. It is agreed that the Equipment values stated in the Lease agreements or other agreements are subject to amendment in accordance with the terms thereof.

3   As to any particular item of Equipment, cover hereunder ceases automatically when the Equipment is sold or otherwise disposed of and for which the Insured has no responsibility to provide War Risks physical damage insurance.

D   The Equipment values set forth in the Schedule shall be used to calculate the final reconciled premiums at the end of the policy period. The values on the Schedule will reflect the Insured's commercial insurance policy schedule of Equipment unless the Administrator determines prior to a loss incident that a valuation under Article II. A. (iii) of this Part is appropriate. The Administrator may alter the Agreed Values at her discretion with ten (10) days notice to the Insured.

III CONDITIONS

A   Subject to the terms, conditions, and exclusions of this Policy, this insurance covers all physical loss or damage to the Equipment described and set forth in the Schedule, while being operated by the Insured (or others approved by the Insured) and engines, navigational instruments, parts and appliances insured under the Insured's commercial hull loss policy.

B   The Insurer will pay, subject to the terms, conditions, and exclusions of this Policy: (1) in respect to total loss, the Sum Insured; and (2) in respect to partial loss:

1   If repairs are made by other than the Insured, the actual cost, as evidenced by bills rendered to the Insured less any discounts granted to the Insured, excluding the cost of overtime and its related overhead unless previously agreed to by the Insurer, to repair the damaged property with material or parts of like kind and quality, plus the reasonable cost of transporting new and/or damaged parts and/or the damaged Equipment to the place of repair and the return of the repaired Equipment to the control of the Insured, plus the reasonable and necessary costs incurred by the Insured in association with these repairs.

2   If repairs are made by the Insured, the total of the following items:

(a)   Actual cost of material or parts of like kind and quality.

(b)   Actual wages paid for direct labor, excluding extra charges for overtime, unless such overtime is consistent with sound business practices and the Insured's obligation to expeditiously and economically repair the damaged Equipment.

(c)   Overhead costs incurred by the Insured which shall be determined by the Administrator as (i) A reasonable percentage of Item 2 in lieu of all overhead, including supervisory services, (ii) actual overhead costs, or (iii) the relevant percentage provided in the Insured's previously effective commercial hull Policy.

(d)   The reasonable cost of transporting new and/or the damaged parts and/or the damaged Equipment to the place of repair and return of the repaired Equipment to the place of accident or home airport.

(e)   If repairs are not made and the Equipment is subsequently disposed of, then the estimated cost by the Administrator (consistent with the usual business practices of the Insured's commercial insurers) of making such repairs to the damaged Equipment with material of like kind and quality or the difference between the Sum Insured of the Equipment before it was damaged and the value of the Equipment in its damaged state.

C   The amount due under this Policy in respect to a partial loss shall not exceed the Sum Insured should the loss payable be for a total loss. When the amount paid hereunder is equal to the Sum Insured, any salvage value remaining shall inure to the benefit of the Insurer. There shall, however, be no abandonment without the consent of the Insurer.

D   The Sum Insured, remaining after loss or damage from a War Risk Occurrence, shall be reduced by the amount of any loss or damage, whether or not covered by this Policy, until repairs have been completed and the value automatically restored in kind.

IV PROMPT NOTICE OF LOSS

A   In the event of any known or suspected War Risk Occurrence which may result in loss, damage, or expense for which the Insurer may become liable, prompt notice thereof, on being known to the Insured, shall be given by the Insured to the Administrator consistent with this section. Failure to give such prompt notice because of uncertainty of the event causation, the occasion of War Risk Occurrences, or intervening regulations shall not necessarily prejudice this insurance.

   1   Notification must be immediate at any time there are fatalities or injuries to crew, passengers or third parties, there is property damage, or damage to the hull is likely to be in excess of $10,000.

   2   Notification may be given as soon as practicable, or as a written claim submission consistent with Article IV., Paragraph B. of this Part, within sixty (60) days to the FAA only when the aircraft suffers superficial damage; there were no fatalities or injuries to crew or passengers or third parties; there was no property damage, the air carrier has documented the aircraft damage with digital photography or by video tape and includes the records with the claim; and the amount of the claim is estimated to be under $10,000.

B   For all claims, the Insured shall render to the Administrator a proof of loss claim signed and sworn to by the Insured no later than 60 days after the determination of a war risk occurrence by the Secretary of Transportation or other cognizant government entity for loss or damage, or expense for which the Insurer may become liable (unless such time is extended in writing by the Insurer), stating the place, time, and cause of the loss or damage, the interest of the Insured and of all others in the Equipment, the Sum Insured at the time of the loss thereof, the amount and nature of the loss or damage, all encumbrances on the Equipment, all changes in title, and all other valid and collectible War Risk hull insurance covering said Equipment.

V   ASSISTANCE AND COOPERATION OF THE INSURED

A   The Insured and/or the Additional Insureds shall not interfere in any negotiations of the Insurer for settlement of any legal proceedings in respect to any War Risk Occurrence for which the Insurer may be liable under this Policy of Insurance. Provided that, in respect to any War Risk Occurrence likely to give rise to a claim under this Policy of Insurance, the Insured is obligated to, and shall take such steps to, protect its and the Insurer's interests as would reasonably be taken in the absence of this or similar insurance. The Insurer shall consult in good faith, and adequately in advance, with the Insured and/or the Additional Insureds regarding its proceeding and settlement strategy and proposed settlements, and ensure that it develops in good faith with the Insured a litigation defense or settlement strategy. The Insured and/or the Additional Insureds shall do nothing after a loss covered by this Policy to the prejudice of such rights of the Insurer. The Insurer and Insured and/or the Additional Insureds will cooperate fully in the investigation of any loss.

B   Whenever required by the Insurer, the Insured and/or the Additional Insureds shall aid in securing information and evidence and in obtaining witnesses and shall cooperate with the Insurer in the defense of any claim or suit or in the appeal from any judgment, in respect of any War Risk Occurrence as herein provided.

VI SUBROGATION RIGHTS

The Insurer shall be subrogated to all the rights which the Insured may have against any other person or entity, in respect of any payment made under this Policy, to the extent of such payment, and the Insured shall, upon the request of the Insurer, execute all documents necessary to secure to the Insurer such rights. The Insured shall do nothing after a loss covered by this Policy to the prejudice of such rights or defenses of the Insurer. The Insurer and Insured will cooperate fully in the investigation of any loss.

VII INSURED AIRCRAFT AND PROPERTY

The insurance provided hereunder covers only loss or damage to the Equipment described in the Schedule (as may be amended from time to time pursuant to Section II, above) while the Equipment is being operated by the Insured (or others approved by the Insured) which shall be deemed to include, but not be limited to stop-overs, ground time, and ferry flights to position or reposition the aircraft and/or maintenance or storage of Equipment.

VIII PAYMENT OF CLAIMS

A   The FAA shall make prompt payment in full of any claim covered under this policy after confirmation of loss. Any subsequent post-incident losses, directly related to the incident shall be covered by this Policy as a loss directly related to the original subject loss incident.

B   The FAA may, at its sole discretion and at any time prior to the final settlement of any claim by the Insured, elect to make a partial payment to the Insured for any loss, damage, or expense covered by this Policy.

IX PREMIUM PAYMENT

A   The actual premium for this Policy of Insurance shall be $0.012 per hundred dollars of the total Sums Insured of the aircraft of the Insured as set forth in the Schedule as in effect of the date of this policy, per year, prorated on an aircraft day basis for the number of days the policy is in effect.

B   The Insured shall estimate a deposit premium as specified in VIII (A) above for each day the policy is in effect and in total for the entire duration of the policy. The Insurer reserves the right to increase or decrease the premium offered by the Insured based on the Insurer's analysis of expected operations by the insured.

C   The total deposit premium will be paid in three installments; each installment representing the fraction of the total duration of coverage provided during each of the periods October 1, 2012 through January 31, 2013; February 1, 2013 through May 31, 2013; and June 1, 2013 through September 30, 2013.

D   Within 30 days of the expiration or termination of this Policy of Insurance, the insured shall calculate and submit data to reconcile the actual premium owed with the deposit premium estimated.

　　1   Final reconciliation will be based on the calculation of aircraft day valuation based upon actual values reported by the Insured to the Insurer.

　　2   If the premium owed is greater than the deposit premium paid by the Insured, the Insured shall pay the premium difference to the Insurer.

　　3   If the premium owed is less than the deposit premium paid by the Insured, the Insurer will refund the premium difference.

THIS ENDS PART I OF THE POLICY OF INSURANCE

UNITED STATES OF AMERICA

DEPARTMENT OF TRANSPORTATION

FEDERAL AVIATION ADMINISTRATION

Policy No. _____

PART II: COMPREHENSIVE

I   COVERAGE

A   The Insurer, represented by the Administrator, acting for the Secretary of Transportation, shall provide by this Policy of Insurance, in accordance with applicable provisions of law and subject to all limitations thereof, comprehensive liability insurance for, (i) the loss of property of others carried by the Insured (or others approved by the Insured) or in the care, custody or control of the Insured or for which the insured has agreed to be responsible (including but not limited to baggage and personal effects, cargo, mail and aircraft and/or aircraft spare parts and equipment), (ii) for the personal injury, bodily injury or death of Passengers or crewmembers of the Insured, and/or (iii) premises liability, products/completed operations liability, and hangarkeepers liability, in the limits shown herein for War Risk Occurrences arising from the Air Transportation Business of _____ (hereinafter, the Insured).

B   Subject to the limits of liability, exclusions, conditions, and other terms of this Policy of Insurance, the Insurer hereby agrees to pay on behalf of the Insured (including Other Insured Parties) and/or its vendors, agents, and subcontractors (additional insureds) whose products and services are required for the operation of its aircraft for which the insured has entered into an indemnity agreement, all sums which the Insured shall be legally liable to pay, or by final judgment be adjudged to pay, to any person or persons, including damages for personal injuries, and/or bodily injuries sustained, including death at any time resulting therefrom, damages for care and loss of services, or by reason of loss or damage to or destruction of property resulting from the occurrence of loss resulting from War Risk Occurrences.

II AMOUNTS

A   The amount of Insurance provided under this Policy of Insurance shall not exceed U.S.$_____ (dollar amount) of liability per occurrence, per aircraft incurred by the Insured and/or the Additional Insureds for losses resulting from a War Risk Occurrence, which is the limit of liability in the Insured's commercial policy in effect on November 25, 2002, or for those who did not hold a commercial all risk insurance policy as of November 25, 2002, the amount equal to the limit contained in the Insured's previous FAA war risk policy, or for those who were not previously insured by the FAA, the limit in the Insured's current commercial all risk insurance policy.

B   The Insured agrees that, if the limits of liability contained in the Insured's commercial insurance policy carried against liabilities arising from risks other than war risks are voluntarily reduced to amounts of insurance less than the limits of liability stated by this Policy, the insurance under this Policy shall be considered to have been automatically amended to the new limits of liability on the Insured's commercial policies at the time of such amendment.

III DEFENSE AND SETTLEMENT OF CLAIMS

A   The Insurer shall have the right and duty to defend any suit or claim against the Insured and/or the Additional Insureds seeking damages on account of any bodily injury, personal injury, or property damage covered under this policy, even if such suit is groundless, false or fraudulent and may make such investigation, negotiation, and settlement of any claim or suit as it deems proper and expedient, but the Insurer shall not be obligated to pay any claim or judgment or to defend any suit or claim after the applicable limit of the Insurer's liability has been exhausted by payment of judgments or settlements.

B   During such time as the Insurer is obligated to defend a claim or claims under the provisions of the preceding paragraph, the Insurer shall pay with respect to such claims:

1   Subject to the applicable limits of liability, all expenses incurred by the Insured and/or the Additional Insureds, all costs taxed against the Insured and/or the Additional Insureds in any suit or claim defended by the Insurer and all interest on the entire amount of any judgment

thereon which accrues after entry of the judgment and before the Insurer has paid or tendered or deposited in court that part of the judgment which does not exceed the limit of the Insurer's liability thereon under this Policy; and

2   In addition to the applicable limits of liability, all reasonable expenses incurred by the Insured and/or the Additional Insureds at the Insurer's request, other than for loss of earnings or for wages or salaries of employees of the Insured and/or the Additional Insureds.

IV PROMPT NOTICE OF LOSS

A   In the event of any known or suspected War Risk Occurrence which may result in loss, damage, or expense for which the Insurer may become liable, prompt notice thereof, on being known to the Insured, shall be given by the Insured to the Administrator consistent with this section. Failure to give such prompt notice because of uncertainty of the event causation, the occasion of War Risk Occurrences, or intervening regulations shall not necessarily prejudice this insurance.

B   If a claim is made or suit is brought against the Insured and/or the Additional Insureds for which the insurer may become liable, the Insured shall immediately notify the Insurer and timely provide a copy of every demand, notice, summons, pleading, motion, document filed with a court, settlement offer, and other process received by the Insured or its representatives.

V ASSISTANCE AND COOPERATION OF THE INSURED

A   The Insured and/or the Additional Insureds shall not interfere in any negotiations of the Insurer for settlement of any legal proceedings in respect to any War Risk Occurrence for which the Insurer may be liable under this Policy of Insurance. Provided that, in respect to any War Risk Occurrence likely to give rise to a claim under this Policy of Insurance, the Insured is obligated to, and shall take such steps to, protect its and the Insurer's interests as would reasonably be taken in the absence of this or similar insurance. The Insurer shall consult in good faith, and adequately in advance, with the Insured and/or the Additional Insureds regarding its proceeding and settlement strategy and proposed settlements, and ensure that it develops in good faith with the Insured a litigation defense or settlement strategy. The Insured and/or the Additional Insureds shall do nothing after a loss covered by this Policy to the prejudice of such rights of the Insurer. The Insurer and Insured and/or the Additional Insureds will cooperate fully in the investigation of any loss.

B   Whenever required by the Insurer, the Insured and/or the Additional Insureds shall aid in securing information and evidence and in obtaining witnesses and shall cooperate with the Insurer in the defense of any claim or suit or in the appeal from any judgment, in respect of any War Risk Occurrence as herein provided.

VI ACTION AGAINST THE INSURER

No action shall lie against the Insurer unless, as a condition precedent thereto, the Insured shall have fully complied with all of the terms of this Policy and until the amount of the Insured's obligations to pay, with respect to the specific legal action or claim in question, shall have been finally determined either by judgment against the Insured after actual trial or by written agreement of the Claimant and the Insurer. Any person or organization or the legal representative thereof who has secured such judgment or written agreement shall thereafter be entitled to recover under this Policy to the extent such judgment or written agreement is not in excess of the remaining insurance afforded by this Policy. Nothing contained in this Policy shall give any person or organization any right to join the Insurer as a codefendant in any action against the Insured and/or Additional Insureds to determine the Insured's liability. Neither the filing nor the adjudication of bankruptcy or insolvency of the Insured or Insured's Estate shall relieve the Insurer of any of its obligations hereunder.

VII SUBROGATION RIGHTS

The Insurer shall be subrogated to all the rights which the Insured and/or Additional Insureds may have against any other person or entity, in respect of any payment made under this Policy, to the extent of such payment, and the Insured shall, upon the request of the Insurer, execute all documents necessary to secure to the Insurer such rights. The Insured shall do nothing after a loss covered by this Policy to the prejudice of such rights or defenses of the Insurer. The Insurer and Insured will cooperate fully in the investigation of any loss.

VIII PAYMENT OF CLAIMS

A   The FAA shall make prompt payment in full, on behalf of the Insured, of any claim covered under this Policy after the Insured and/or the Additional Insureds shall become legally liable to pay, or by final judgment be adjudged to pay. Any subsequent post-incident losses, directly related to the incident, incurred shall be covered by this Policy as a loss directly related to the original subject loss incident.

B   The FAA may at its discretion, and at any time prior to the final settlement of any claim by the Insured and/or the Additional Insureds, elect to make a partial payment to the Insured for any loss, damage, or expense covered by this Policy.

IX PREMIUM PAYMENT

A   The actual premium for this Part II of the Policy of Insurance shall be based upon whether the Insured conducts passenger or air freight operations, or a combination of the two. The premium calculations are set into four Classes based upon the amount of coverage set forth in Article II. Class I calculations address Article II coverage amounts that are less than Five Hundred Million Dollars ($500,000,000). Class II calculations address Article II coverage amounts that range from Five Hundred Million Dollars ($500,000,000) but less than One Billion U.S. Dollars ($1,000,000,000). Class III calculations address Article II coverage amounts that range from One Billion U.S. Dollars ($1,000,000,000) or more, but less than One and One-Half Billion U.S. Dollars ($1,500,000,000). Class IV calculations address Article II coverage amounts that are One and One-Half Billion U.S. Dollars ($1,500,000,000) or more. The premium for each Class shall be calculated as follows:

Class I. If the amount of coverage in Article II amounts to less than Five Hundred Million U.S. Dollars ($500,000,000), the total premium shall be calculated as the sum of premiums for passenger and freight operations as set forth below:

Formulae for Class I Premium for Passenger Operations = ($0.14 × number of enplanements) + ($0.14 × RPM/1000)

Premium for Freight Operations = $0.01 × RTM/1000 Class II. If the amount of coverage in Article II amounts to Five Hundred Million U.S. Dollars ($500,000,000) or more, but less than One Billion U.S. Dollars ($1,000,000,000), the total premium shall be calculated as the sum of premiums for passenger and freight operations as set forth below:

Formulae for Class II Premium for Passenger Operations = ($0.18 × number of enplanements) + ($0.18 × RPM/1000)

Premium for Freight Operations = $0.01 × RTM/1000 Class III. If the amount of coverage in Article II amounts to One Billion U.S. Dollars ($1,000,000,000) or more, but less than One and One-Half Billion U.S. Dollars ($1,500,000,000), the total premium shall be calculated as the sum of premiums for passenger and freight operations as set forth below:

Formulae for Class III Premium for Passenger Operations = ($0.23 × number of enplanements) + ($0.23 × RPM/1000)

Premium for Freight Operations = $0.02 × RTM/1000 Class IV. If the amount of coverage in Article II amounts to One and One-Half Billion U.S. Dollars ($1,500,000,000) or more, the total premium shall be calculated as the sum of premiums for passenger and freight operations as set forth below:

Formulae for Class IV Premium for Passenger Operations = ($0.23 × number of enplanements) + ($0.23 × RPM/1000)

Premium for Freight Operations = $0.02 × RTM/1000

B   The Insured shall estimate a deposit premium as specified in IX (A) above for each day the policy is in effect and in total for the entire duration of the policy. The Insurer reserves the right to increase or decrease the premium offered by the Insured based on its analysis of expected operations by the insured.

C   The total deposit premium will be paid in three installments; each installment representing the fraction of the total duration of coverage provided during each of the periods October 1, 2012 through January 31, 2013; February 1, 2013 through May 31, 2013; and June 1, 2013 through September 30, 2013.

D   Within 30 days of the expiration or termination of this Policy of Insurance, the insured shall calculate and submit data to reconcile the actual premium owed with the deposit premium estimated.

    1   If the premium owed is greater than the deposit premium paid by the Insured, the Insured shall pay the premium difference to the Insurer.

    2   If the premium owed is less than the deposit premium paid by the Insured, the Insurer must refund the premium difference.

THIS ENDS PART II OF THE POLICY OF INSURANCE

UNITED STATES OF AMERICA

DEPARTMENT OF TRANSPORTATION

FEDERAL AVIATION ADMINISTRATION

Policy No. _____

PART III: THIRD PARTY WAR RISK LIABILITY INSURANCE

I   COVERAGE

A   The Insurer, represented by the Administrator, acting for the Secretary of Transportation, shall provide by this Policy of Insurance, in accordance with applicable provisions of law and subject to all limitations thereof, and upon the payment of a premium, pursuant to the provisions of chapter 443, Third Party War Risk Liability Insurance (hereinafter, the Insurance) of the type indicated and in the limits shown herein for War Risk Occurrences arising from the Air Transportation Business of _____ (hereinafter, the Insured, including Other Insured Parties), and, through indemnification agreements entered into by the Insured with its vendors, agents, and subcontractors for goods or services related to the Insured's Air Transportation Business.

B   Subject to the limits of liability, exclusions, conditions, and other terms of this Policy of Insurance, the Insurer hereby agrees to pay on behalf of the Insured all sums which the Insured (including Other Insured Parties) and/or its vendors, agents, and subcontractors whose products and services are required for the operation of its aircraft for which the insured has entered into an indemnity agreement shall be legally liable to pay to any person or persons who are not Passengers or employees who are on active duty in the course of their employment of the Insured, or by final judgment be adjudged to pay to any such person or persons, including damages for personal injuries and/or bodily injuries sustained, including death at any time resulting

306

therefrom, damages for care and loss of services, or by reason of loss or damage to or destruction of property, including the loss of use thereof, resulting from a loss resulting from a War Risk Occurrence arising from the Insured's Air Transportation Business.

II AMOUNTS

The amount of Insurance provided under this Policy of Insurance shall not exceed U.S.$_____ (dollar amount equal to the limit of third-party liability in the Insured's previous FAA war risk policy, or for those who were not previously insured by the FAA, two times the per-occurrence liability limit in the Insured's current commercial all risk insurance policy) incurred by the Insured and/or Additional Insureds for losses resulting from a War Risk Occurrence.

III DEFENSE AND SETTLEMENT OF CLAIMS

A   The Insurer shall have the right and duty to defend any suit or claim against the Insured and/or Additional Insureds seeking damages on account of any bodily injury, personal injury, or property damage covered under this Policy of Insurance, even if such suit is groundless, false or fraudulent and may make such investigation, negotiation, and settlement of any claim or suit as it deems proper and expedient, but the Insurer shall not be obligated to pay any claim or judgment or to defend any suit or claim after the applicable limit of the Insurer's liability has been exhausted by payment of judgments or settlements.

B   During such time as the Insurer is obligated to defend a claim or claims under the provisions of the preceding paragraph, the Insurer shall pay with respect to such claims:

1   Subject to the applicable limits of liability, all expenses incurred by the Insured, all costs taxed against the Insured in any suit or claim defended by the Insurer and all interest on the entire amount of any judgment thereon which accrues after entry of the judgment and before the Insurer has paid or tendered or deposited in court that part of the judgment which does not exceed the limit of the Insurer's liability thereon under this Policy; and

2   In addition to the applicable limits of liability, all reasonable expenses incurred by the Insured at the Insurer's request, other than for loss of earnings or for wages or salaries of employees of the Insured.

IV PROMPT NOTICE OF LOSS

A   In the event of any known or suspected War Risk Occurrence which may result in loss, damage, or expense for which the Insurer may become liable, prompt notice thereof, on being known to the Insured, shall be given by the Insured to the Administrator consistent with this section. Failure to give such prompt notice because of uncertainty of the event causation, the occasion of War Risk Occurrences, or intervening regulations shall not necessarily prejudice this insurance.

B   If a claim is made or suit is brought against the Insured and/or the Additional Insureds for which the insurer may become liable, the Insured shall immediately notify the Insurer and timely provide a copy of every demand, notice, summons, pleading, motion, document filed with a court, settlement offer, and other process received by the Insured or its representatives.

V   ASSISTANCE AND COOPERATION OF THE INSURED

A   The Insured shall not interfere in any negotiations by the Insurer for settlement of any legal proceedings in respect of any War Risk Occurrence for which the Insurer may be liable under this Policy of Insurance. Provided, that in respect of any War Risk Occurrence likely to give rise to a claim under this Policy of Insurance, the Insured is obligated to, and shall take such steps to protect its and the Insurer's interests as would reasonably be taken in the absence of this or similar insurance. The Insurer shall consult in good faith, and adequately in advance, with the Insured regarding its proceeding and settlement strategy and proposed settlements, and ensure that it develops in good faith with the Insured and/or Additional Insureds a litigation defense or settlement strategy.

B   Whenever required by the Insurer, the Insured shall aid in securing information and evidence and in obtaining witnesses and shall cooperate with the Insurer in the defense of any claim or suit or in the appeal from any judgment, in respect of any War Risk Occurrence as herein provided.

VI   ACTION AGAINST THE INSURER

No action shall lie against the Insurer unless, as a condition precedent thereto, the Insured shall have fully complied with all of the terms of this Policy of Insurance and until the amount of the Insured's obligations to pay, with respect to the specific legal action or claim in question, shall have been finally determined either by judgment against the Insured after actual trial or by written agreement of the claimant and the Insurer. Any person or organization or the legal representative thereof who has secured such judgment or written agreement shall thereafter be entitled to recover under this Policy of Insurance to the extent such judgment or written agreement is not in excess of the remaining insurance afforded by this Policy of Insurance. Nothing contained in this Policy of Insurance shall give any person or organization any right to join the Insurer as a co-defendant in any action against the Insured and/or Additional Insureds to determine the liability of the Insured. Neither the filing nor the adjudication of bankruptcy or insolvency of the Insured or the Estate of the Insured shall relieve the Insurer of any of its obligations hereunder.

VII   SUBROGATION RIGHTS

The Insurer shall be subrogated to all the rights which the Insured may have against any other person or entity, in respect of any payment made under this Policy of Insurance, to the extent of such payment, and the Insured shall, upon the request of the Insurer, execute all documents necessary to secure to the Insurer such rights.

The Insured shall do nothing after a loss covered by this Policy to the prejudice of such rights or defenses of the Insurer. The Insurer and Insured will cooperate fully in the investigation of any loss.

VIII   PAYMENT OF CLAIMS

A   The Insurer shall make prompt payment in full, on behalf of the Insured and/or Additional Insureds of any claim covered under this Policy of Insurance after the Insured and/or Additional Insureds becomes legally liable to pay, or by final judgment be adjudged to pay. Any subsequent post-incident losses incurred that are directly related to the incident shall be covered by this Policy of Insurance as a loss directly related to the original subject loss incident.

B   The Insurer may at its discretion, and at any time prior to final settlement of any claim by the Insured, elect to make a partial payment to the insured for any loss, damage, or expense covered by this Policy of Insurance.

IX PREMIUM PAYMENT

A   The actual premium for this Part III of the Policy of Insurance shall be based upon whether the Insured conducts passenger or air freight operations, or a combination of the two. The premium calculations are set into four Classes based upon the amount of coverage set forth in Article II of this Part III. Class I calculations address Article II coverage amounts that are lessthan One Billion U.S. Dollars ($1,000,000,000). Class II calculations address Article II coverage amounts that range from One Billion Dollars ($1,000,000,000) or more, to less than Two Billion U.S. Dollars ($2,000,000,000). Class III calculations address Article II coverage amounts that range from Two Billion U.S. Dollars ($2,000,000,000) or more, but less than Three Billion U.S. Dollars ($3,000,000,000). Class IV calculations address Article II coverage amounts that range from Three Billion U.S. Dollars ($3,000,000,000) or more. The premium for each Class shall be calculated as follows:

Class I. If the amount of coverage in Article II amounts to less than One Billion U.S. Dollars ($1,000,000,000), the total premium shall be calculated as the sum of premiums for passenger and freight operations as set forth below:

Formulae for Class I Premium for Passenger Operations $==$ ($0.03 × number of enplanements)+($0.03 × RPM/1000)

Premium for Freight Operations = $0.17 × RTM/1000 Class II. If the amount of coverage in Article II amounts to One Billion U.S. Dollars ($1,000,000,000) or more, but less than Two Billion U.S. Dollars ($2,000,000,000), the total premium shall be calculated as the sum of premiums for passenger and freight operations as set forth below:

Formulae for Class II Premium for Passenger Operations = ($0.04 × number of enplanements)+($0.04 × RPM/1000)

Premium for Freight Operations = $0.25 × RTM/1000 Class III. If the amount of coverage in Article II amounts to Two Billion U.S. Dollars ($2,000,000,000) or more, but less than Three Billion U.S. Dollars ($3,000,000,000), the total premium shall be calculated as the sum of premiums for passenger and freight operations as set forth below:

Formulae for Class III Premium for Passenger Operations = ($0.05 × number of enplanements)+($0.05 × RPM/1000)

Premium for Freight Operations = $0.30 × RTM/1000 Class IV. If the amount of coverage in Article II amounts to Three Billion U.S. Dollars ($3,000,000,000) or more, the total premium shall be calculated as the sum of premiums for passenger and freight operations as set forth below:

Formulae for Class IV Premium for Passenger Operations = ($0.05 × number of enplanements)+($0.05 × RPM/1000)

Premium for Freight Operations = $0.33 × RTM/1000

B   The Insured shall estimate a deposit premium as specified in IX (A) above for each day the policy is in effect and in total for the entire duration of the policy. The Insurer reserves the right to increase or decrease the premium offered by the Insured based on its analysis of expected operations by the insured.

C   The total deposit premium will be paid in three installments; each install-ment representing the fraction of the total duration of coverage provided during each of the periods October 1, 2012 through January 31, 2013; February 1, 2013 through May 31, 2013; and June 1, 2013 through September 30, 2013. The Insured shall pay the first estimated premium installment within ten (10) days of the beginning of the policy period, and subsequent estimated insurance premium installments within ten (10) days of the first day of each of the installment periods.

D   Within 30 days of the expiration or termination of this Policy of Insurance, the insured shall calculate and submit data to reconcile the actual premium owed with the deposit premium estimated.
   1   If the premium owed is greater than the deposit premium paid by the Insured, the Insured shall pay the premium difference to the Insurer.
   2   If the premium owed is less than the deposit premium paid by the Insured, the Insurer will refund the premium difference.

THIS ENDS PART III OF THE POLICY OF INSURANCE

INSURANCE PROVISIONS AVN48B

WAR, HI-JACKING AND OTHER PERILS EXCLUSION CLAUSE (AVIATION)

This Policy does not cover claims caused by

(a) War, invasion, acts of foreign enemies, hostilities (whether war be declared or not), civil war, rebellion, revolution, insurrection, martial law, military or usurped power or attempts at usurpation of power.

(b) Any hostile detonation of any weapon of war employing atomic or nuclear fission and/or fusion or other like reaction or radioactive force or matter.

(c) Strikes, riots, civil commotions or labour disturbances.

(d) Any act of one or more persons, whether or not agents of a sovereign Power, for political or terrorist purposes and whether the loss or damage resulting therefrom is accidental or intentional.

(e) Any malicious act or act of sabotage.

(f) Confiscation, nationalisation, seizure, restraint, detention, appropriation, requisition for title or use by or under the order of any Government (whether civil, military or de facto) or public or local authority.

(g) Hi-jacking or any unlawful seizure or wrongful exercise of control of the Aircraft or crew in Flight (including any attempt at such seizure or control) made by any person or persons on board the Aircraft acting without the consent of the Insured.

Furthermore this Policy does not cover claims arising while the Aircraft is outside the control of the Insured by reason of any of the above perils. The Aircraft shall be deemed to have been restored to the control of the Insured on the safe return of the Aircraft to the Insured at an airfield not excluded by the geographical limits of this Policy, and entirely suitable for the operation of the Aircraft (such safe return shall require that the Aircraft be parked with engines shut down and under no duress).

AVN 48B 1.10.96

INSURANCE PROVISIONS CUT THROUGH CLAUSE

CUT-THROUGH ENDORSEMENT

The Reinsurers hereby agree, at the request and with the agreement of the Reinsured, that if a valid hull or aircraft spares claim arises hereunder the Reinsurers shall pay to the order of the party(/ies) entitled to indemnity under the original insurance effected by the Insured that portion of any loss which the Reinsurers would otherwise be liable to pay to the Reinsured, subject to the following provisions:

(1) such loss payment shall be in lieu of payment to the Reinsured or its successors in interest and assigns, and shall fully discharge and release the Reinsurers from any and all liability in connection with such a claim under the hull and aircraft spares insurances;

(2) such loss payment shall be made notwithstanding non-payment of the Reinsured's portion under the original insurance;

(3) the Reinsurers reserve the right to set off against such payment any outstanding premiums due on the subject hull or aircraft spares;

(4) if the Reinsured is declared insolvent, bankrupt, in liquidation, in dissolution or in administration by a court of competent jurisdiction to which the Reinsured is subject, the Reinsurers shall only be obliged to make payment under this Endorsement if the court consents to such payment and confirms that such payment fully discharges and releases Reinsurers from further liability in relation to such a claim under the hull or aircraft spares insurances, such consent and confirmation being in a form satisfactory to the Reinsurers. The Reinsurers shall take reasonable steps to obtain such consent and confirmation at Reinsurers' cost. If there is a dispute as to such matters, then the Reinsurers' liability shall be determined by such court at Reinsurers' cost, prior to payment;

(5) Reinsurers shall not be obliged to make a payment under this Endorsement if such payment would contravene the laws of the jurisdiction to which the Reinsured is subject. The Reinsurers and the Reinsured shall each take all reasonable steps at their own cost to obtain any necessary governmental consent or licence in order to permit such payment to be lawfully made.

AVN 109 24.9.09

SELECTED TREATIES CAPE TOWN CONVENTION

CONVENTION ON INTERNATIONAL INTERESTS IN MOBILE EQUIPMENT

THE STATES PARTIES TO THIS CONVENTION,

AWARE of the need to acquire and use mobile equipment of high value or particular economic significance and to facilitate the financing of the acquisition and use of such equipment in an efficient manner,

RECOGNISING the advantages of asset-based financing and leasing for this purpose and desiring to facilitate these types of transaction by establishing clear rules to govern them,

MINDFUL of the need to ensure that interests in such equipment are recognised and protected universally,

DESIRING to provide broad and mutual economic benefits for all interested parties,

BELIEVING that such rules must reflect the principles underlying asset-based financing and leasing and promote the autonomy of the parties necessary in these transactions,

CONSCIOUS of the need to establish a legal framework for international interests in such equipment and for that purpose to create an international registration system for their protection,

TAKING INTO CONSIDERATION the objectives and principles enunciated in existing Conventions relating to such equipment,

HAVE AGREED upon the following provisions:

Chapter I
Sphere of application and general provisions

Article 1 — Definitions

In this Convention, except where the context otherwise requires, the following terms are employed with the meanings set out below:

    (a)   'agreement' means a security agreement, a title reservation agreement or a leasing agreement;

(b) 'assignment' means a contract which, whether by way of security or otherwise, confers on the assignee associated rights with or without a transfer of the related international interest;

(c) 'associated rights' means all rights to payment or other performance by a debtor under an agreement which are secured by or associated with the object;

(d) 'commencement of the insolvency proceedings' means the time at which the insolvency proceedings are deemed to commence under the applicable insolvency law;

(e) 'conditional buyer' means a buyer under a title reservation agreement;

(f) 'conditional seller' means a seller under a title reservation agreement;

(g) 'contract of sale' means a contract for the sale of an object by a seller to a buyer which is not an agreement as defined in (a) above;

(h) 'court' means a court of law or an administrative or arbitral tribunal established by a Contracting State;

(i) 'creditor' means a chargee under a security agreement, a conditional seller under a title reservation agreement or a lessor under a leasing agreement;

(j) 'debtor' means a chargor under a security agreement, a conditional buyer under a title reservation agreement, a lessee under a leasing agreement or a person whose interest in an object is burdened by a registrable non-consensual right or interest;

(k) 'insolvency administrator' means a person authorised to administer the reorganisation or liquidation, including one authorised on an interim basis, and includes a debtor in possession if permitted by the applicable insolvency law;

(l) 'insolvency proceedings' means bankruptcy, liquidation or other collective judicial or administrative proceedings, including interim proceedings, in which the assets and affairs of the debtor are subject to control or supervision by a court for the purposes of reorganisation or liquidation;

(m) 'interested persons' means:
(i) the debtor;
(ii) any person who, for the purpose of assuring performance of any of the obligations in favour of the creditor, gives or issues a suretyship or demand guarantee or a standby letter of credit or any other form of credit insurance;
(iii) any other person having rights in or over the object;

(n) 'internal transaction' means a transaction of a type listed in Article 2(2)(a) to (c) where the centre of the main interests of all parties to such transaction is situated, and the relevant object located (as specified in the Protocol), in the same Contracting State at the time of the conclusion of the contract and where the interest created by the transaction has been registered in a national registry in that Contracting State which has made a declaration under Article 50(1);

(o) 'international interest' means an interest held by a creditor to which Article 2 applies;

(p) 'International Registry' means the international registration facilities established for the purposes of this Convention or the Protocol;

(q) 'leasing agreement' means an agreement by which one person (the lessor) grants a right to possession or control of an object (with or without an option to purchase) to another person (the lessee) in return for a rental or other payment;

(r) 'national interest' means an interest held by a creditor in an object and created by an internal transaction covered by a declaration under Article 50(1);

(s) 'non-consensual right or interest' means a right or interest conferred under the law of a Contracting State which has made a declaration under Article 39 to secure the performance of an obligation, including an obligation to a State, State entity or an intergovernmental or private organisation;

(t) 'notice of a national interest' means notice registered or to be registered in the International Registry that a national interest has been created;

(u) 'object' means an object of a category to which Article 2 applies;

(v) 'pre-existing right or interest' means a right or interest of any kind in or over an object created or arising before the effective date of this Convention as defined by Article 0(2)(a);

(w) 'proceeds' means money or non-money proceeds of an object arising from the total or partial loss or physical destruction of the object or its total or partial confiscation, condemnation or requisition;

(x) 'prospective assignment' means an assignment that is intended to be made in the future, upon the occurrence of a stated event, whether or not the occurrence of the event is certain;

(y) 'prospective international interest' means an interest that is intended to be created or provided for in an object as an international interest in the future, upon the occurrence of a stated event (which may include the debtor's acquisition of an interest in the object), whether or not the occurrence of the event is certain;

(z) 'prospective sale' means a sale which is intended to be made in the future, upon the occurrence of a stated event, whether or not the occurrence of the event is certain;

(aa) 'Protocol' means, in respect of any category of object and associated rights to which this Convention applies, the Protocol in respect of that category of object and associated rights;

(bb) 'registered' means registered in the International Registry pursuant to Chapter V;

(cc) 'registered interest' means an international interest, a registrable non-consensual right or interest or a national interest specified in a notice of a national interest registered pursuant to Chapter V;

(dd) 'registrable non-consensual right or interest' means a non-consensual right or interest registrable pursuant to a declaration deposited under Article 40;

(ee) 'Registrar' means, in respect of the Protocol, the person or body designated by that Protocol or appointed under Article 17(2)(b);

(ff) 'regulations' means regulations made or approved by the Supervisory Authority pursuant to the Protocol;

(gg) 'sale' means a transfer of ownership of an object pursuant to a contract of sale;

(hh) 'secured obligation' means an obligation secured by a security interest;

(ii) 'security agreement' means an agreement by which a chargor grants or agrees to grant to a chargee an interest (including an ownership interest) in or over an object to secure the performance of any existing or future obligation of the chargor or a third person;

(jj) 'security interest' means an interest created by a security agreement;

(kk) 'Supervisory Authority' means, in respect of the Protocol, the Supervisory Authority referred to in Article 17(1);

(ll) 'title reservation agreement' means an agreement for the sale of an object on terms that ownership does not pass until fulfilment of the condition or conditions stated in the agreement;

(mm) 'unregistered interest' means a consensual interest or non-consensual right or interest other than an interest to which Article 39 applies) which has not been registered, whether or not it is registrable under this Convention; and

(nn) 'writing' means a record of information (including information communicated by teletransmission) which is in tangible or other form and is capable of being reproduced in tangible form on a subsequent occasion and which indicates by reasonable means a person's approval of the record.

Article 2 — The international interest

1   This Convention provides for the constitution and effects of an international interest in certain categories of mobile equipment and associated rights.

2   For the purposes of this Convention, an international interest in mobile equipment is an interest, constituted under Article 7, in a uniquely identifiable object of a category of such objects listed in paragraph 3 and designated in the Protocol:

(a)   granted by the chargor under a security agreement;

(b)   vested in a person who is the conditional seller under a title reservation agreement; or

(c)   vested in a person who is the lessor under a leasing agreement.

An interest falling within sub-paragraph (a) does not also fall within sub-paragraph (b) or (c).

3   The categories referred to in the preceding paragraphs are:

(a)   airframes, aircraft engines and helicopters;

(b)   railway rolling stock; and

(c)   space assets.

4   The applicable law determines whether an interest to which paragraph 2 applies falls within subparagraph (a), (b) or (c) of that paragraph.

5   An international interest in an object extends to proceeds of that object.

Article 3 — Sphere of application

1   This Convention applies when, at the time of the conclusion of the agreement creating or providing for the international interest, the debtor is situated in a Contracting State.

2   The fact that the creditor is situated in a non-Contracting State does not affect the applicability of this Convention.

Article 4 — Where debtor is situated

1   For the purposes of Article 3(1), the debtor is situated in any Contracting State:
     (a)   under the law of which it is incorporated or formed;
     (b)   where it has its registered office or statutory seat;
     (c)   where it has its centre of administration; or
     (d)   where it has its place of business.

2   A reference in sub-paragraph (d) of the preceding paragraph to the debtor's place of business shall, if it has more than one place of business, mean its principal place of business or, if it has no place of business, its habitual residence.

Article 5 — Interpretation and applicable law

1   In the interpretation of this Convention, regard is to be had to its purposes as set forth in the preamble, to its international character and to the need to promote uniformity and predictability in its application.

2   Questions concerning matters governed by this Convention which are not expressly settled in it are to be settled in conformity with the general principles on which it is based or, in the absence of such principles, in conformity with the applicable law.

3   References to the applicable law are to the domestic rules of the law applicable by virtue of the rules of private international law of the forum State.

4   Where a State comprises several territorial units, each of which has its own rules of law in respect of the matter to be decided, and where there is no indication of the relevant territorial unit, the law of that State decides which is the territorial unit whose rules shall govern. In the absence of any such rule, the law of the territorial unit with which the case is most closely connected shall apply.

Article 6 — Relationship between the Convention and the Protocol

1   This Convention and the Protocol shall be read and interpreted together as a single instrument.

2   To the extent of any inconsistency between this Convention and the Protocol, the Protocol shall prevail.

Chapter II
Constitution of an international interest

Article 7 — Formal requirements

An interest is constituted as an international interest under this Convention where the agreement creating or providing for the interest:

(a) is in writing;
(b) relates to an object of which the chargor, conditional seller or lessor has power to dispose;
(c) enables the object to be identified in conformity with the Protocol; and
(d) in the case of a security agreement, enables the secured obligations to be determined, but without the need to state a sum or maximum sum secured.

Chapter III
Default remedies

Article 8 — Remedies of chargee

1 In the event of default as provided in Article 11, the chargee may, to the extent that the chargor has at any time so agreed and subject to any declaration that may be made by a Contracting State under Article 54, exercise any one or more of the following remedies:
   (a) take possession or control of any object charged to it;
   (b) sell or grant a lease of any such object;
   (c) collect or receive any income or profits arising from the management or use of any such object.

2 The chargee may alternatively apply for a court order authorising or directing any of the acts referred to in the preceding paragraph.

3 Any remedy set out in sub-paragraph (a), (b) or (c) of paragraph 1 or by Article 13 shall be exercised in a commercially reasonable manner. A remedy shall be deemed to be exercised in a commercially reasonable manner where it is exercised in conformity with a provision of the security agreement except where such a provision is manifestly unreasonable.

4 A chargee proposing to sell or grant a lease of an object under paragraph 1 shall give reasonable prior notice in writing of the proposed sale or lease to:
   (a) interested persons specified in Article 1(m)(i) and (ii); and
   (b) interested persons specified in Article 1(m)(iii) who have given notice of their rights to the chargee within a reasonable time prior to the sale or lease.

5 Any sum collected or received by the chargee as a result of exercise of any of the remedies set out in paragraph 1 or 2 shall be applied towards discharge of the amount of the secured obligations.

6 Where the sums collected or received by the chargee as a result of the exercise of any remedy set out in paragraph 1 or 2 exceed the amount

318

secured by the security interest and any reasonable costs incurred in the exercise of any such remedy, then unless otherwise ordered by the court the chargee shall distribute the surplus among holders of subsequently ranking interests which have been registered or of which the chargee has been given notice, in order of priority, and pay any remaining balance to the chargor.

Article 9 — Vesting of object in satisfaction; redemption

1　At any time after default as provided in Article 11, the chargee and all the interested persons may agree that ownership of (or any other interest of the chargor in) any object covered by the security interest shall vest in the chargee in or towards satisfaction of the secured obligations.

2　The court may on the application of the chargee order that ownership of (or any other interest of the chargor in) any object covered by the security interest shall vest in the chargee in or towards satisfaction of the secured obligations.

3　The court shall grant an application under the preceding paragraph only if the amount of the secured obligations to be satisfied by such vesting is commensurate with the value of the object after taking account of any payment to be made by the chargee to any of the interested persons.

4　At any time after default as provided in Article 11 and before sale of the charged object or the making of an order under paragraph 2, the chargor or any interested person may discharge the security interest by paying in full the amount secured, subject to any lease granted by the chargee under Article 8(1)(b) or ordered under Article 8(2). Where, after such default, the payment of the amount secured is made in full by an interested person other than the debtor, that person is subrogated to the rights of the chargee.

5　Ownership or any other interest of the chargor passing on a sale under Article 8(1)(b) or passing under paragraph 1 or 2 of this Article is free from any other interest over which the chargee's security interest has priority under the provisions of Article 29.

Article 10 — Remedies of conditional seller or lessor

In the event of default under a title reservation agreement or under a leasing agreement as provided in Article 11, the conditional seller or the lessor, as the case may be, may:

(a)　subject to any declaration that may be made by a Contracting State under Article 54, terminate the agreement and take possession or control of any object to which the agreement relates; or

(b)　apply for a court order authorising or directing either of these acts.

Article 11 — Meaning of default

1   The debtor and the creditor may at any time agree in writing as to the events that constitute a default or otherwise give rise to the rights and remedies specified in Articles 8 to 10 and 13.

2   Where the debtor and the creditor have not so agreed, "default" for the purposes of Articles 8 to 10 and 13 means a default which substantially deprives the creditor of what it is entitled to expect under the agreement.

Article 12 — Additional remedies

Any additional remedies permitted by the applicable law, including any remedies agreed upon by the parties, may be exercised to the extent that they are not inconsistent with the mandatory provisions of this Chapter as set out in Article 15.

Article 13 — Relief pending final determination

1   Subject to any declaration that it may make under Article 55, a Contracting State shall ensure that a creditor who adduces evidence of default by the debtor may, pending final determination of its claim and to the extent that the debtor has at any time so agreed, obtain from a court speedy relief in the form of such one or more of the following orders as the creditor requests:
(a)   preservation of the object and its value;
(b)   possession, control or custody of the object;
(c)   immobilisation of the object; and
(d)   lease or, except where covered by sub-paragraphs (a) to (c), management of the object and the income therefrom.

2   In making any order under the preceding paragraph, the court may impose such terms as it considers necessary to protect the interested persons in the event that the creditor:
(a)   in implementing any order granting such relief, fails to perform any of its obligations to the debtor under this Convention or the Protocol; or
(b)   fails to establish its claim, wholly or in part, on the final determination of that claim.

3   Before making any order under paragraph 1, the court may require notice of the request to be given to any of the interested persons.

4   Nothing in this Article affects the application of Article 8(3) or limits the availability of forms of interim relief other than those set out in paragraph 1.

Article 14 — Procedural requirements

Subject to Article 54(2), any remedy provided by this Chapter shall be exercised in conformity with the procedure prescribed by the law of the place where the remedy is to be exercised.

Article 15 — Derogation

In their relations with each other, any two or more of the parties referred to in this Chapter may at any time, by agreement in writing, derogate from or vary the effect of any of the preceding provisions of this Chapter except Articles 8(3) to (6), 9(3) and (4), 13(2) and 14.

Chapter IV
The international registration system

Article 16 — The International Registry

1   An International Registry shall be established for registrations of:
    (a)   international interests, prospective international interests and registrable non-consensual rights and interests;
    (b)   assignments and prospective assignments of international interests;
    (c)   acquisitions of international interests by legal or contractual subrogations under the applicable law;
    (d)   notices of national interests; and
    (e)   subordinations of interests referred to in any of the preceding subparagraphs.

2   Different international registries may be established for different categories of object and associated rights.

3   For the purposes of this Chapter and Chapter V, the term 'registration' includes, where appropriate, an amendment, extension or discharge of a registration.

Article 17 — The Supervisory Authority and the Registrar

1   There shall be a Supervisory Authority as provided by the Protocol.
2   The Supervisory Authority shall:
    (a)   establish or provide for the establishment of the International Registry;
    (b)   except as otherwise provided by the Protocol, appoint and dismiss the Registrar;
    (c)   ensure that any rights required for the continued effective operation of the International Registry in the event of a change of Registrar will vest in or be assignable to the new Registrar;
    (d)   after consultation with the Contracting States, make or approve and ensure the publication of regulations pursuant to the Protocol dealing with the operation of the International Registry;
    (e)   establish administrative procedures through which complaints concerning the operation of the International Registry can be made to the Supervisory Authority;
    (f)   supervise the Registrar and the operation of the International Registry;
    (g)   at the request of the Registrar, provide such guidance to the Registrar as the Supervisory Authority thinks fit;

(h) set and periodically review the structure of fees to be charged for the services and facilities of the International Registry;

(i) do all things necessary to ensure that an efficient notice-based electronic registration system exists to implement the objectives of this Convention and the Protocol; and

(j) report periodically to Contracting States concerning the discharge of its obligations under this Convention and the Protocol.

3 The Supervisory Authority may enter into any agreement requisite for the performance of its functions, including any agreement referred to in Article 27(3).

4 The Supervisory Authority shall own all proprietary rights in the data bases and archives of the International Registry.

5 The Registrar shall ensure the efficient operation of the International Registry and perform the functions assigned to it by this Convention, the Protocol and the regulations.

Chapter V
Other matters relating to registration

Article 18 — Registration requirements

1 The Protocol and regulations shall specify the requirements, including the criteria for the identification of the object:

(a) for effecting a registration (which shall include provision for prior electronic transmission of any consent from any person whose consent is required under Article 20);

(b) for making searches and issuing search certificates, and, subject thereto;

(c) for ensuring the confidentiality of information and documents of the International Registry other than information and documents relating to a registration.

2 The Registrar shall not be under a duty to enquire whether a consent to registration under Article 20 has in fact been given or is valid.

3 Where an interest registered as a prospective international interest becomes an international interest, no further registration shall be required provided that the registration information is sufficient for a registration of an international interest.

4 The Registrar shall arrange for registrations to be entered into the International Registry data base and made searchable in chronological order of receipt, and the file shall record the date and time of receipt.

5 The Protocol may provide that a Contracting State may designate an entity or entities in its territory as the entry point or entry points through which the information required for registration shall or may be transmitted to the International Registry. A Contracting State making such a designation may specify the requirements, if any, to be satisfied before such information is transmitted to the International Registry.

Article 19 — Validity and time of registration

1 A registration shall be valid only if made in conformity with Article 20.
2 A registration, if valid, shall be complete upon entry of the required information into the International Registry data base so as to be searchable.
3 A registration shall be searchable for the purposes of the preceding paragraph at the time when:
   (a) the International Registry has assigned to it a sequentially ordered file number; and
   (b) the registration information, including the file number, is stored in durable form and may be accessed at the International Registry.
4 If an interest first registered as a prospective international interest becomes an international interest, that international interest shall be treated as registered from the time of registration of the prospective international interest provided that the registration was still current immediately before the international interest was constituted as provided by Article 7.
5 The preceding paragraph applies with necessary modifications to the registration of a prospective assignment of an international interest.
6 A registration shall be searchable in the International Registry data base according to the criteria prescribed by the Protocol.

Article 20 — Consent to registration

1 An international interest, a prospective international interest or an assignment or prospective assignment of an international interest may be registered, and any such registration amended or extended prior to its expiry, by either party with the consent in writing of the other.
2 The subordination of an international interest to another international interest may be registered by or with the consent in writing at any time of the person whose interest has been subordinated.
3 A registration may be discharged by or with the consent in writing of the party in whose favour it was made.
4 The acquisition of an international interest by legal or contractual subrogation may be registered by the subrogee.
5 A registrable non-consensual right or interest may be registered by the holder thereof.
6 A notice of a national interest may be registered by the holder thereof.

Article 21 — Duration of registration

Registration of an international interest remains effective until discharged or until expiry of the period specified in the registration.

Article 22 — Searches

1   Any person may, in the manner prescribed by the Protocol and regulations, make or request a search of the International Registry by electronic means concerning interests or prospective international interests registered therein.

2   Upon receipt of a request therefore, the Registrar, in the manner prescribed by the Protocol and regulations, shall issue a registry search certificate by electronic means with respect to any object:

    (a)   stating all registered information relating thereto, together with a statement indicating the date and time of registration of such information; or

    (b)   stating that there is no information in the International Registry relating thereto.

3   A search certificate issued under the preceding paragraph shall indicate that the creditor named in the registration information has acquired or intends to acquire an international interest in the object but shall not indicate whether what is registered is an international interest or a prospective international interest, even if this is ascertainable from the relevant registration information.

Article 23 — List of declarations and declared non-consensual rights or interests

The Registrar shall maintain a list of declarations, withdrawals of declaration and of the categories of nonconsensual right or interest communicated to the Registrar by the Depositary as having been declared by Contracting States in conformity with Articles 39 and 40 and the date of each such declaration or withdrawal of declaration. Such list shall be recorded and searchable in the name of the declaring State and shall be made available as provided in the Protocol and regulations to any person requesting it.

Article 24 — Evidentiary value of certificates

A document in the form prescribed by the regulations which purports to be a certificate issued by the International Registry is prima facie proof:

    (a)   that it has been so issued; and

    (b)   of the facts recited in it, including the date and time of a registration.

Article 25 — Discharge of registration

1   Where the obligations secured by a registered security interest or the obligations giving rise to a registered non-consensual right or interest have been discharged, or where the conditions of transfer of title under a registered title reservation agreement have been fulfilled, the holder of such interest shall, without undue delay, procure the discharge of the registration

after written demand by the debtor delivered to or received at its address stated in the registration.

2   Where a prospective international interest or a prospective assignment of an international interest has been registered, the intending creditor or intending assignee shall, without undue delay, procure the discharge of the registration after written demand by the intending debtor or assignor which is delivered to or received at its address stated in the registration before the intending creditor or assignee has given value or incurred a commitment to give value.

3   Where the obligations secured by a national interest specified in a registered notice of a national interest have been discharged, the holder of such interest shall, without undue delay, procure the discharge of the registration after written demand by the debtor delivered to or received at its address stated in the registration.

4   Where a registration ought not to have been made or is incorrect, the person in whose favour the registration was made shall, without undue delay, procure its discharge or amendment after written demand by the debtor delivered to or received at its address stated in the registration.

Article 26 — Access to the international registration facilities

No person shall be denied access to the registration and search facilities of the International Registry on any ground other than its failure to comply with the procedures prescribed by this Chapter.

Chapter VI
Privileges and immunities of the Supervisory Authority and the Registrar

Article 27 — Legal personality; immunity

[Omitted]

Chapter VII
Liability of the Registrar

Article 28 — Liability and financial assurances

[Omitted]

Chapter VIII
Effects of an international interest as against third parties

Article 29 — Priority of competing interests

1   A registered interest has priority over any other interest subsequently registered and over an unregistered interest.

2   The priority of the first-mentioned interest under the preceding paragraph applies:

    (a)   even if the first-mentioned interest was acquired or registered with actual knowledge of the other interest; and

    (b)   even as regards value given by the holder of the first-mentioned interest with such knowledge.

3   The buyer of an object acquires its interest in it:

    (a)   subject to an interest registered at the time of its acquisition of that interest; and

    (b)   free from an unregistered interest even if it has actual knowledge of such an interest.

4   The conditional buyer or lessee acquires its interest in or right over that object:

    (a)   subject to an interest registered prior to the registration of the international interest held by its conditional seller or lessor; and

    (b)   free from an interest not so registered at that time even if it has actual knowledge of that interest.

5   The priority of competing interests or rights under this Article may be varied by agreement between the holders of those interests, but an assignee of a subordinated interest is not bound by an agreement to subordinate that interest unless at the time of the assignment a subordination had been registered relating to that agreement.

6   Any priority given by this Article to an interest in an object extends to proceeds.

7   This Convention:

    (a)   does not affect the rights of a person in an item, other than an object, held prior to its installation on an object if under the applicable law those rights continue to exist after the installation; and

    (b)   does not prevent the creation of rights in an item, other than an object, which has previously been installed on an object where under the applicable law those rights are created.

Article 30 — Effects of insolvency

1   In insolvency proceedings against the debtor an international interest is effective if prior to the commencement of the insolvency proceedings that interest was registered in conformity with this Convention.

2   Nothing in this Article impairs the effectiveness of an international interest in the insolvency proceedings where that interest is effective under the applicable law.

3   Nothing in this Article affects:

    (a)   any rules of law applicable in insolvency proceedings relating to the avoidance of a transaction as a preference or a transfer in fraud of creditors; or

(b)   any rules of procedure relating to the enforcement of rights to property which is under the control or supervision of the insolvency administrator.

Chapter IX
Assignments of associated rights and international interests; rights of subrogation

Article 31 — Effects of assignment

1   Except as otherwise agreed by the parties, an assignment of associated rights made in conformity with Article 32 also transfers to the assignee:
(a)   the related international interest; and
(b)   all the interests and priorities of the assignor under this Convention.

2   Nothing in this Convention prevents a partial assignment of the assignor's associated rights. In the case of such a partial assignment the assignor and assignee may agree as to their respective rights concerning the related international interest assigned under the preceding paragraph but not so as adversely to affect the debtor without its consent.

3   Subject to paragraph 4, the applicable law shall determine the defences and rights of set-off available to the debtor against the assignee.

4   The debtor may at any time by agreement in writing waive all or any of the defences and rights of set-off referred to in the preceding paragraph other than defences arising from fraudulent acts on the part of the assignee.

5   In the case of an assignment by way of security, the assigned associated rights revest in the assignor, to the extent that they are still subsisting, when the obligations secured by the assignment have been discharged.

Article 32 — Formal requirements of assignment

1   An assignment of associated rights transfers the related international interest only if it:
(a)   is in writing;
(b)   enables the associated rights to be identified under the contract from which they arise; and
(c)   in the case of an assignment by way of security, enables the obligations secured by the assignment to be determined in accordance with the Protocol but without the need to state a sum or maximum sum secured.

2   An assignment of an international interest created or provided for by a security agreement is not valid unless some or all related associated rights also are assigned.

3   This Convention does not apply to an assignment of associated rights which is not effective to transfer the related international interest.

Article 33 — Debtor's duty to assignee

1   To the extent that associated rights and the related international interest have been transferred in accordance with Articles 31 and 32, the debtor in relation to those rights and that interest is bound by the assignment and has a duty to make payment or give other performance to the assignee, if but only if:

    (a)   the debtor has been given notice of the assignment in writing by or with the authority of the assignor; and

    (b)   the notice identifies the associated rights.

2   Irrespective of any other ground on which payment or performance by the debtor discharges the latter from liability, payment or performance shall be effective for this purpose if made in accordance with the preceding paragraph.

3   Nothing in this Article shall affect the priority of competing assignments.

Article 34 — Default remedies in respect of assignment by way of security

In the event of default by the assignor under the assignment of associated rights and the related international interest made by way of security, Articles 8, 9 and 11 to 14 apply in the relations between the assignor and the assignee (and, in relation to associated rights, apply in so far as those provisions are capable of application to intangible property) as if references:

(a)   to the secured obligation and the security interest were references to the obligation secured by the assignment of the associated rights and the related international interest and the security interest created by that assignment;

(b)   to the chargee or creditor and chargor or debtor were references to the assignee and assignor;

(c)   to the holder of the international interest were references to the assignee; and

(d)   to the object were references to the assigned associated rights and the related international interest.

Article 35 — Priority of competing assignments

1   Where there are competing assignments of associated rights and at least one of the assignments includes the related international interest and is registered, the provisions of Article 29 apply as if the references to a registered interest were references to an assignment of the associated rights and the related registered interest and as if references to a registered or unregistered interest were references to a registered or unregistered assignment.

2   Article 30 applies to an assignment of associated rights as if the references to an international interest were references to an assignment of the associated rights and the related international interest.

Article 36 — Assignee's priority with respect to associated rights

1   The assignee of associated rights and the related international interest whose assignment has been registered only has priority under Article 35(1) over another assignee of the associated rights:

    (a)   if the contract under which the associated rights arise states that they are secured by or associated with the object; and

    (b)   to the extent that the associated rights are related to an object.

2   For the purposes of sub-paragraph (b) of the preceding paragraph, associated rights are related to an object only to the extent that they consist of rights to payment or performance that relate to:

    (a)   a sum advanced and utilised for the purchase of the object;

    (b)   a sum advanced and utilised for the purchase of another object in which the assignor held another international interest if the assignor transferred that interest to the assignee and the assignment has been registered;

    (c)   the price payable for the object;

    (d)   the rentals payable in respect of the object; or

    (e)   other obligations arising from a transaction referred to in any of the preceding subparagraphs.

3   In all other cases, the priority of the competing assignments of the associated rights shall be determined by the applicable law.

Article 37 — Effects of assignor's insolvency

The provisions of Article 30 apply to insolvency proceedings against the assignor as if references to the debtor were references to the assignor.

Article 38 — Subrogation

1   Subject to paragraph 2, nothing in this Convention affects the acquisition of associated rights and the related international interest by legal or contractual subrogation under the applicable law.

2   The priority between any interest within the preceding paragraph and a competing interest may be varied by agreement in writing between the holders of the respective interests but an assignee of a subordinated interest is not bound by an agreement to subordinate that interest unless at the time of the assignment a subordination had been registered relating to that agreement.

Chapter X
Rights or interests subject to declarations by Contracting States

Article 39 — Rights having priority without registration

1   A Contracting State may at any time, in a declaration deposited with the Depositary of the Protocol declare, generally or specifically:

    (a)   those categories of non-consensual right or interest (other than a right or interest to which Article 40 applies) which under that State's law have priority over an interest in an object equivalent to that of the holder of a registered international interest and which shall have priority over a registered international interest, whether in or outside insolvency proceedings; and

    (b)   that nothing in this Convention shall affect the right of a State or State entity, intergovernmental organisation or other private provider of public services to arrest or detain an object under the laws of that State for payment of amounts owed to such entity, organisation or provider directly relating to those services in respect of that object or another object.

2   A declaration made under the preceding paragraph may be expressed to cover categories that are created after the deposit of that declaration.

3   A non-consensual right or interest has priority over an international interest if and only if the former is of a category covered by a declaration deposited prior to the registration of the international interest.

4   Notwithstanding the preceding paragraph, a Contracting State may, at the time of ratification, acceptance, approval of, or accession to the Protocol, declare that a right or interest of a category covered by a declaration made under sub-paragraph (a) of paragraph 1 shall have priority over an international interest registered prior to the date of such ratification, acceptance, approval or accession.

Article 40 — Registrable non-consensual rights or interests

A Contracting State may at any time in a declaration deposited with the Depositary of the Protocol list the categories of non-consensual right or interest which shall be registrable under this Convention as regards any category of object as if the right or interest were an international interest and shall be regulated accordingly. Such a declaration may be modified from time to time.

Chapter XI
Application of the Convention to sales

Article 41 — Sale and prospective sale

This Convention shall apply to the sale or prospective sale of an object as provided for in the Protocol with any modifications therein.

Chapter XII
Jurisdiction

Article 42 — Choice of forum

1   Subject to Articles 43 and 44, the courts of a Contracting State chosen by the parties to a transaction have jurisdiction in respect of any claim brought under this Convention, whether or not the chosen forum has a connection with the parties or the transaction. Such jurisdiction shall be exclusive unless otherwise agreed between the parties.

2   Any such agreement shall be in writing or otherwise concluded in accordance with the formal requirements of the law of the chosen forum.

Article 43 — Jurisdiction under Article 13

1   The courts of a Contracting State chosen by the parties and the courts of the Contracting State on the territory of which the object is situated have jurisdiction to grant relief under Article 13(1)(a), (b), (c) and Article 13(4) in respect of that object.

2   Jurisdiction to grant relief under Article 13(1)(d) or other interim relief by virtue of Article 13(4) may be exercised either:

    (a)   by the courts chosen by the parties; or

    (b)   by the courts of a Contracting State on the territory of which the debtor is situated, being relief which, by the terms of the order granting it, is enforceable only in the territory of that Contracting State.

3   A court has jurisdiction under the preceding paragraphs even if the final determination of the claim referred to in Article 13(1) will or may take place in a court of another Contracting State or by arbitration.

Article 44 — Jurisdiction to make orders against the Registrar

1   The courts of the place in which the Registrar has its centre of administration shall have exclusive jurisdiction to award damages or make orders against the Registrar.

2   Where a person fails to respond to a demand made under Article 25 and that person has ceased to exist or cannot be found for the purpose of enabling an order to be made against it requiring it to procure discharge of the registration, the courts referred to in the preceding paragraph shall have exclusive jurisdiction, on the application of the debtor or intending debtor, to make an order directed to the Registrar requiring the Registrar to discharge the registration.

3   Where a person fails to comply with an order of a court having jurisdiction under this Convention or, in the case of a national interest, an order of a court of competent jurisdiction requiring that person to procure the amendment or discharge of a registration, the courts referred to in paragraph 1 may direct the Registrar to take such steps as will give effect to that order.

4   Except as otherwise provided by the preceding paragraphs, no court may make orders or give judgments or rulings against or purporting to bind the Registrar.

Article 45 — Jurisdiction in respect of insolvency proceedings

The provisions of this Chapter are not applicable to insolvency proceedings.

Chapter XIII
Relationship with other Conventions

Article 45 bis — Relationship with the *United Nations Convention on the Assignment of Receivables in International Trade*

This Convention shall prevail over the *United Nations Convention on the Assignment of Receivables in International Trade*, opened for signature in New York on 12 December 2001, as it relates to the assignment of receivables which are associated rights related to international interests in aircraft objects, railway rolling stock and space assets.

Article 46 — Relationship with the *UNIDROIT Convention on International Financial Leasing*

The Protocol may determine the relationship between this Convention and the *UNIDROIT Convention on International Financial Leasing*, signed at Ottawa on 28 May 1988.

Chapter XIV
Final provisions

Article 47 — Signature, ratification, acceptance, approval or accession

1   This Convention shall be open for signature in Cape Town on 16 November 2001 by States participating in the Diplomatic Conference to Adopt a Mobile Equipment Convention and an Aircraft Protocol held at Cape Town from 29 October to 16 November 2001. After 16 November 2001, the Convention shall be open to all States for signature at the Headquarters of the International Institute for the Unification of Private Law (UNIDROIT) in Rome until it enters into force in accordance with Article 49.

2   This Convention shall be subject to ratification, acceptance or approval by States which have signed it.

3   Any State which does not sign this Convention may accede to it at any time.

4   Ratification, acceptance, approval or accession is effected by the deposit of a formal instrument to that effect with the Depositary.

Article 48 — Regional Economic Integration Organisations

1   A Regional Economic Integration Organisation which is constituted by sovereign States and has competence over certain matters governed by this Convention may similarly sign, accept, approve or accede to this Convention. The Regional Economic Integration Organisation shall in that case have the rights and obligations of a Contracting State, to the extent that that Organisation has competence over matters governed by this Convention. Where the number of Contracting States is relevant in this Convention, the Regional Economic Integration Organisation shall not count as a Contracting State in addition to its Member States which are Contracting States.

2   The Regional Economic Integration Organisation shall, at the time of signature, acceptance, approval or accession, make a declaration to the Depositary specifying the matters governed by this Convention in respect of which competence has been transferred to that Organisation by its Member States. The Regional Economic Integration Organisation shall promptly notify the Depositary of any changes to the distribution of competence, including new transfers of competence, specified in the declaration under this paragraph.

3   Any reference to a 'Contracting State' or 'Contracting States' or 'State Party' or 'States Parties' in this Convention applies equally to a Regional Economic Integration Organisation where the context so requires.

Article 49 — Entry into force

1   This Convention enters into force on the first day of the month following the expiration of three months after the date of the deposit of the third instrument of ratification, acceptance, approval or accession but only as regards a category of objects to which a Protocol applies:
    (a)   as from the time of entry into force of that Protocol;
    (b)   subject to the terms of that Protocol; and
    (c)   as between States Parties to this Convention and that Protocol.

2   For other States this Convention enters into force on the first day of the month following the expiration of three months after the date of the deposit of their instrument of ratification, acceptance, approval or accession but only as regards a category of objects to which a Protocol applies and subject, in relation to such Protocol, to the requirements of sub-paragraphs (a), (b) and (c) of the preceding paragraph.

Article 50 — Internal transactions

1   A Contracting State may, at the time of ratification, acceptance, approval of, or accession to the Protocol, declare that this Convention shall not apply to a transaction which is an internal transaction in relation to that State with regard to all types of objects or some of them.

2　Notwithstanding the preceding paragraph, the provisions of Articles 8(4), 9(1), 16, Chapter V, Article 29, and any provisions of this Convention relating to registered interests shall apply to an internal transaction.

3　Where notice of a national interest has been registered in the International Registry, the priority of the holder of that interest under Article 29 shall not be affected by the fact that such interest has become vested in another person by assignment or subrogation under the applicable law.

Article 51 — Future Protocols

1　The Depositary may create working groups, in co-operation with such relevant nongovernmental organisations as the Depositary considers appropriate, to assess the feasibility of extending the application of this Convention, through one or more Protocols, to objects of any category of high-value mobile equipment, other than a category referred to in Article 2(3), each member of which is uniquely identifiable, and associated rights relating to such objects.

2　The Depositary shall communicate the text of any preliminary draft Protocol relating to a category of objects prepared by such a working group to all States Parties to this Convention, all member States of the Depositary, member States of the United Nations which are not members of the Depositary and the relevant intergovernmental organisations, and shall invite such States and organisations to participate in intergovernmental negotiations for the completion of a draft Protocol on the basis of such a preliminary draft Protocol.

3　The Depositary shall also communicate the text of any preliminary draft Protocol prepared by such a working group to such relevant non-governmental organisations as the Depositary considers appropriate. Such non-governmental organisations shall be invited promptly to submit comments on the text of the preliminary draft Protocol to the Depositary and to participate as observers in the preparation of a draft Protocol.

4　When the competent bodies of the Depositary adjudge such a draft Protocol ripe for adoption, the Depositary shall convene a diplomatic conference for its adoption.

5　Once such a Protocol has been adopted, subject to paragraph 6, this Convention shall apply to the category of objects covered thereby.

6　Article 45 bis of this Convention applies to such a Protocol only if specifically provided for in that Protocol.

Article 52 — Territorial units

1　If a Contracting State has territorial units in which different systems of law are applicable in relation to the matters dealt with in this Convention, it may, at the time of ratification, acceptance, approval or accession, declare that this Convention is to extend to all its territorial units or only to one or more of them and may modify its declaration by submitting another declaration at any time.

2   Any such declaration shall state expressly the territorial units to which this Convention applies.

3   If a Contracting State has not made any declaration under paragraph 1, this Convention shall apply to all territorial units of that State.

4   Where a Contracting State extends this Convention to one or more of its territorial units, declarations permitted under this Convention may be made in respect of each such territorial unit, and the declarations made in respect of one territorial unit may be different from those made in respect of another territorial unit.

5   If by virtue of a declaration under paragraph 1, this Convention extends to one or more territorial units of a Contracting State:

    (a)   the debtor is considered to be situated in a Contracting State only if it is incorporated or formed under a law in force in a territorial unit to which this Convention applies or if it has its registered office or statutory seat, centre of administration, place of business or habitual residence in a territorial unit to which this Convention applies;

    (b)   any reference to the location of the object in a Contracting State refers to the location of the object in a territorial unit to which this Convention applies; and

    (c)   any reference to the administrative authorities in that Contracting State shall be construed as referring to the administrative authorities having jurisdiction in a territorial unit to which this Convention applies.

Article 53 — Determination of courts

A Contracting State may, at the time of ratification, acceptance, approval of, or accession to the Protocol, declare the relevant 'court' or 'courts' for the purposes of Article 1 and Chapter XII of this Convention.

Article 54 — Declarations regarding remedies

1   A Contracting State may, at the time of ratification, acceptance, approval of, or accession to the Protocol, declare that while the charged object is situated within, or controlled from its territory the chargee shall not grant a lease of the object in that territory.

2   A Contracting State shall, at the time of ratification, acceptance, approval of, or accession to the Protocol, declare whether or not any remedy available to the creditor under any provision of this Convention which is not there expressed to require application to the court may be exercised only with leave of the court.

Article 55 — Declarations regarding relief pending final determination

A Contracting State may, at the time of ratification, acceptance, approval of, or accession to the Protocol, declare that it will not apply the provisions of Article 13 or Article 43, or both, wholly or in part. The declaration shall specify under which conditions the relevant Article will be applied, in case it will be applied partly, or otherwise which other forms of interim relief will be applied.

Article 56 — Reservations and declarations

1   No reservations may be made to this Convention but declarations authorised by Articles 39, 40, 50, 52, 53, 54, 55, 57, 58 and 60 may be made in accordance with these provisions.

2   Any declaration or subsequent declaration or any withdrawal of a declaration made under this Convention shall be notified in writing to the Depositary.

Article 57 — Subsequent declarations

1   A State Party may make a subsequent declaration, other than a declaration authorised under Article 60, at any time after the date on which this Convention has entered into force for it, by notifying the Depositary to that effect.

2   Any such subsequent declaration shall take effect on the first day of the month following the expiration of six months after the date of receipt of the notification by the Depositary. Where a longer period for that declaration to take effect is specified in the notification, it shall take effect upon the expiration of such longer period after receipt of the notification by the Depositary.

3   Notwithstanding the previous paragraphs, this Convention shall continue to apply, as if no such subsequent declarations had been made, in respect of all rights and interests arising prior to the effective date of any such subsequent declaration.

Article 58 — Withdrawal of declarations

1   Any State Party having made a declaration under this Convention, other than a declaration authorised under Article 60, may withdraw it at any time by notifying the Depositary. Such withdrawal is to take effect on the first day of the month following the expiration of six months after the date of receipt of the notification by the Depositary.

2   Notwithstanding the previous paragraph, this Convention shall continue to apply, as if no such withdrawal of declaration had been made, in respect of all rights and interests arising prior to the effective date of any such withdrawal.

Article 59 — Denunciations

1   Any State Party may denounce this Convention by notification in writing to the Depositary.

2   Any such denunciation shall take effect on the first day of the month following the expiration of twelve months after the date on which notification is received by the Depositary.

3   Notwithstanding the previous paragraphs, this Convention shall continue to apply, as if no such denunciation had been made, in respect of all rights and interests arising prior to the effective date of any such denunciation.

Article 60 — Transitional provisions

1.   Unless otherwise declared by a Contracting State at any time, the Convention does not apply to a preexisting right or interest, which retains the priority it enjoyed under the applicable law before the effective date of this Convention.

2.   For the purposes of Article 1 (v) and of determining priority under this Convention:

   (a)   'effective date of this Convention' means in relation to a debtor the time when this Convention enters into force or the time when the State in which the debtor is situated becomes a Contracting State, whichever is the later; and

   (b)   the debtor is situated in a State where it has its centre of administration or, if it has no centre of administration, its place of business or, if it has more than one place of business, its principal place of business or, if it has no place of business, its habitual residence.

3.   A Contracting State may in its declaration under paragraph 1 specify a date, not earlier than three years after the date on which the declaration becomes effective, when this Convention and the Protocol will become applicable, for the purpose of determining priority, including the protection of any existing priority, to pre-existing rights or interests arising under an agreement made at a time when the debtor was situated in a State referred to in sub-paragraph (b) of the preceding paragraph but only to the extent and in the manner specified in its declaration.

Article 61 — Review Conferences, amendments and related matters

1.   The Depositary shall prepare reports yearly or at such other time as the circumstances may require for the States Parties as to the manner in which the international regimen established in this Convention has operated in practice. In preparing such reports, the Depositary shall take into account the reports of the Supervisory Authority concerning the functioning of the international registration system.

2. At the request of not less than twenty-five percent of the States Parties, Review Conferences of States Parties shall be convened from time to time by the Depositary, in consultation with the Supervisory Authority, to consider:

    (a) the practical operation of this Convention and its effectiveness in facilitating the asset-based financing and leasing of the objects covered by its terms;

    (b) the judicial interpretation given to, and the application made of the terms of this Convention and the regulations;

    (c) the functioning of the international registration system, the perform-ance of the Registrar and its oversight by the Supervisory Authority, taking into account the reports of the Supervisory Authority; and

    (d) whether any modifications to this Convention or the arrangements relating to the International Registry are desirable.

3. Subject to paragraph 4, any amendment to this Convention shall be approved by at least a two-thirds majority of States Parties participating in the Conference referred to in the preceding paragraph and shall then enter into force in respect of States which have ratified, accepted or approved such amendment when ratified, accepted, or approved by three States in accordance with the provisions of Article 49 relating to its entry into force.

4. Where the proposed amendment to this Convention is intended to apply to more than one category of equipment, such amendment shall also be approved by at least a two-thirds majority of States Parties to each Protocol that are participating in the Conference referred to in paragraph 2.

Article 62 — Depositary and its functions

1. Instruments of ratification, acceptance, approval or accession shall be deposited with the International Institute for the Unification of Private Law (UNIDROIT), which is hereby designated the Depositary.

2. The Depositary shall:

    (a) inform all Contracting States of:

    (i) each new signature or deposit of an instrument of ratification, acceptance, approval or accession, together with the date thereof;

    (ii) the date of entry into force of this Convention;

    (iii) each declaration made in accordance with this Convention, together with the date thereof;

    (iv) the withdrawal or amendment of any declaration, together with the date thereof; and

    (v) the notification of any denunciation of this Convention together with the date thereof and the date on which it takes effect;

    (b) transmit certified true copies of this Convention to all Contracting States;

    (c) provide the Supervisory Authority and the Registrar with a copy of each instrument of ratification, acceptance, approval or accession, together with the date of deposit thereof, of each declaration or withdrawal or amendment of a declaration and of each notification of denunciation, together with the date of notification thereof, so that the information contained therein is easily and fully available; and

(d) perform such other functions customary for depositaries.

IN WITNESS WHEREOF the undersigned Plenipotentiaries, having been duly authorised, have signed this Convention.

DONE at Cape Town, this sixteenth day of November, two thousand and one, in a single original in the English, Arabic, Chinese, French, Russian and Spanish languages, all texts being equally authentic, such authenticity to take effect upon verification by the Joint Secretariat of the Conference under the authority of the President of the Conference within ninety days hereof as to the conformity of the texts with one another.

SELECTED TREATIES PROTOCOL
TO THE CONVENTION

<u>Protocol</u>

PROTOCOL TO THE CONVENTION ON INTERNATIONAL INTERESTS IN MOBILE EQUIPMENT ON MATTERS SPECIFIC TO AIRCRAFT EQUIPMENT

THE STATES PARTIES TO THIS PROTOCOL,

CONSIDERING it necessary to implement the *Convention on International Interests in Mobile Equipment* (hereinafter referred to as 'the Convention') as it relates to aircraft equipment, in the light of the purposes set out in the preamble to the Convention,

MINDFUL of the need to adapt the Convention to meet the particular requirements of aircraft finance and to extend the sphere of application of the Convention to include contracts of sale of aircraft equipment,

MINDFUL of the principles and objectives of the *Convention on International Civil Aviation*, signed at Chicago on 7 December 1944,

HAVE AGREED upon the following provisions relating to aircraft equipment:

Chapter I
Sphere of application and general provisions

Article I — Defined terms

1   In this Protocol, except where the context otherwise requires, terms used in it have the meanings set out in the Convention.

2   In this Protocol the following terms are employed with the meanings set out below:

(a)   'aircraft' means aircraft as defined for the purposes of the Chicago Convention which are either airframes with aircraft engines installed thereon or helicopters;

(b)   'aircraft engines' means aircraft engines (other than those used in military, customs or police services) powered by jet propulsion or turbine or piston technology and:

(i)   in the case of jet propulsion aircraft engines, have at least 1750 lb of thrust or its equivalent; and

(ii)   in the case of turbine-powered or piston-powered aircraft engines, have at least 550 rated take-off shaft horsepower or its equivalent, together with all modules and other installed, incorporated or attached accessories, parts and equipment and all data, manuals and records relating thereto;

(c) 'aircraft objects' means airframes, aircraft engines and helicopters;

(d) 'aircraft register' means a register maintained by a State or a common mark registering authority for the purposes of the Chicago Convention;

(e) 'airframes' means airframes (other than those used in military, customs or police services) that, when appropriate aircraft engines are installed thereon, are type certified by the competent aviation authority to transport:

(i) at least eight (8) persons including crew; or

(ii) goods in excess of 2750 kilograms,

together with all installed, incorporated or attached accessories, parts and equipment (other than aircraft engines), and all data, manuals and records relating thereto;

(f) 'authorised party' means the party referred to in Article XIII(3);

(g) 'Chicago Convention' means the Convention on International Civil Aviation, signed at Chicago on 7 December 1944, as amended, and its Annexes;

(h) 'common mark registering authority' means the authority maintaining a register in accordance with Article 77 of the Chicago Convention as implemented by the Resolution adopted on 14 December 1967 by the Council of the International Civil Aviation Organization on nationality and registration of aircraft operated by international operating agencies;

(i) 'de-registration of the aircraft' means deletion or removal of the registration of the aircraft from its aircraft register in accordance with the Chicago Convention;

(j) 'guarantee contract' means a contract entered into by a person as guarantor;

(k) 'guarantor' means a person who, for the purpose of assuring performance of any obligations in favour of a creditor secured by a security agreement or under an agreement, gives or issues a suretyship or demand guarantee or a standby letter of credit or any other form of credit insurance;

(l) 'helicopters' means heavier-than-air machines (other than those used in military, customs or police services) supported in flight chiefly by the reactions of the air on one or more power-driven rotors on substantially vertical axes and which are type certified by the competent aviation authority to transport:

(i) at least five (5) persons including crew; or

(ii) goods in excess of 450 kilograms, together with all installed, incorporated or attached accessories, parts and equipment (including rotors), and all data, manuals and records relating thereto;

(m) 'insolvency-related event' means:

(i)   the commencement of the insolvency proceedings; or

(ii)   the declared intention to suspend or actual suspension of payments by the debtor where the creditor's right to institute insolvency proceedings against the debtor or to exercise remedies under the Convention is prevented or suspended by law or State action;

(n)   'primary insolvency jurisdiction' means the Contracting State in which the centre of the debtor's main interests is situated, which for this purpose shall be deemed to be the place of the debtor's statutory seat or, if there is none, the place where the debtor is incorporated or formed, unless proved otherwise;

(o)   'registry authority' means the national authority or the common mark registering authority, maintaining an aircraft register in a Contracting State and responsible for the registration and deregistration of an aircraft in accordance with the Chicago Convention; and

(p)   'State of registry' means, in respect of an aircraft, the State on the national register of which an aircraft is entered or the State of location of the common mark registering authority maintaining the aircraft register.

Article II — Application of Convention as regards aircraft objects

1   The Convention shall apply in relation to aircraft objects as provided by the terms of this Protocol.

2   The Convention and this Protocol shall be known as the Convention on International Interests in Mobile Equipment as applied to aircraft objects.

Article III — Application of Convention to sales

The following provisions of the Convention apply as if references to an agreement creating or providing for an international interest were references to a contract of sale and as if references to an international interest, a prospective international interest, the debtor and the creditor were references to a sale, a prospective sale, the seller and the buyer respectively:

Articles 3 and 4;

Article 16(1)(a);

Article 19(4);

Article 20(1) (with regard to registration of a contract of sale or a prospective sale);

Article 25(2) (with regard to a prospective sale); and

Article 30.

In addition, the general provisions of Article 1, Article 5, Chapters IV to VII, Article 29 (other than Article 29(3) which is replaced by Article XIV(1) and (2)), Chapter X, Chapter XII (other than Article 43), Chapter XIII and Chapter XIV (other than Article 60) shall apply to contracts of sale and prospective sales.

Article IV — Sphere of application

1   Without prejudice to Article 3(1) of the Convention, the Convention shall also apply in relation to a helicopter, or to an airframe pertaining to an aircraft, registered in an aircraft register of a Contracting State which is the State of registry, and where such registration is made pursuant to an agreement for registration of the aircraft it is deemed to have been effected at the time of the agreement.

2   For the purposes of the definition of 'internal transaction' in Article 1 of the Convention:
    (a)   an airframe is located in the State of registry of the aircraft of which it is a part;
    (b)   an aircraft engine is located in the State of registry of the aircraft on which it is installed or, if it is not installed on an aircraft, where it is physically located; and
    (c)   a helicopter is located in its State of registry, at the time of the conclusion of the agreement creating or providing for the interest.

3   The parties may, by agreement in writing, exclude the application of Article XI and, in their relations with each other, derogate from or vary the effect of any of the provisions of this Protocol except Article IX (2)(4).

Article V — Formalities, effects and registration of contracts of sale

1   For the purposes of this Protocol, a contract of sale is one which:
    (a)   is in writing;
    (b)   relates to an aircraft object of which the seller has power to dispose; and
    (c)   enables the aircraft object to be identified in conformity with this Protocol.

2   A contract of sale transfers the interest of the seller in the aircraft object to the buyer according to its terms.

3   Registration of a contract of sale remains effective indefinitely. Registration of a prospective sale remains effective unless discharged or until expiry of the period, if any, specified in the registration.

Article VI — Representative capacities

A person may enter into an agreement or a sale, and register an international interest in, or a sale of, an aircraft object, in an agency, trust or other representative capacity. In such case, that person is entitled to assert rights and interests under the Convention.

Article VII — Description of aircraft objects

A description of an aircraft object that contains its manufacturer's serial number, the name of the manufacturer and its model designation is necessary and sufficient to identify the object for the purposes of Article 7(c) of the Convention and Article V(1)(c) of this Protocol.

Article VIII — Choice of law

1   This Article applies only where a Contracting State has made a declaration pursuant to Article XXX(1).

2   The parties to an agreement, or a contract of sale, or a related guarantee contract or subordination agreement may agree on the law which is to govern their contractual rights and obligations, wholly or in part.

3   Unless otherwise agreed, the reference in the preceding paragraph to the law chosen by the parties is to the domestic rules of law of the designated State or, where that State comprises several territorial units, to the domestic law of the designated territorial unit.

Chapter II
Default remedies, priorities and assignments

Article IX — Modification of default remedies provisions

1   In addition to the remedies specified in Chapter III of the Convention, the creditor may, to the extent that the debtor has at any time so agreed and in the circumstances specified in that Chapter:
     (a)   procure the de-registration of the aircraft; and
     (b)   procure the export and physical transfer of the aircraft object from the territory in which it is situated.

2   The creditor shall not exercise the remedies specified in the preceding paragraph without the prior consent in writing of the holder of any registered interest ranking in priority to that of the creditor.

3   Article 8(3) of the Convention shall not apply to aircraft objects. Any remedy given by the Convention in relation to an aircraft object shall be exercised in a commercially reasonable manner. A remedy shall be deemed to be exercised in a commercially reasonable manner where it is exercised in conformity with a provision of the agreement except where such a provision is manifestly unreasonable.

4   A chargee giving ten or more working days' prior written notice of a proposed sale or lease to interested persons shall be deemed to satisfy the requirement of providing 'reasonable prior notice' specified in Article 8(4) of the Convention. The foregoing shall not prevent a chargee and a chargor or a guarantor from agreeing to a longer period of prior notice.

5   The registry authority in a Contracting State shall, subject to any applicable safety laws and regulations, honour a request for de-registration and export if:
     (a)   the request is properly submitted by the authorised party under a recorded irrevocable deregistration and export request authorisation; and
     (b)   the authorised party certifies to the registry authority, if required by that authority, that all registered interests ranking in priority to that of the creditor in whose favour the authorisation has been issued have

been discharged or that the holders of such interests have consented to the deregistration and export.

6   A chargee proposing to procure the de-registration and export of an aircraft under paragraph 1 otherwise than pursuant to a court order shall give reasonable prior notice in writing of the proposed deregistration and export to:

(a)   interested persons specified in Article 1(m)(i) and (ii) of the Convention; and

(b)   interested persons specified in Article 1(m)(iii) of the Convention who have given notice of their rights to the chargee within a reasonable time prior to the de-registration and export.

Article X — Modification of provisions regarding relief pending final determination

1   This Article applies only where a Contracting State has made a declaration under Article XXX(2) and to the extent stated in such declaration.

2   For the purposes of Article 13(1) of the Convention, 'speedy' in the context of obtaining relief means within such number of working days from the date of filing of the application for relief as is specified in a declaration made by the Contracting State in which the application is made.

3   Article 13(1) of the Convention applies with the following being added immediately after subparagraph (d):

(e)   if at any time the debtor and the creditor specifically agree, sale and application of proceeds therefrom, and Article 43(2) applies with the insertion after the words 'Article 13(1)(d)' of the words 'and (e)'.

4   Ownership or any other interest of the debtor passing on a sale under the preceding paragraph is free from any other interest over which the creditor's international interest has priority under the provisions of Article 29 of the Convention.

5   The creditor and the debtor or any other interested person may agree in writing to exclude the application of Article 13(2) of the Convention.

6   With regard to the remedies in Article IX(1):

(a)   they shall be made available by the registry authority and other administrative authorities, as applicable, in a Contracting State no later than five working days after the creditor notifies such authorities that the relief specified in Article IX(1) is granted or, in the case of relief granted by a foreign court, recognised by a court of that Contracting State, and that the creditor is entitled to procure those remedies in accordance with the Convention; and

(b)   the applicable authorities shall expeditiously co-operate with and assist the creditor in the exercise of such remedies in conformity with the applicable aviation safety laws and regulations.

7   Paragraphs 2 and 6 shall not affect any applicable aviation safety laws and regulations.

Article XI — Remedies on insolvency

1   This Article applies only where a Contracting State that is the primary insolvency jurisdiction has made a declaration pursuant to Article XXX(3).

Alternative A

2   Upon the occurrence of an insolvency-related event, the insolvency administrator or the debtor, as applicable, shall, subject to paragraph 7, give possession of the aircraft object to the creditor no later than the earlier of:
   (a)   the end of the waiting period; and
   (b)   the date on which the creditor would be entitled to possession of the aircraft object if this Article did not apply.

3   For the purposes of this Article, the 'waiting period' shall be the period specified in a declaration of the Contracting State which is the primary insolvency jurisdiction.

4   References in this Article to the 'insolvency administrator' shall be to that person in its official, not in its personal, capacity.

5   Unless and until the creditor is given the opportunity to take possession under paragraph 2:
   (a)   the insolvency administrator or the debtor, as applicable, shall preserve the aircraft object and maintain it and its value in accordance with the agreement; and
   (b)   the creditor shall be entitled to apply for any other forms of interim relief available under the applicable law.

6   Sub-paragraph (a) of the preceding paragraph shall not preclude the use of the aircraft object under arrangements designed to preserve the aircraft object and maintain it and its value.

7   The insolvency administrator or the debtor, as applicable, may retain possession of the aircraft object where, by the time specified in paragraph 2, it has cured all defaults other than a default constituted by the opening of insolvency proceedings and has agreed to perform all future obligations under the agreement. A second waiting period shall not apply in respect of a default in the performance of such future obligations.

8   With regard to the remedies in Article IX(1):
   (a)   they shall be made available by the registry authority and the administrative authorities in a Contracting State, as applicable, no later than five working days after the date on which the creditor notifies such authorities that it is entitled to procure those remedies in accordance with the Convention; and
   (b)   the applicable authorities shall expeditiously co-operate with and assist the creditor in the exercise of such remedies in conformity with the applicable aviation safety laws and regulations.

9   No exercise of remedies permitted by the Convention or this Protocol may be prevented or delayed after the date specified in paragraph 2.

10 No obligations of the debtor under the agreement may be modified without the consent of the creditor.

11 Nothing in the preceding paragraph shall be construed to affect the authority, if any, of the insolvency administrator under the applicable law to terminate the agreement.

12 No rights or interests, except for non-consensual rights or interests of a category covered by a declaration pursuant to Article 39(1), shall have priority in insolvency proceedings over registered interests.

13 The Convention as modified by Article IX of this Protocol shall apply to the exercise of any remedies under this Article.

Alternative B

2 Upon the occurrence of an insolvency-related event, the insolvency administrator or the debtor, as applicable, upon the request of the creditor, shall give notice to the creditor within the time specified in a declaration of a Contracting State pursuant to Article XXX(3) whether it will:
   (a) cure all defaults other than a default constituted by the opening of insolvency proceedings and agree to perform all future obligations, under the agreement and related transaction documents; or
   (b) give the creditor the opportunity to take possession of the aircraft object, in accordance with the applicable law.

3 The applicable law referred to in sub-paragraph (b) of the preceding paragraph may permit the court to require the taking of any additional step or the provision of any additional guarantee.

4 The creditor shall provide evidence of its claims and proof that its international interest has been registered.

5 If the insolvency administrator or the debtor, as applicable, does not give notice in conformity with paragraph 2, or when the insolvency administrator or the debtor has declared that it will give the creditor the opportunity to take possession of the aircraft object but fails to do so, the court may permit the creditor to take possession of the aircraft object upon such terms as the court may order and may require the taking of any additional step or the provision of any additional guarantee.

6 The aircraft object shall not be sold pending a decision by a court regarding the claim and the international interest.

Article XII — Insolvency assistance

1 This Article applies only where a Contracting State has made a declaration pursuant to Article XXX(1).

2 The courts of a Contracting State in which an aircraft object is situated shall, in accordance with the law of the Contracting State, co-operate to the maximum extent possible with foreign courts and foreign insolvency administrators in carrying out the provisions of Article XI.

Article XIII — De-registration and export request authorisation

1   This Article applies only where a Contracting State has made a declaration pursuant to Article XXX(1).

2   Where the debtor has issued an irrevocable de-registration and export request authorisation substantially in the form annexed to this Protocol and has submitted such authorisation for recordation to the registry authority, that authorisation shall be so recorded.

3   The person in whose favour the authorisation has been issued (the 'authorised party') or its certified designee shall be the sole person entitled to exercise the remedies specified in Article IX(1) and may do so only in accordance with the authorisation and applicable aviation safety laws and regulations. Such authorisation may not be revoked by the debtor without the consent in writing of the authorised party. The registry authority shall remove an authorisation from the registry at the request of the authorised party.

4   The registry authority and other administrative authorities in Contracting States shall expeditiously cooperate with and assist the authorised party in the exercise of the remedies specified in Article IX.

Article XIV — Modification of priority provisions

1   A buyer of an aircraft object under a registered sale acquires its interest in that object free from an interest subsequently registered and from an unregistered interest, even if the buyer has actual knowledge of the unregistered interest.

2   A buyer of an aircraft object acquires its interest in that object subject to an interest registered at the time of its acquisition.

3   Ownership of or another right or interest in an aircraft engine shall not be affected by its installation on or removal from an aircraft.

4   Article 29(7) of the Convention applies to an item, other than an object, installed on an airframe, aircraft engine or helicopter.

Article XV — Modification of assignment provisions

Article 33(1) of the Convention applies as if the following were added immediately after sub-paragraph (b):

and (c) the debtor has consented in writing, whether or not the consent is given in advance of the assignment or identifies the assignee.

Article XVI — Debtor provisions

1   In the absence of a default within the meaning of Article 11 of the Convention, the debtor shall be entitled to the quiet possession and use of the object in accordance with the agreement as against:

(a)   its creditor and the holder of any interest from which the debtor takes free pursuant to Article 29(4) of the Convention or, in the capacity

348

of buyer, Article XIV(1) of this Protocol, unless and to the extent that the debtor has otherwise agreed; and

(b) the holder of any interest to which the debtor's right or interest is subject pursuant to Article 29(4) of the Convention or, in the capacity of buyer, Article XIV(2) of this Protocol, but only to the extent, if any, that such holder has agreed.

2 Nothing in the Convention or this Protocol affects the liability of a creditor for any breach of the agreement under the applicable law in so far as that agreement relates to an aircraft object.

Chapter III
Registry provisions relating to international interests in aircraft objects

Article XVII — The Supervisory Authority and the Registrar

1 The Supervisory Authority shall be the international entity designated by a Resolution adopted by the Diplomatic Conference to Adopt a Mobile Equipment Convention and an Aircraft Protocol.

2 Where the international entity referred to in the preceding paragraph is not able and willing to act as Supervisory Authority, a Conference of Signatory and Contracting States shall be convened to designate another Supervisory Authority.

3 The Supervisory Authority and its officers and employees shall enjoy such immunity from legal and administrative process as is provided under the rules applicable to them as an international entity or otherwise.

4 The Supervisory Authority may establish a commission of experts, from among persons nominated by Signatory and Contracting States and having the necessary qualifications and experience, and entrust it with the task of assisting the Supervisory Authority in the discharge of its functions.

5 The first Registrar shall operate the International Registry for a period of five years from the date of entry into force of this Protocol. Thereafter, the Registrar shall be appointed or reappointed at regular five-yearly intervals by the Supervisory Authority.

Article XVIII — First regulations

The first regulations shall be made by the Supervisory Authority so as to take effect upon the entry into force of this Protocol.

Article XIX — Designated entry points

1 Subject to paragraph 2, a Contracting State may at any time designate an entity or entities in its territory as the entry point or entry points through which there shall or may be transmitted to the International Registry information required for registration other than registration of a notice of

a national interest or a right or interest under Article 40 in either case arising under the laws of another State.

2   A designation made under the preceding paragraph may permit, but not compel, use of a designated entry point or entry points for information required for registrations in respect of aircraft engines.

Article XX — Additional modifications to Registry provisions

1   For the purposes of Article 19(6) of the Convention, the search criteria for an aircraft object shall be the name of its manufacturer, its manufacturer's serial number and its model designation, supplemented as necessary to ensure uniqueness. Such supplementary information shall be specified in the regulations.

2   For the purposes of Article 25(2) of the Convention and in the circumstances there described, the holder of a registered prospective international interest or a registered prospective assignment of an international interest or the person in whose favour a prospective sale has been registered shall take such steps as are within its power to procure the discharge of the registration no later than five working days after the receipt of the demand described in such paragraph.

3   The fees referred to in Article 17(2)(h) of the Convention shall be determined so as to recover the reasonable costs of establishing, operating and regulating the International Registry and the reasonable costs of the Supervisory Authority associated with the performance of the functions, exercise of the powers, and discharge of the duties contemplated by Article 17(2) of the Convention.

4   The centralised functions of the International Registry shall be operated and administered by the Registrar on a twenty-four hour basis. The various entry points shall be operated at least during working hours in their respective territories.

5   The amount of the insurance or financial guarantee referred to in Article 28(4) of the Convention shall, in respect of each event, not be less than the maximum value of an aircraft object as determined by the Supervisory Authority.

6   Nothing in the Convention shall preclude the Registrar from procuring insurance or a financial guarantee covering events for which the Registrar is not liable under Article 28 of the Convention.

Chapter IV
Jurisdiction

Article XXI — Modification of jurisdiction provisions

For the purposes of Article 43 of the Convention and subject to Article 42 of the Convention, a court of a Contracting State also has jurisdiction where the object is a helicopter, or an airframe pertaining to an aircraft, for which that State is the State of registry.

Article XXII — Waivers of sovereign immunity

1   Subject to paragraph 2, a waiver of sovereign immunity from jurisdiction of the courts specified in Article 42 or Article 43 of the Convention or relating to enforcement of rights and interests relating to an aircraft object under the Convention shall be binding and, if the other conditions to such jurisdiction or enforcement have been satisfied, shall be effective to confer jurisdiction and permit enforcement, as the case may be.

2   A waiver under the preceding paragraph must be in writing and contain a description of the aircraft object.

Chapter V
Relationship with other conventions

Article XXIII — Relationship with the Convention on the International Recognition of Rights in Aircraft

The Convention shall, for a Contracting State that is a party to the *Convention on the International Recognition of Rights in Aircraft*, signed at Geneva on 19 June 1948, supersede that Convention as it relates to aircraft, as defined in this Protocol, and to aircraft objects. However, with respect to rights or interests not covered or affected by the present Convention, the Geneva Convention shall not be superseded.

Article XXIV — Relationship with the Convention for the Unification of Certain Rules Relating to the Precautionary Attachment of Aircraft

1   The Convention shall, for a Contracting State that is a Party to the *Convention for the Unification of Certain Rules Relating to the Precautionary Attachment of Aircraft*, signed at Rome on 29 May 1933, supersede that Convention as it relates to aircraft, as defined in this Protocol.

2   A Contracting State Party to the above Convention may declare, at the time of ratification, acceptance, approval of, or accession to this Protocol, that it will not apply this Article.

Article XXV — Relationship with the UNIDROIT Convention on International Financial Leasing

The Convention shall supersede the *UNIDROIT Convention on International Financial Leasing*, signed at Ottawa on 28 May 1988, as it relates to aircraft objects.

Chapter VI
Final provisions

Article XXVI — Signature, ratification, acceptance, approval or accession

1   This Protocol shall be open for signature in Cape Town on 16 November 2001 by States participating in the Diplomatic Conference to Adopt a Mobile Equipment Convention and an Aircraft Protocol held at Cape Town from 29 October to 16 November 2001. After 16 November 2001, this Protocol shall be open to all States for signature at the Headquarters of the International Institute for the Unification of Private Law (UNIDROIT) in Rome until it enters into force in accordance with Article XXVIII.

2   This Protocol shall be subject to ratification, acceptance or approval by States which have signed it.

3   Any State which does not sign this Protocol may accede to it at any time.

4   Ratification, acceptance, approval or accession is effected by the deposit of a formal instrument to that effect with the Depositary.

5   A State may not become a Party to this Protocol unless it is or becomes also a Party to the Convention.

Article XXVII — Regional Economic Integration Organisations

1   A Regional Economic Integration Organisation which is constituted by sovereign States and has competence over certain matters governed by this Protocol may similarly sign, accept, approve or accede to this Protocol. The Regional Economic Integration Organisation shall in that case have the rights and obligations of a Contracting State, to the extent that that Organisation has competence over matters governed by this Protocol. Where the number of Contracting States is relevant in this Protocol, the Regional Economic Integration Organisation shall not count as a Contracting State in addition to its Member States which are Contracting States.

2   The Regional Economic Integration Organisation shall, at the time of signature, acceptance, approval or accession, make a declaration to the Depositary specifying the matters governed by this Protocol in respect of which competence has been transferred to that Organisation by its Member States. The Regional Economic Integration Organisation shall promptly notify the Depositary of any changes to the distribution of competence, including new transfers of competence, specified in the declaration under this paragraph.

3   Any reference to a 'Contracting State' or 'Contracting States' or 'State Party' or 'States Parties' in this Protocol applies equally to a Regional Economic Integration Organisation where the context so requires.

Article XXVIII — Entry into force

1   This Protocol enters into force on the first day of the month following the expiration of three months after the date of the deposit of the eighth instrument of ratification, acceptance, approval or accession, between the States which have deposited such instruments.

2   For other States this Protocol enters into force on the first day of the month following the expiration of three months after the date of the deposit of its instrument of ratification, acceptance, approval or accession.

Article XXIX — Territorial units

1   If a Contracting State has territorial units in which different systems of law are applicable in relation to the matters dealt with in this Protocol, it may, at the time of ratification, acceptance, approval or accession, declare that this Protocol is to extend to all its territorial units or only to one or more of them and may modify its declaration by submitting another declaration at any time.

2   Any such declaration shall state expressly the territorial units to which this Protocol applies.

3   If a Contracting State has not made any declaration under paragraph 1, this Protocol shall apply to all territorial units of that State.

4   Where a Contracting State extends this Protocol to one or more of its territorial units, declarations permitted under this Protocol may be made in respect of each such territorial unit, and the declarations made in respect of one territorial unit may be different from those made in respect of another territorial unit.

5   If by virtue of a declaration under paragraph 1, this Protocol extends to one or more territorial units of a Contracting State:

    (a)   the debtor is considered to be situated in a Contracting State only if it is incorporated or formed under a law in force in a territorial unit to which the Convention and this Protocol apply or if it has its registered office or statutory seat, centre of administration, place of business or habitual residence in a territorial unit to which the Convention and this Protocol apply;

    (b)   any reference to the location of the object in a Contracting State refers to the location of the object in a territorial unit to which the Convention and this Protocol apply; and

    (c)   any reference to the administrative authorities in that Contracting State shall be construed as referring to the administrative authorities having jurisdiction in a territorial unit to which the Convention and this Protocol apply and any reference to the national register or to the registry authority in that Contracting State shall be construed as referring to the aircraft register in force or to the registry authority having jurisdiction in the territorial unit or units to which the Convention and this Protocol apply.

Article XXX — Declarations relating to certain provisions

1   A Contracting State may, at the time of ratification, acceptance, approval of, or accession to this Protocol, declare that it will apply any one or more of Articles VIII, XII and XIII of this Protocol.

2   A Contracting State may, at the time of ratification, acceptance, approval of, or accession to this Protocol, declare that it will apply Article X of this Protocol, wholly or in part. If it so declares with respect to Article X(2), it shall specify the time-period required thereby.

3   A Contracting State may, at the time of ratification, acceptance, approval of, or accession to this Protocol, declare that it will apply the entirety of Alternative A, or the entirety of Alternative B of Article XI and, if so, shall specify the types of insolvency proceeding, if any, to which it will apply Alternative A and the types of insolvency proceeding, if any, to which it will apply Alternative B. A Contracting State making a declaration pursuant to this paragraph shall specify the time-period required by Article XI.

4   The courts of Contracting States shall apply Article XI in conformity with the declaration made by the Contracting State which is the primary insolvency jurisdiction.

5   A Contracting State may, at the time of ratification, acceptance, approval of, or accession to this Protocol, declare that it will not apply the provisions of Article XXI, wholly or in part. The declaration shall specify under which conditions the relevant Article will be applied, in case it will be applied partly, or otherwise which other forms of interim relief will be applied.

Article XXXI — Declarations under the Convention

Declarations made under the Convention, including those made under Articles 39, 40, 50, 53, 54, 55, 57, 58 and 60 of the Convention, shall be deemed to have also been made under this Protocol unless stated otherwise.

Article XXXII — Reservations and declarations

1   No reservations may be made to this Protocol but declarations authorised by Articles XXIV, XXIX, XXX, XXXI, XXXIII and XXXIV may be made in accordance with these provisions.

2   Any declaration or subsequent declaration or any withdrawal of a declaration made under this Protocol shall be notified in writing to the Depositary.

Article XXXIII — Subsequent declarations

1   A State Party may make a subsequent declaration, other than a declaration made in accordance with Article XXXI under Article 60 of the Convention, at any time after the date on which this Protocol has entered into force for it, by notifying the Depositary to that effect.

2   Any such subsequent declaration shall take effect on the first day of the month following the expiration of six months after the date of receipt of the

notification by the Depositary. Where a longer period for that declaration to take effect is specified in the notification, it shall take effect upon the expiration of such longer period after receipt of the notification by the Depositary.

3   Notwithstanding the previous paragraphs, this Protocol shall continue to apply, as if no such subsequent declarations had been made, in respect of all rights and interests arising prior to the effective date of any such subsequent declaration.

Article XXXIV — Withdrawal of declarations

1   Any State Party having made a declaration under this Protocol, other than a declaration made in accordance with Article XXXI under Article 60 of the Convention, may withdraw it at any time by notifying the Depositary. Such withdrawal is to take effect on the first day of the month following the expiration of six months after the date of receipt of the notification by the Depositary.

2   Notwithstanding the previous paragraph, this Protocol shall continue to apply, as if no such withdrawal of declaration had been made, in respect of all rights and interests arising prior to the effective date of any such withdrawal.

Article XXXV — Denunciations

1   Any State Party may denounce this Protocol by notification in writing to the Depositary.

2   Any such denunciation shall take effect on the first day of the month following the expiration of twelve months after the date of receipt of the notification by the Depositary.

3   Notwithstanding the previous paragraphs, this Protocol shall continue to apply, as if no such denunciation had been made, in respect of all rights and interests arising prior to the effective date of any such denunciation.

Article XXXVI — Review Conferences, amendments and related matters

1   The Depositary, in consultation with the Supervisory Authority, shall prepare reports yearly, or at such other time as the circumstances may require, for the States Parties as to the manner in which the international regime established in the Convention as amended by this Protocol has operated in practice. In preparing such reports, the Depositary shall take into account the reports of the Supervisory Authority concerning the functioning of the international registration system.

2   At the request of not less than twenty-five percent of the States Parties, Review Conferences of the States Parties shall be convened from time to

time by the Depositary, in consultation with the Supervisory Authority, to consider:

(a) the practical operation of the Convention as amended by this Protocol and its effectiveness in facilitating the asset-based financing and leasing of the objects covered by its terms;

(b) the judicial interpretation given to, and the application made of the terms of this Protocol and the regulations;

(c) the functioning of the international registration system, the performance of the Registrar and its oversight by the Supervisory Authority, taking into account the reports of the Supervisory Authority; and

(d) whether any modifications to this Protocol or the arrangements relating to the International Registry are desirable.

3   Any amendment to this Protocol shall be approved by at least a two-thirds majority of States Parties participating in the Conference referred to in the preceding paragraph and shall then enter into force in respect of States which have ratified, accepted or approved such amendment when it has been ratified, accepted or approved by eight States in accordance with the provisions of Article XXVIII relating to its entry into force.

Article XXXVII — Depositary and its functions

1   Instruments of ratification, acceptance, approval or accession shall be deposited with the International Institute for the Unification of Private Law (UNIDROIT), which is hereby designated the Depositary.

2   The Depositary shall:

(a) inform all Contracting States of:

(i) each new signature or deposit of an instrument of ratification, acceptance, approval or accession, together with the date thereof;

(ii) the date of entry into force of this Protocol;

(iii) each declaration made in accordance with this Protocol, together with the date thereof;

(iv) the withdrawal or amendment of any declaration, together with the date thereof; and

(v) the notification of any denunciation of this Protocol together with the date thereof and the date on which it takes effect;

(b) transmit certified true copies of this Protocol to all Contracting States;

(c) provide the Supervisory Authority and the Registrar with a copy of each instrument of ratification, acceptance, approval or accession, together with the date of deposit thereof, of each declaration or withdrawal or amendment of a declaration and of each notification of denunciation, together with the date of notification thereof, so that the information contained therein is easily and fully available; and

(d) perform such other functions customary for depositaries.

IN WITNESS WHEREOF the undersigned Plenipotentiaries, having been duly authorised, have signed this Protocol.

DONE at Cape Town, this sixteenth day of November, two thousand and one, in a single original in the English, Arabic, Chinese, French, Russian and Spanish languages, all texts being equally authentic, such authenticity to take effect upon verification by the Joint Secretariat of the Conference under the authority of the President of the Conference within ninety days hereof as to the conformity of the texts with one another.

Annex

FORM OF IRREVOCABLE DE-REGISTRATION AND EXPORT REQUEST AUTHORISATION

Annex referred to in Article XIII

See Appendix I

SELECTED STATUTES SECTION 1110

11 USC § 1110 Aircraft equipment and vessels

(a) (1) Except as provided in paragraph (2) and subject to subsection (b), the right of a secured party with a security interest in equipment described in paragraph (3), or of a lessor or conditional vendor of such equipment, to take possession of such equipment in compliance with a security agreement, lease, or conditional sale contract, and to enforce any of its other rights or remedies, under such security agreement, lease, or conditional sale contract, to sell, lease, or otherwise retain or dispose of such equipment, is not limited or otherwise affected by any other provision of this title or by any power of the court.

(2) The right to take possession and to enforce the other rights and remedies described in paragraph (1) shall be subject to section 362 if—

(A) before the date that is 60 days after the date of the order for relief under this chapter, the trustee, subject to the approval of the court, agrees to perform all obligations of the debtor under such security agreement, lease, or conditional sale contract; and

(B) any default, other than a default of a kind specified in section 365(b)(2), under such security agreement, lease, or conditional sale contract—

(i) that occurs before the date of the order is cured before the expiration of such 60-day period;

(ii) that occurs after the date of the order and before the expiration of such 60-day period is cured before the later of—

(I) the date that is 30 days after the date of the default; or

(II) the expiration of such 60-day period; and

(iii) that occurs on or after the expiration of such 60-day period is cured in compliance with the terms of such security agreement, lease, or conditional sale contract, if a cure is permitted under that agreement, lease, or contract.

(3) The equipment described in this paragraph—

(A) is—

(i) an aircraft, aircraft engine, propeller, appliance, or spare part (as defined in section 40102 of title 49) that is subject to a security interest granted by, leased to, or conditionally sold to a debtor that, at the time such transaction is entered into, holds an air carrier operating certificate issued pursuant to chapter 447 of title 49 for Appendix C-1 Selected Statutes Section 1110 aircraft capable of carrying 10 or more individuals or 6,000 pounds or more of cargo; or

    (ii)   a vessel documented under chapter 121 of title 46 that is subject to a security interest granted by, leased to, or conditionally sold to a debtor that is a water carrier that, at the time such transaction is entered into, holds a certificate of public convenience and necessity or permit issued by the Department of Transportation; and

(B)   includes all records and documents relating to such equipment that are required, under the terms of the security agreement, lease, or conditional sale contract, to be surrendered or returned by the debtor in connection with the surrender or return of such equipment.

(4)   Paragraph (1) applies to a secured party, lessor, or conditional vendor acting in its own behalf or acting as trustee or otherwise in behalf of another party.

(b)   The trustee and the secured party, lessor, or conditional vendor whose right to take possession is protected under subsection (a) may agree, subject to the approval of the court, to extend the 60-day period specified in subsection (a)(1).

(c)   (1)   In any case under this chapter, the trustee shall immediately surrender and return to a secured party, lessor, or conditional vendor, described in subsection (a)(1), equipment described in subsection (a)(3), if at any time after the date of the order for relief under this chapter such secured party, lessor, or conditional vendor is entitled pursuant to subsection (a)(1) to take possession of such equipment and makes a written demand for such possession to the trustee.

    (2)   At such time as the trustee is required under paragraph (1) to surrender and return equipment described in subsection (a)(3), any lease of such equipment, and any security agreement or conditional sale contract relating to such equipment, if such security agreement or conditional sale contract is an executory contract, shall be deemed rejected.

(d)   With respect to equipment first placed in service on or before October 22, 1994, for purposes of this section—

    (1)   the term 'lease' includes any written agreement with respect to which the lessor and the debtor, as lessee, have expressed in the agreement or in a substantially contemporaneous writing that the agreement is to be treated as a lease for Federal income tax purposes; and

    (2)   the term 'security interest' means a purchase-money equipment security interest.

SELECTED STATUTES NEW YORK UCC ARTICLE 9 (AND RELATED SECTIONS) – SECURITY INTERESTS AND PERFECTION

New York UCC Section 1–201(37)

(37) 'Security interest' means an interest in personal property or fixtures which secures payment or performance of an obligation. The term also includes any interest of a consignor and a buyer of accounts, chattel paper, a payment intangible, or a promissory note in a transaction that is subject to Article 9. The special property interest of a buyer of goods on identification of those goods to a contract for sale under Section 2 401 is not a 'security interest', but a buyer may also acquire a 'security interest' by complying with Article 9. Except as otherwise provided in Section 2 505, the right of a seller or lessor of goods under Article 2 or 2 A to retain or acquire possession of the goods is not a 'security interest', but a seller or lessor may also acquire a 'security interest' by complying with Article 9. The retention or reservation of title by a seller of goods notwithstanding shipment or delivery to the buyer (Section 2 401) is limited in effect to a reservation of a 'security interest'.

(a) Whether a transaction creates a lease or security interest is determined by the facts of each case; however, a transaction creates a security interest if the consideration the lessee is to pay the lessor for the right to possession and use of the goods is an obligation for the term of the lease not subject to termination by the lessee, and:
  (i) the original term of the lease is equal to or greater than the remaining economic life of the goods,
  (ii) the lessee is bound to renew the lease for the remaining economic life of the goods or is bound to become the owner of the goods,
  (iii) the lessee has an option to renew the lease for the remaining economic life of the goods for no additional consideration or nominal additional consideration upon compliance with the lease agreement, or
  (iv) the lessee has an option to become the owner of the goods for no additional consideration or nominal additional consideration upon compliance with the lease agreement.
(b) A transaction does not create a security interest merely because it provides that:
  (i) the present value of the consideration the lessee is obligated to pay the lessor for the right to possession and use of the goods is substantially equal to or is greater than the fair market value of the goods at the time the lease is entered into,

(ii)   the lessee assumes risk of loss of the goods, or agrees to pay taxes, insurance, filing, recording, or registration fees, or service or maintenance costs with respect to the goods,

(iii) the lessee has an option to renew the lease or to become the owner of the goods,

(iv)   the lessee has an option to renew the lease for a fixed rent that is equal to or greater than the reasonably predictable fair market rent for the use of the goods for the term of the renewal at the time the option is to be performed, or

(v)   the lessee has an option to become the owner of the goods for a fixed price that is equal to or greater than the reasonably predictable fair market value of the goods at the time the option is to be performed.

(c)   For purposes of this subsection (37):

(i)   Additional consideration is not nominal if (A) when the option to renew the lease is granted to the lessee the rent is stated to be the fair market rent for the use of the goods for the term of the renewal determined at the time the option is to be performed, or (B) when the option to become the owner of the goods is granted to the lessee the price is stated to be the fair market value of the goods determined at the time the option is to be performed. Additional consideration is nominal if it is less than the lessee's reasonably predictable cost of performing under the lease agreement if the option is not exercised;

(ii)   'Reasonably predictable' and 'remaining economic life of the goods' are to be determined with reference to the facts and circumstances at the time the transaction is entered into; and

(iii) 'Present value' means the amount as of a date certain of one or more sums payable in the future, discounted to the date certain. The discount is determined by the interest rate specified by the parties if the rate is not manifestly unreasonable at the time the transaction is entered into; otherwise, the discount is determined by a commercially reasonable rate that takes into account the facts and circumstances of each case at the time the transaction was entered into.

§ 9–203: Attachment and Enforceability of Security Interest; Proceeds; Supporting Obligations; Formal Requisites

(a)   **Attachment**. A security interest attaches to collateral when it becomes enforceable against the debtor with respect to the collateral, unless an agreement expressly postpones the time of attachment.

(b)   **Enforceability**. Except as otherwise provided in subsections (c) through (i), a security interest is enforceable against the debtor and third parties with respect to the collateral only if:

(1)   value has been given;

(2)   the debtor has rights in the collateral or the power to transfer rights in the collateral to a secured party; and

(3) one of the following conditions is met:

    (A) the debtor has authenticated a security agreement that provides a description of the collateral and, if the security interest covers timber to be cut, a description of the land concerned;

    (B) the collateral is not a certificated security and is in the possession of the secured party under Section 9–313 pursuant to the debtor's security agreement;

    (C) the collateral is a certificated security in registered form and the security certificate has been delivered to the secured party under Section 8–301 pursuant to the debtor's security agreement; or

    (D) the collateral is deposit accounts, electronic chattel paper, investment property, or letter-of-credit rights, and the secured party has control under Section 9–104, 9–105, 9–106, or 9–107 pursuant to the debtor's security agreement.

(c) **Other UCC provisions**. Subsection (b) is subject to Section 4–210 on the security interest of a collecting bank, Section 5–118 on the security interest of a letter-of-credit issuer or nominated person, Section 9–110 on a security interest arising under Article 2 or 2-A, and Section 9–206 on security interests in investment property.

(d) **When a person becomes bound by another person's security agreement**. A person becomes bound as debtor by a security agreement entered into by another person if, by operation of law other than this article or by contract:

(1) the security agreement becomes effective to create a security interest in the person's property; or

(2) the person becomes generally obligated for the obligations of the other person, including the obligation secured under the security agreement, and acquires or succeeds to all or substantially all of the assets of the other person.

(e) **Effect of new debtor becoming bound**. If a new debtor becomes bound as debtor by a security agreement entered into by another person:

(1) the agreement satisfies subsection (b)(3) with respect to existing or after-acquired property of the new debtor to the extent the property is described in the agreement; and

(2) another agreement is not necessary to make a security interest in the property enforceable.

(f) **Proceeds and supporting obligations**. The attachment of a security interest in collateral gives the secured party the rights to proceeds provided by Section 9–315 and is also attachment of a security interest in a supporting obligation for the collateral.

(g) **Lien securing right to payment**. The attachment of a security interest in a right to payment or performance secured by a security interest or other lien on personal or real property is also attachment of a security interest in the security interest, mortgage, or other lien.

(h)   **Security entitlement carried in securities account**. The attachment of a security interest in a securities account is also attachment of a security interest in the security entitlements carried in the securities account.

(i)   **Commodity contracts carried in commodity account**. The attachment of a security interest in a commodity account is also attachment of a security interest in the commodity contracts carried in the commodity account.

SELECTED STATUTES NEW YORK UCC
ARTICLE 9 – ENFORCEMENT OF REMEDIES

§ 9–601: Rights after Default; Judicial Enforcement; Consignor or Buyer of Accounts, Chattel Paper, Payment Intangibles, or Promissory Notes

(a) **Rights of secured party after default**. After default, a secured party has the rights provided in this part and, except as otherwise provided in Section 9–602, those provided by agreement of the parties. A secured party:

    (1) may reduce a claim to judgment, foreclose, or otherwise enforce the claim, security interest, or agricultural lien by any available judicial procedure; and

    (2) if the collateral is documents, may proceed either as to the documents or as to the goods they cover.

(b) **Rights and duties of secured party in possession or control**. A secured party in possession of collateral or control of collateral under Section 9–104, 9–105, 9–106, or 9–107 has the rights and duties provided in Section 9–207.

(c) **Rights cumulative; simultaneous exercise**. The rights under subsections (a) and (b) are cumulative and may be exercised simultaneously.

(d) **Rights of debtor and obligor**. Except as otherwise provided in subsection (g) and Section 9–605, after default, a debtor and an obligor have the rights provided in this part and by agreement of the parties.

(e) **Lien of levy after judgment**. If a secured party has reduced its claim to judgment, the lien of any levy that may be made upon the collateral by virtue of an execution based upon the judgment relates back to the earliest of:

    (1) the date of perfection of the security interest or agricultural lien in the collateral;

    (2) the date of filing a financing statement covering the collateral; or

    (3) any date specified in a statute under which the agricultural lien was created.

(f) **Execution sale**. A sale pursuant to an execution is a foreclosure of the security interest or agricultural lien by judicial procedure within the meaning of this section. A secured party may purchase at the sale and thereafter hold the collateral free of any other requirements of this article.

(g) **Consignor or buyer of certain rights to payment**. Except as otherwise provided in Section 9–607(c), this part imposes no duties upon a secured party that is a consignor or is a buyer of accounts, chattel paper, payment intangibles, or promissory notes.

§ 9–602: Waiver and Variance of Rights and Duties

Except as otherwise provided in Section 9–624, to the extent that they give rights to a debtor or obligor and impose duties on a secured party, the debtor or obligor may not waive or vary the rules stated in the following listed sections:

(a) Section 9–207(b)(4)(C), which deals with use and operation of the collateral by the secured party;

(b) Section 9–210, which deals with requests for an accounting and requests concerning a list of collateral and statement of account;

(c) Section 9–607(c), which deals with collection and enforcement of collateral;

(d) Sections 9–608(a) and 9–615(c) to the extent that they deal with application or payment of noncash proceeds of collection, enforcement, or disposition;

(e) Sections 9–608(a) and 9–615(d) to the extent that they require accounting for or payment of surplus proceeds of collateral;

(f) Section 9–609 to the extent that it imposes upon a secured party that takes possession of collateral without judicial process the duty to do so without breach of the peace;

(g) Sections 9–610(b), 9–611, 9–613, and 9–614, which deal with disposition of collateral;

(h) Section 9–615(f), which deals with calculation of a deficiency or surplus when a disposition is made to the secured party, a person related to the secured party, or a secondary obligor;

(i) Section 9–616, which deals with explanation of the calculation of a surplus or deficiency;

(j) Sections 9–620, 9–621, and 9–622, which deal with acceptance of collateral in satisfaction of obligation;

(k) Section 9–623, which deals with redemption of collateral;

(l) Section 9–624, which deals with permissible waivers; and

(m) Sections 9–625 and 9–626, which deal with the secured party's liability for failure to comply with this article.

§ 9–603: Agreement on Standards Concerning Rights and Duties

(a) **Agreed standards.** The parties may determine by agreement the standards measuring the fulfillment of the rights of a debtor or obligor and the duties of a secured party under a rule stated in Section 9–602 if the standards are not manifestly unreasonable.

(b) **Agreed standards inapplicable to breach of peace.** Subsection (a) does not apply to the duty under Section 9–609 to refrain from breaching the peace.

§ 9–604: Procedure If Security Agreement Covers Real Property, Fixtures, or Cooperative Interests

[Omitted]

§ 9–605: Unknown Debtor or Secondary Obligor

A secured party does not owe a duty based on its status as secured party:

(a) to a person that is a debtor or obligor, unless the secured party knows:
  (1) that the person is a debtor or obligor;
  (2) the identity of the person; and
  (3) how to communicate with the person; or
(b) to a secured party or lienholder that has filed a financing statement against a person, unless the secured party knows:
  (1) that the person is a debtor; and
  (2) the identity of the person.

§ 9–606: Time of Default for Agricultural Lien

[Omitted]

§ 9–607: Collection and Enforcement by Secured Party

(a) **Collection and enforcement generally.** If so agreed, and in any event after default, a secured party:
  (1) may notify an account debtor or other person obligated on collateral to make payment or otherwise render performance to or for the benefit of the secured party;
  (2) may take any proceeds to which the secured party is entitled under Section 9–315;
  (3) may enforce the obligations of an account debtor or other person obligated on collateral and exercise the rights of the debtor with respect to the obligation of the account debtor or other person obligated on collateral to make payment or otherwise render performance to the debtor, and with respect to any property that secures the obligations of the account debtor or other person obligated on the collateral;
  (4) if it holds a security interest in a deposit account perfected by control under Section 9–104(a)(1), may apply the balance of the deposit account to the obligation secured by the deposit account; and
  (5) if it holds a security interest in a deposit account perfected by control under Section 9–104(a) (2) or (3), may instruct the bank to pay the balance of the deposit account to or for the benefit of the secured party.
(b) **Nonjudicial enforcement of mortgage.** If necessary to enable a secured party to exercise under subsection (a)(3) the right of a debtor to enforce a mortgage nonjudicially, the secured party may record in the office in which a record of the mortgage is recorded:
  (1) a copy of the security agreement that creates or provides for a security interest in the obligation secured by the mortgage; and

(2)   the secured party's sworn affidavit in recordable form stating that:
  (A)   a default has occurred; and
  (B)   the secured party is entitled to enforce the mortgage non-judicially.

(c)   **Commercially reasonable collection and enforcement**. A secured party shall proceed in a commercially reasonable manner if the secured party:
  (1)   undertakes to collect from or enforce an obligation of an account debtor or other person obligated on collateral; and
  (2)   is entitled to charge back uncollected collateral or otherwise to full or limited recourse against the debtor or a secondary obligor.

(d)   **Expenses of collection and enforcement**. A secured party may deduct from the collections made pursuant to subsection (c) reasonable expenses of collection and enforcement, including reasonable attorney's fees and legal expenses incurred by the secured party.

(e)   **Duties to secured party not affected**. This section does not determine whether an account debtor, bank, or other person obligated on collateral owes a duty to a secured party.

§ 9–608: Application of Proceeds of Collection or Enforcement; Liability for Deficiency and Right to Surplus

(a)   **Application of proceeds, surplus, and deficiency if obligation secured**. If a security interest or agricultural lien secures payment or performance of an obligation, the following rules apply:
  (1)   A secured party shall apply or pay over for application the cash proceeds of collection or enforcement under Section 9–607 in the following order to:
    (A)   the reasonable expenses of collection and enforcement and, to the extent provided for by agreement and not prohibited by law, reasonable attorney's fees and legal expenses incurred by the secured party;
    (B)   the satisfaction of obligations secured by the security interest or agricultural lien under which the collection or enforcement is made; and
    (C)   the satisfaction of obligations secured by any subordinate security interest in or other lien on the collateral subject to the security interest or agricultural lien under which the collection or enforcement is made if the secured party receives an authenticated demand for proceeds before distribution of the proceeds is completed.
  (2)   If requested by a secured party, a holder of a subordinate security interest or other lien shall furnish reasonable proof of the interest or lien within a reasonable time. Unless the holder complies, the secured party need not comply with the holder's demand under paragraph (1)(C).

(3) A secured party need not apply or pay over for application noncash proceeds of collection and enforcement under Section 9–607 unless the failure to do so would be commercially unreasonable. A secured party that applies or pays over for application noncash proceeds shall do so in a commercially reasonable manner.

(4) A secured party shall account to and pay a debtor for any surplus, and the obligor is liable for any deficiency.

(b) No surplus or deficiency in sales of certain rights to payment. If the underlying transaction is a sale of accounts, chattel paper, payment intangibles, or promissory notes, the debtor is not entitled to any surplus, and the obligor is not liable for any deficiency.

§ 9–609: Secured Party's Right to Take Possession after Default

(a) Possession; rendering equipment unusable; disposition on debtor's premises. After default, a secured party:

(1) may take possession of the collateral; and

(2) without removal, may render equipment unusable and dispose of collateral on a debtor's premises under Section 9–610.

(b) **Judicial and nonjudicial process**. A secured party may proceed under subsection (a):

(1) pursuant to judicial process; or

(2) without judicial process, if it proceeds without breach of the peace.

(c) **Assembly of collateral**. If so agreed, and in any event after default, a secured party may require the debtor to assemble the collateral and make it available to the secured party at a place to be designated by the secured party which is reasonably convenient to both parties.

§ 9–610: Disposition of Collateral after Default

(a) **Disposition after default**. After default, a secured party may sell, lease, license, or otherwise dispose of any or all of the collateral in its present condition or following any commercially reasonable preparation or processing.

(b) **Commercially reasonable disposition**. Every aspect of a disposition of collateral, including the method, manner, time, place, and other terms, must be commercially reasonable. If commercially reasonable, a secured party may dispose of collateral by public or private proceedings, by one or more contracts, as a unit or in parcels, and at any time and place and on any terms.

(c) **Purchase by secured party**. A secured party may purchase collateral:

(1) at a public disposition; or

(2) at a private disposition only if the collateral is of a kind that is customarily sold on a recognized market or the subject of widely distributed standard price quotations.

(d) **Warranties on disposition**. A contract for sale, lease, license, or other disposition includes the warranties relating to title, possession, quiet enjoyment, and the like which by operation of law accompany a voluntary disposition of property of the kind subject to the contract.

(e) **Disclaimer of warranties**. A secured party may disclaim or modify warranties under subsection (d):

(1) in a manner that would be effective to disclaim or modify the warranties in a voluntary disposition of property of the kind subject to the contract of disposition; or

(2) by communicating to the purchaser a record evidencing the contract for disposition and including an express disclaimer or modification of the warranties.

(f) **Record sufficient to disclaim warranties**. A record is sufficient to disclaim warranties under subsection (e) if it indicates 'There is no warranty relating to title, possession, quiet enjoyment, or the like in this disposition' or uses words of similar import.

§ 9–611: Notification Before Disposition of Collateral

(a) **'Notification date.'** In this section, 'notification date' means the earlier of the date on which:

(1) a secured party sends to the debtor and any secondary obligor an authenticated notification of disposition; or

(2) the debtor and any secondary obligor waive the right to notification.

(b) **Notification of disposition required**. Except as otherwise provided in subsection (d), a secured party that disposes of collateral under Section 9–610 shall send to the persons specified in subsection (c) a reasonable authenticated notification of disposition.

(c) **Persons to be notified**. To comply with subsection (b), the secured party shall send an authenticated notification of disposition to:

(1) the debtor;

(2) any secondary obligor; and

(3) if the collateral is other than consumer goods:

(A) any other person from which the secured party has received, before the notification date, an authenticated notification of a claim of an interest in the collateral;

(B) any other secured party or lienholder that, 10 days before the notification date, held a security interest in or other lien on the collateral perfected by the filing of a financing statement that:

(i) identified the collateral;

(ii) was indexed under the debtor's name as of that date; and

(iii) was filed in the office in which to file a financing statement against the debtor covering the collateral as of that date; and

(C) any other secured party that, 10 days before the notification date, held a security interest in the collateral perfected by compliance with a statute, regulation, or treaty described in Section 9–311(a).

(d) **Subsection (b) inapplicable: perishable collateral; recognized market.** Subsection (b) does not apply if the collateral is perishable or threatens to decline speedily in value or is of a type customarily sold on a recognized market.

(e) **Compliance with subsection (c)(3)(B).** A secured party complies with the requirement for notification prescribed by subsection (c)(3)(B) if:

(1) not later than twenty days or earlier than thirty days before the notification date, the secured party requests, in a commercially reasonable manner, information concerning financing statements indexed under the debtor's name in the office indicated in subsection (c)(3)(B); and

(2) before the notification date, the secured party:

(A) did not receive a response to the request for information; or

(B) received a response to the request for information and sent an authenticated notification of disposition to each secured party or other lienholder named in that response whose financing statement covered the collateral.

§ 9–612: Timeliness of Notification Before Disposition of Collateral

(a) **Reasonable time is question of fact.** Except as otherwise provided in subsection (b), whether a notification is sent within a reasonable time is a question of fact.

(b) **10-day period sufficient in non-consumer transaction.** In a transaction other than a consumer transaction, a notification of disposition sent after default and 10 days or more before the earliest time of disposition set forth in the notification is sent within a reasonable time before the disposition.

§ 9–613: Contents and Form of Notification Before Disposition of Collateral: General

Except in a consumer-goods transaction, the following rules apply:

(a) The contents of a notification of disposition are sufficient if the notification:
(1) describes the debtor and the secured party;
(2) describes the collateral that is the subject of the intended disposition;
(3) states the method of intended disposition;
(4) states that the debtor is entitled to an accounting of the unpaid indebtedness and states the charge, if any, for an accounting; and
(5) states the time and place of a public disposition or the time after which any other disposition is to be made.

(b) Whether the contents of a notification that lacks any of the information specified in subsection (a) are nevertheless sufficient is a question of fact.

(c)   The contents of a notification providing substantially the information specified in subsection (a) are sufficient, even if the notification includes:

    (1)   information not specified by subsection (a); or

    (2)   minor errors that are not seriously misleading.

(d)   A particular phrasing of the notification is not required.

(e)   The following form of notification and the form appearing in Section 9–614(c), when completed, each provides sufficient information:

NOTIFICATION OF DISPOSITION OF COLLATERAL

To: (Name of debtor, obligor, or other person to which the notification is sent)

From: (Name, address, and telephone number of secured party)

Name of Debtor(s): (Include only if debtor(s) are not an addressee)

(For a public disposition:)

We will sell (or lease or license, as applicable) the (describe collateral) (to the highest qualified bidder) in public as follows:

Day and Date: _____

Time: _____

Place: _____

(For a private disposition:)

We will sell (or lease or license, as applicable) the (describe collateral) privately sometime after (day and date).

You are entitled to an accounting of the unpaid indebtedness secured by the property that we intend to sell (or lease or license, as applicable) (for a charge of $_____). You may request an accounting by calling us at (telephone number).

§ 9–614: Contents and Form of Notification Before Disposition of Collateral: Consumer-goods Transaction

[Omitted]

§ 9–615: Application of Proceeds of Disposition; Liability for Deficiency and Right to Surplus

(a)   **Application of proceeds.** A secured party shall apply or pay over for application the cash proceeds of disposition under Section 9–610 in the following order to:

    (1)   the reasonable expenses of retaking, holding, preparing for disposition, processing, and disposing, and, to the extent provided for by agreement and not prohibited by law, reasonable attorney's fees and legal expenses incurred by the secured party;

371

(1-a) in the case of a cooperative organization security interest, the holder thereof in the amount secured thereby;

(2) the satisfaction of obligations secured by the security interest or agricultural lien under which the disposition is made;

(3) the satisfaction of obligations secured by any subordinate security interest in or other subordinate lien on the collateral if:

(A) the secured party receives from the holder of the subordinate security interest or other lien an authenticated demand for proceeds before distribution of the proceeds is completed; and

(B) in a case in which a consignor has an interest in the collateral, the subordinate security interest or other lien is senior to the interest of the consignor; and

(4) a secured party that is a consignor of the collateral if the secured party receives from the consignor an authenticated demand for proceeds before distribution of the proceeds is completed.

(b) **Proof of subordinate interest**. If requested by a secured party, a holder of a subordinate security interest or other lien shall furnish reasonable proof of the interest or lien within a reasonable time. Unless the holder does so, the secured party need not comply with the holder's demand under subsection (a)(3).

(c) **Application of noncash proceeds**. A secured party need not apply or pay over for application noncash proceeds of disposition under Section 9–610 unless the failure to do so would be commercially unreasonable. A secured party that applies or pays over for application noncash proceeds shall do so in a commercially reasonable manner.

(d) **Surplus or deficiency if obligation secured**. If the security interest under which a disposition is made secures payment or performance of an obligation, after making the payments and applications required by subsection (a) and permitted by subsection (c):

(1) unless subsection (a)(4) requires the secured party to apply or pay over cash proceeds to a consignor, the secured party shall account to and pay a debtor for any surplus; and

(2) the obligor is liable for any deficiency.

(e) **No surplus or deficiency in sales of certain rights to payment**. If the underlying transaction is a sale of accounts, chattel paper, payment intangibles, or promissory notes:

(1) the debtor is not entitled to any surplus; and

(2) the obligor is not liable for any deficiency.

(f) **Calculation of surplus or deficiency in disposition to person related to secured party**. The surplus or deficiency following a disposition is calculated based on the amount of proceeds that would have been realized in a disposition complying with this part to a transferee other than the secured party, a person related to the secured party, or a secondary obligor if:

(1) the transferee in the disposition is the secured party, a person related to the secured party, or a secondary obligor; and

(2)   the amount of proceeds of the disposition is significantly below the range of proceeds that a complying disposition to a person other than the secured party, a person related to the secured party, or a secondary obligor would have brought.

(g) **Cash proceeds received by junior secured party**. A secured party that receives cash proceeds of a disposition in good faith and without knowledge that the receipt violates the rights of the holder of a security interest or other lien that is not subordinate to the security interest or agricultural lien under which the disposition is made:

(1)   takes the cash proceeds free of the security interest or other lien;

(2)   is not obligated to apply the proceeds of the disposition to the satisfaction of obligations secured by the security interest or other lien; and

(3)   is not obligated to account to or pay the holder of the security interest or other lien for any surplus.

§ 9–616: Explanation of Calculation of Surplus or Deficiency

(a) **Definitions**. In this section:

(1)   'Explanation' means a writing that:

    (A)   states the amount of the surplus or deficiency;

    (B)   provides an explanation in accordance with subsection (c) of how the secured party calculated the surplus or deficiency;

    (C)   states, if applicable, that future debits, credits, charges, including additional credit service charges or interest, rebates, and expenses may affect the amount of the surplus or deficiency; and

    (D)   provides a telephone number or mailing address from which additional information concerning the transaction is available.

(2)   'Request' means a record:

    (A)   authenticated by a debtor or consumer obligor;

    (B)   requesting that the recipient provide an explanation; and

    (C)   sent after disposition of the collateral under Section 9–610.

(b) **Explanation of calculation**. In a consumer-goods transaction in which the debtor is entitled to a surplus or a consumer obligor is liable for a deficiency under Section 9–615, the secured party shall:

(1)   send an explanation to the debtor or consumer obligor, as applicable, after the disposition and:

    (A)   before or when the secured party accounts to the debtor and pays any surplus or first makes written demand on the consumer obligor after the disposition for payment of the deficiency; and

    (B)   within fourteen days after receipt of a request; or

(2)   in the case of a consumer obligor who is liable for a deficiency, within fourteen days after receipt of a request, send to the consumer obligor a record waiving the secured party's right to a deficiency.

(c) **Required information**. To comply with subsection (a)(1)(B), a writing must provide the following information in the following order:

(1) the aggregate amount of obligations secured by the security interest under which the disposition was made, and, if the amount reflects a rebate of unearned interest or credit service charge, an indication of that fact, calculated as of a specified date:

(A) if the secured party takes or receives possession of the collateral after default, not more than thirty-five days before the secured party takes or receives possession; or

(B) if the secured party takes or receives possession of the collateral before default or does not take possession of the collateral, not more than thirty-five days before the disposition;

(2) the amount of proceeds of the disposition;

(3) the aggregate amount of the obligations after deducting the amount of proceeds;

(4) the amount, in the aggregate or by type, and types of expenses, including expenses of retaking, holding, preparing for disposition, processing, and disposing of the collateral, and attorney's fees secured by the collateral which are known to the secured party and relate to the current disposition;

(5) the amount, in the aggregate or by type, and types of credits, including rebates of interest or credit service charges, to which the obligor is known to be entitled and which are not reflected in the amount in paragraph (1); and

(6) the amount of the surplus or deficiency.

(d) **Substantial compliance.** A particular phrasing of the explanation is not required. An explanation complying substantially with the requirements of subsection (a) is sufficient, even if it includes minor errors that are not seriously misleading.

(e) **Charges for responses**. A debtor or consumer obligor is entitled without charge to one response to a request under this section during any six-month period in which the secured party did not send to the debtor or consumer obligor an explanation pursuant to subsection (b)(1). The secured party may require payment of a charge not exceeding 25 dollars for each additional response.

§ 9–617: Rights of Transferee of Collateral

(a) **Effects of disposition**. A secured party's disposition of collateral after default:

(1) transfers to a transferee for value all of the debtor's rights in the collateral;

(2) discharges the security interest under which the disposition is made; and

(3) discharges any subordinate security interest or other subordinate lien other than liens created under any law of this state that are not to be discharged.

(b) **Rights of good-faith transferee**. A transferee that acts in good faith takes free of the rights and interests described in subsection (a), even if the secured party fails to comply with this article or the requirements of any judicial proceeding.

(c) **Rights of other transferee**. If a transferee does not take free of the rights and interests described in subsection (a), the transferee takes the collateral subject to:

(1) the debtor's rights in the collateral;

(2) the security interest or agricultural lien under which the disposition is made; and

(3) any other security interest or other lien.

§ 9–618: Rights and Duties of Certain Secondary Obligors

(a) **Rights and duties of secondary obligor**. A secondary obligor acquires the rights and becomes obligated to perform the duties of the secured party after the secondary obligor:

(1) receives an assignment of a secured obligation from the secured party;

(2) receives a transfer of collateral from the secured party and agrees to accept the rights and assume the duties of the secured party; or

(3) is subrogated to the rights of a secured party with respect to collateral.

(b) **Effect of assignment, transfer, or subrogation**. An assignment, transfer, or subrogation described in subsection (a):

(1) is not a disposition of collateral under Section 9–610; and

(2) relieves the secured party of further duties under this article.

§ 9–619: Transfer of Record or Legal Title

(a) **'Transfer statement.'** In this section, 'transfer statement' means a record authenticated by a secured party stating:

(1) that the debtor has defaulted in connection with an obligation secured by specified collateral;

(2) that the secured party has exercised its post-default remedies with respect to the collateral;

(3) that, by reason of the exercise, a transferee has acquired the rights of the debtor in the collateral; and

(4) the name and mailing address of the secured party, debtor, and transferee.

(b) **Effect of transfer statement**. A transfer statement entitles the transferee to the transfer of record of all rights of the debtor in the collateral specified in the statement in any official filing, recording, registration, or certificate-of-title system covering the collateral. If a transfer statement is presented with the applicable fee and request form to the official or office responsible for maintaining the system, the official or office shall:

(1) accept the transfer statement;

(2) promptly amend its records to reflect the transfer; and

(3) if applicable, issue a new appropriate certificate of title in the name of the transferee.

(c) **Transfer not a disposition; no relief of secured party's duties**. A transfer of the record or legal title to collateral to a secured party under subsection (b) or otherwise is not of itself a disposition of collateral under this article and does not of itself relieve the secured party of its duties under this article.

§ 9–620: Acceptance of Collateral in Full or Partial Satisfaction of Obligation; Compulsory Disposition of Collateral

(a) **Conditions to acceptance in satisfaction**. Except as otherwise provided in subsection (g), a secured party may accept collateral in full or partial satisfaction of the obligation it secures only if:
(1) the debtor consents to the acceptance under subsection (c);
(2) the secured party does not receive, within the time set forth in subsection (d), a notification of objection to the proposal authenticated by:
(A) a person to which the secured party was required to send a proposal under Section 9–621; or
(B) any other person, other than the debtor, holding an interest in the collateral subordinate to the security interest that is the subject of the proposal;
(3) if the collateral is consumer goods, the collateral is not in the possession of the debtor when the debtor consents to the acceptance; and
(4) subsection (e) does not require the secured party to dispose of the collateral or the debtor waives the requirement pursuant to Section 9–624.

(b) **Purported acceptance ineffective**. A purported or apparent acceptance of collateral under this section is ineffective unless:
(1) the secured party consents to the acceptance in an authenticated record or sends a proposal to the debtor; and
(2) the conditions of subsection (a) are met.

(c) **Debtor's consent**. For purposes of this section:
(1) a debtor consents to an acceptance of collateral in partial satisfaction of the obligation it secures only if the debtor agrees to the terms of the acceptance in a record authenticated after default; and
(2) a debtor consents to an acceptance of collateral in full satisfaction of the obligation it secures only if the debtor agrees to the terms of the acceptance in a record authenticated after default or the secured party:
(A) sends to the debtor after default a proposal that is unconditional or subject only to a condition that collateral not in the possession of the secured party be preserved or maintained;
(B) in the proposal, proposes to accept collateral in full satisfaction of the obligation it secures; and

376

(C) does not receive a notification of objection authenticated by the debtor within twenty days after the proposal is sent.

(d) **Effectiveness of notification**. To be effective under subsection (a)(2), a notification of objection must be received by the secured party:

(1) in the case of a person to which the proposal was sent pursuant to Section 9–621, within 20 days after notification was sent to that person; and

(2) in other cases:

(A) within 20 days after the last notification was sent pursuant to Section 9–621; or

(B) if a notification was not sent, before the debtor consents to the acceptance under subsection (c).

[Paragraphs (e), (f) and (g) omitted as 'consumer' provisions]

§ 9–621: Notification of Proposal to Accept Collateral

(a) **Persons to which proposal to be sent**. A secured party that desires to accept collateral in full or partial satisfaction of the obligation it secures shall send its proposal to:

(1) any person from which the secured party has received, before the debtor consented to the acceptance, an authenticated notification of a claim of an interest in the collateral;

(2) any other secured party or lienholder that, 10 days before the debtor consented to the acceptance, held a security interest in or other lien on the collateral perfected by the filing of a financing statement that:

(A) identified the collateral;

(B) was indexed under the debtor's name as of that date; and

(C) was filed in the office or offices in which to file a financing statement against the debtor covering the collateral as of that date; and

(3) any other secured party that, 10 days before the debtor consented to the acceptance, held a security interest in the collateral perfected by compliance with a statute, regulation, or treaty described in Section 9–311(a).

(b) Proposal to be sent to secondary obligor in partial satisfaction. A secured party that desires to accept collateral in partial satisfaction of the obligation it secures shall send its proposal to any secondary obligor in addition to the persons described in subsection (a).

§ 9–622: Effect of Acceptance of Collateral

(a) **Effect of acceptance**. A secured party's acceptance of collateral in full or partial satisfaction of the obligation it secures:

(1) discharges the obligation to the extent consented to by the debtor;

(2) transfers to the secured party all of a debtor's rights in the collateral;

(3)   discharges the security interest or agricultural lien that is the subject of the debtor's consent and any subordinate security interest or other subordinate lien; and

(4)   terminates any other subordinate interest.

(b)   **Discharge of subordinate interest notwithstanding noncompliance.** A subordinate interest is discharged or terminated under subsection (a), even if the secured party fails to comply with this article.

§ 9–623: Right to Redeem Collateral

(a)   **Persons that may redeem.** A debtor, any secondary obligor, or any other secured party or lienholder may redeem collateral.

(b)   **Requirements for redemption.** To redeem collateral, a person shall tender:

(1)   fulfillment of all obligations secured by the collateral; and

(2)   the reasonable expenses and attorney's fees described in Section 9–615(a)(1).

(c)   **When redemption may occur.** A redemption may occur at any time before a secured party:

(1)   has collected collateral under Section 9–607;

(2)   has disposed of collateral or entered into a contract for its disposition under Section 9–610; or

(3)   has accepted collateral in full or partial satisfaction of the obligation it secures under Section 9–622.

§ 9–624: Waiver

(a)   **Waiver of disposition notification.** A debtor or secondary obligor may waive the right to notification of disposition of collateral under Section 9–611 only by an agreement to that effect entered into and authenticated after default.

(b)   **Waiver of mandatory disposition.** A debtor may waive the right to require disposition of collateral under Section 9–620 (e) only by an agreement to that effect entered into and authenticated after default.

(c)   **Waiver of redemption right.** Except in a consumer–goods transaction, a debtor or secondary obligor may waive the right to redeem collateral under Section 9–623 only by an agreement to that effect entered into and authenticated after default.

SELECTED STATUTES U.S. LESSOR LIABILITY STATUTE 49 USCS § 44112

§ 44112: Limitation of liability

(a) **Definitions**. In this section—

(1) 'lessor' means a person leasing for at least 30 days a civil aircraft, aircraft engine, or propeller.

(2) 'owner' means a person that owns a civil aircraft, aircraft engine, or propeller.

(3) 'secured party' means a person having a security interest in, or security title to, a civil aircraft, aircraft engine, or propeller under a conditional sales contract, equipment trust contract, chattel or corporate mortgage, or similar instrument.

(b) **Liability**. A lessor, owner, or secured party is liable for personal injury, death, or property loss or damage on land or water only when a civil aircraft, aircraft engine, or propeller is in the actual possession or control of the lessor, owner, or secured party, and the personal injury, death, or property loss or damage occurs because of—

(1) the aircraft, engine, or propeller; or

(2) the flight of, or an object falling from, the aircraft, engine, or propeller.

APPENDIX D

CREDIT RATINGS EQUIVALENCY GUIDE

Moody's	S&P	Fitch	Kroll	NAIC*
Aaa	AAA	AAA	AAA	1
Aa1	AA+	AA+	AA+	1
Aa2	AA	AA	AA	1
Aa3	AA-	AA-	AA-	1
A1	A+	A+	A+	1
A2	A	A	A	1
A3	A-	A-	A-1	1
Baa1	BBB+	BBB+	BBB+	2
Baa2	BBB	BBB	BBB	2
Baa3	BBB-	BBB-	BBB-	2
Ba1	BB+	BB+	BB+	3
Ba2	BB	BB	BB	3
Ba3	BB-	BB-	BB-	3
B1	B+	B+	B+	4
B2	B	B	B	4
B3	B-	B-	B-	4

* National Association of Insurance Commissioners

CONTRACTING SALES

CONVENTION ON INTERNATIONAL INTERESTS IN MOBILE EQUIPMENT

CONVENTION RELATIVE AUX GARANTIES INTERNATIONALES

PORTANT SUR DES MATÉRIELS D'ÉQUIPEMENT MOBILES

Adoption:	Place: Cape Town
	Date: 16.11.2001
Entry into force:	01.03.2006 (Art. 49(1))
Contracting States:	73
Regional economic integration organizations:	1
Depositary:	UNIDROIT

State	Signature		Ratification/ Accession	(Entry into force	Declaration or reservation (art.)
Afghanistan	-	AS	25.07.2006	01.11.2006	39(1)(a)-(b), 40, 52, 53, 54(2)
Albania	-	AS	30.10.2007	01.02.2008	39(1)(a)-(b),54(2)
Angola	-	AS	30.04.2006	01.08.2006	39(1)(a), 40, 54(2)
Australia	-	AS	26.05.2015	01.09.2015	39(1)(a), 53, 54(2), 55
Bahrain	-	AS	27.11.2012	01.03.2013	39(1)(a)-(b), 40, 54(2)
Bangladesh	-	AS	15.12.2008	01.04.2009	39(1)(a)-(b), 40, 52, 53, 54(2)
Belarus	-	AS	28.06.2011	01.01.2012	54(2)
Bhutan	-	AS	04.07.2014	01.11.2014	39(1)(a), 54(2)
Brazil	-	AS	30.11.2011	01.03.2012	39(1)(a)-(b), 39(4), 53, 54(2)
Burkina Faso	-	AS	12.12.2014	- [1]	-
Burundi	16.11.2001	-	-	-	-
Cameroon	-	AS	19.04.2011	01.08.2011	39(1)(a), 40, 52, 53, 54(2)
Canada	31.03.2004	RT	21.12.2012	...	39(1)(a)-(b), 39(4), 52*, 53*, 54(2), 60
Cape Verde	-	AS	26.09.2007	01.01.2008	39(1)(a), 40, 53, 54(2)
Chile	16.11.2001	-	-	-	-
China	16.11.2001	RT	03.02.2009	01.06.2009	39(1)(a)-(b), 39(4), 40, 50, 53, 54(1),54(2), 55
Colombia	-	AS	19.02.2007	01.06.2007	39(1)(a), 54(2)

State	Signature		Ratification/ Accession	(Entry into force	Declaration or reservation (art.)
Congo	16.11.2001	AC	25.01.2013	01.05.2013	39(1)(a)-(b), 40, 52, 53, 54(2)
Costa Rica	-	AS	26.08.2011	- [1]	53
Côte d'Ivoire	-	AS	09.02.2015	01.07.2016	54(2)*
Cuba	16.11.2001	RT	28.01.2009	01.05.2009	54(2)
Democratic Republic of the Congo	-	AS	06.05.2016	01.09.2016	39(1)(a)-(b), 40, 53, 54(2)
Denmark		AS	26.10.2015	01.02.2016	39(1)(a)-(b), 40, 52, 54(2), 55
Egypt	-	AS	10.12.2014	01.04.2015	39(1)(a), 54(2)
Ethiopia	16.11.2001	RT	21.11.2003	01.03.2006	39(1)(a), 40, 54(2)
Fiji	-	AS	05.09.2011	01.09.2012	54(2)*
France	16.11.2001	-	-	-	-
Gabon	-	AS	16.04.2010	01.08.2017	54(2)
Germany	17.09.2002	-	-	-	D
Ghana	16.11.2001	-	-	-	
India	-	AS	31.03.2008	01.07.2008	39(1)(a)-(b), 40, 52, 53, 54(2)
Indonesia	-	AS	16.03.2007	01.07.2007	39(1)(a)-(b), 40, 53, 54(2)
Ireland	-	AS	29.07.2005	01.03.2006	39(1)(a)-(b), 54(2)
Italy	06.12.2001	-	-	-	-
Jamaica	16.11.2001	-	-	-	-
Jordan	16.11.2001	RT	31.08.2010	01.12.2010	39(1)(a), 54(2)
Kazakhstan	-	AS	21.01.2009	01.10.2011	39(1)(a)*-(b)*, 39(4)*, 40*, 53*, 54(2)*
Kenya	16.11.2001	RT	13.10.2006	01.02.2007	39(1)(a)*-(b), 40, 53, 54(2)
Kuwait	-	AS	30.10.2013	01.02.2014	54(2)
Latvia	-	AS	08.02.2011	01.06.2011	54(2)
Lesotho	16.11.2001	-	-	-	-
Luxembourg	-	AS	27.06.2008	01.10.2008	53, 54(2)
Madagascar	-	SA	10.04.2013	01.08.2013	39(1)(a)-(b), 40, 53, 54(2)
Malawi	-	AC	16.01.2014	01.05.2014	39(1)(a)-(b), 40, 53, 54(2)
Malaysia	-	AS	02.11.2005	01.03.2006	39(1)(a)-(b), 40, 53, 54(2)
Malta	-	AS	01.10.2010	01.02.2011	39(1)(a), 39(4), 40, 53, 54(2)
Mexico	-	AS	31.07.2007	01.11.2007	39(1)(a)-(b), 50, 53, 54(2), 60

State	Signature		Ratification/ Accession	(Entry into force	Declaration or reservation (art.)
Moldova	-	AS	26.06.2015	- [1]	-
Mongolia	-	AS	19.10.2006	01.02.2007	39(1)(a)-(b), 53, 54(2)
Mozambique	-	AS	30.01.2012	01.11.2013	39(1)(a), 40, 54(2)
Myanmar	-	AS	03.12.2012	01.04.2013	39(1)(a), 40, 52, 53, 54(2)
Netherlands	-	AS	17.05.2010	. . .	39(1)(a)-(b), 52, 53, 54(2)
New Zealand	-	AS	20.07.2010	01.11.2010	39(1)(a), 52, 53, 54(2), 55
Nigeria	16.11.2001	RT	16.12.2003	01.03.2006	39(1)(a)*, 40*, 53*, 54(2)*
Norway	-	AS	20.12.2010	01.04.2011	39(1)(a)-(b), 40, 54(2), 55
Oman	-	AS	21.03.2005	01.03.2006	39(1)(a)-(b), 40, 52, 53, 54(2)
Pakistan	-	AS	22.01.2004	01.03.2006	39(1)(a)-(b), 39(4), 40, 52, 53, 54(2)
Panama	11.09.2002	RT	28.07.2003	01.03.2006	39(1)(a)-(b), 39(4), 50, 53, 54(2)
Russian Federation	-	AS	25.05.2011	01.09.2011	39(1)(a)-(b), 53, 54(2)
Rwanda	-	AS	28.01.2010	01.05.2010	39(1)(a), 40, 52, 53, 54(2)
San Marino	-	AS	09.09.2014	01.01.2015	39(1)(a)-(b), 39(4), 40, 53, 54(2)
Saudi Arabia	12.03.2003	RT	27.06.2008	01.10.2008	54(2)
Senegal	02.04.2002	RT	09.01.2006	01.05.2006	39(1)(a)-(b), 40, 52, 53, 54(2)
Seychelles	-	AS	13.09.2010	- [1]	-
Sierra Leone	-	AS	26.07.2016	01.11.2016	39(1)(a)-(b), 40, 53, 54(2)
Singapore	-	-	28.01.2009	01.05.2009	39(1)(a)-(b), 39(4), 53, 54(2)
South Africa	16.11.2001	RT	18.01.2007	01.05.2007	39(1)(a)-(b), 40, 54(2)
Spain	-	AS	28.06.2013	01.03.2016	39(1) (a)-(b)*, 40*, 52, 53*, 54(2)
Sudan	16.11.2001	-	-	-	-
Swaziland	-	AS	17.11.2016	- [1]	-
Sweden	-	AS	30.12.2015	01.04.2016	39(1)(a)-(b), 39(4), 40, 54(2)
Switzerland	16.11.2001	-	-	-	-
Syrian Arab. Rep.	-	AS	07.08.2007	- [1]	-
Tajikistan	-	AS	31.05.2011	01.09.2011	54(2)
Togo	-	AS	27.01.2010	01.04.2012	39(1)(a)*-(b)*, 40*, 53*, 54(2)*

State	Signature		Ratification/ Accession	(Entry into force	Declaration or reservation (art.)
Tonga	16.11.2001	-	-	-	-
Turkey	16.11.2001	RT	23.08.2011	01.12.2011	39(1)(a)-(b), 40, 50, 54(2)
Ukraine	09.03.2004	RT	31.07.2012	01.11.2012	50, 53, 54(2)
Un. Arab Emirates	-	AS	29.04.2008	01.08.2008	39(1)(a)-(b), 40, 52, 53, 54(2)
United Kingdom	16.11.2001	RT	27.07.2015	01.11.2015	39(1)(a)-(b), 39(4), 52, 53, 54(2)
Un. Rep. of Tanzania	16.11.2001		30.01.2009	01.05.2009	54(2)
United States of America	09.05.2003	RT	28.10.2004	01.03.2006	39(1)(a)-(b), 54(2)
Viet Nam	-	AS	17.09.2014	01.01.2015	39(1)(a)-(b), 40, 53, 54(2)
Zimbabwe	-	AS	13.05.2008	- [1]	-

[1]   Subject to: Convention Article 49(1)
[2]   This State has provided UNIDROIT with information about its laws and policies in relation to the Convention
[3]   The Kingdom of the Netherlands deposited its instrument of accession to the Convention on 20 July 2010 for the Netherlands Antilles (Curaçao, Sint Maarten, Bonaire, Sint Eustatius and Saba) and Aruba. As from 10 October 2010, following a modification of the internal constitutional relations within the Kingdom of the Netherlands, the reference to the 'Netherlands Antilles' is to be replaced by 'Curaçao, Sint Maarten and the Caribbean part of the Netherlands (the islands of Bonaire, Saba and Sint Eustatius)'. See www.unidroit.org/nationalinfo-2001capetown.
*   Affected by withdrawal and/or subsequent declaration.

INSOLVENCY DECLARATION (2012)

COUNTRY	Protocol Article XI Alternative A	Protocol Article XI Alternative B	Waiting Period Declared for Purposes of Protocol Article XI(3) (days unless otherwise stated)	Amended National Insolvency Law
Afghanistan	X		60	
Angola	X		60	
Bangladesh	X		60	
Belarus				
Brazil	X		30	
Cape Verde	X		60	
China	X		60	
Colombia	X		60	
Cuba				
Ethiopia	X		30 working days	
Fiji	X		60	
India	X		2 calendar months	
Indonesia	X		60	
Ireland				No
Jordan	X		60	
Kazakhstan	X		60	
Kenya	X		60	
Luxembourg	X		60	
Malaysia	X		40 working days	
Malta				Yes
Mexico		X	the time period expressed by the parties	
Mongolia	X		60 working days	
Kingdom of The Netherlands				
New Zealand	X		60	
Nigeria	X		30	
Norway	X		60	
Oman	X		60	
Pakistan	X		60	
Panama	X		60	
Russia	X		60	
Rwanda	X		60	
Saudi Arabia				
Senegal	X		30	
Singapore	X		30	
South Africa	X		30	

COUNTRY	Protocol Article XI Alternative A	Protocol Article XI Alternative B	Waiting Period Declared for Purposes of Protocol Article XI(3) (days unless otherwise stated)	Amended National Insolvency Law
Tajikistan	X		60	
Togo	X		30	
Turkey	X		60	
UAE	X		60	
Ukraine	X		60	
United Republic of Tanzania	X		30	
United States of America				No (Section 1110 Applicable; 60 days)

The declarations made by the European Union (EU) under the Convention and the Protocol prevent Contracting States who are EU Member States from making the insolvency declaration but do not prevent an EU Member State from amending its national insolvency laws to have the same effect as if the declaration had been made.

This Chart was modified from the Chart in Part II of The Practitioner's Guide to the Cape Town Convention and the Aircraft Protocol referred to in the introduction to Part VII of the Handbook.

APPENDIX F

COMMERCIAL AIRCRAFT

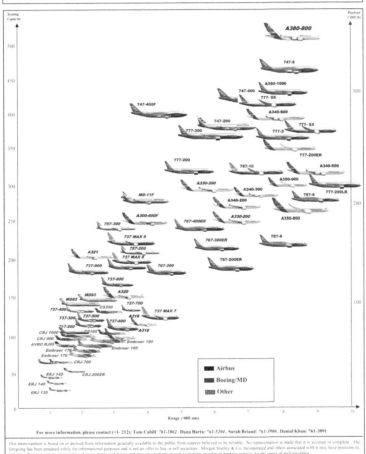

Airbus Aircraft

Airbus A310
Airbus A320 Family
Airbus A330
Airbus A340
Airbus A350
Airbus A380

Boeing Aircraft

Boeing 717
Boeing 737
Boeing 747
Boeing 757
Boeing 767
Boeing 777
Boeing 787

Russian Aircraft

Ilyushin Il-96
Tupolev Tu-204 Family

Regional Aircraft

Antonov An-140
ATR
BAe Avro RJ
Bombardier Dash 8
Bombardier CRJ
Embraer EMB-120
Embraer ERJ
Embraer-170/190
Fokker 50
Fokker 100
Saab
Xian MA60

CERTIFICATE OF REPOSSESSION

U.S. DEPARTMENT OF TRANSPORTATION FEDERAL AVIATION ADMINISTRATION

Aircraft Registration Branch
PO Box 25504
Oklahoma City, Oklahoma 73125–0504

CERTIFICATE OF REPOSSESSION OF ENCUMBERED AIRCRAFT AND TERMINATION OF LEASE

The undersigned hereby certifies that it is the Loan Trustee under the Security Agreement (as defined below), acting on behalf of the true and lawful holder of the loan certificate that evidences indebtedness secured by that certain Loan and Security Agreement (_____) (the 'Security Agreement'), on the following aircraft:

Airframe Manufacturer and Model:
Airframe serial number:
FAA registration number:
Engine Manufacturer and Model:
Engine Serial Numbers:

(the airframe and the _____ (___) engines described above are hereinafter collectively referred to as the 'Aircraft'). Each of the above described engines has _____ pounds of thrust or more (or the equivalent thereof). Said Security Agreement on the above Aircraft bears the date of _____ __, ____ and was executed by between _____, as mortgagor, and _____, as mortgagee. This Security Agreement was recorded under Title 49 U.S.C. § 44107 *et seq.*, as more particularly described on Exhibit A attached hereto.

The Aircraft was leased pursuant to that certain Lease Agreement _____, dated as of _____ __, ____ between the Owner trustee, as lessor, and _____, as lessee (the 'Lessee'), as supplemented and as more particularly described in Exhibit A attached hereto (collectively, the 'Lease').

Pursuant to the terms and conditions of the Security Agreement, the Lease was collaterally assigned by the Owner Trustee to the Loan Trustee.

The Lessee filed a voluntary petition in bankruptcy in _____ and rejected the Lease in _____ in its chapter 11 case.

From at least _____, and continuing thereafter, the aforesaid Owner Trustee breached the obligations and promises contained in the Security Agreement and the Loan Certificates (as defined in the Security Agreement) secured thereby, which breaches constitute events of default under the Security Agreement.

From at least _____ and continuing thereafter, the Lessee breached its obligations and promises contained in the Lease, which breaches constitute events of default under the Lease and the Security Agreement.

The undersigned certifies that it has performed all obligations imposed on it by the Security Agreement and the Lease, as collaterally assigned by the Security Agreement, and applicable local laws; that in accordance with the granting clauses and Section (Mortgage Events of Default; Remedies) of the Security Agreement, and in accordance with the terms of the Lease, as collaterally assigned by the Security Agreement, and pursuant to pertinent laws of the state of New York, the undersigned repossessed the Aircraft described above and foreclosed on the day of _____; and that pursuant to local law and the remedial terms of the Security Agreement, divested the Owner Trustee, the Lessee, and any and all persons claiming by, through or under them, of any and all right, title, claim or interest they had or may have had in and/or to the Aircraft and in, to and/or under the Lease, and the Loan Trustee or its nominee now owns the aforesaid Aircraft free and clear of all rights and claims of any person or entity whatsoever.

As to the Aircraft and the Aircraft alone, the Lease is hereby terminated and any encumbrances on the Aircraft that exists by virtue of the Lease is hereby released.

Neither the execution of this Certificate nor the termination of the Lease with respect to and only with respect to the Aircraft, shall (i) affect or limit, in any way whatever, the right or remedy of the Loan Trustee or any beneficiary thereof, (ii) be construed to be a waiver of any right or remedy of the Loan Trustee or any beneficiary thereof, under the Lease or applicable law that the Loan Trustee or any beneficiary thereof may have against any other person or entity, all of which rights and remedies are hereby expressly reserved, or (iii) release the Lessee from its obligations pursuant to the Lease or any other terms and conditions applicable to Lessee under the Lease. This Certificate shall not be, or be deemed to be, an acknowledgment or admission by the Loan Trustee of the release of the Lessee from its obligations and liabilities under the Lease. This Certificate is intended only to vest title in the name of the Loan Trustee or its nominee and to remove the lien of the Lease from the Aircraft Registry so that the related Aircraft may be disposed of, transferred or conveyed by the Loan Trustee, as the foreclosing secured party, or its nominee free from any encumbrance of the Lease.

_____

By:
Title:

EXHIBIT A
TO
CERTIFICATE OF REPOSSESSION DESCRIPTION OF
SECURITY AGREEMENT

Loan and Security Agreement _____ dated as of _____ between
_____, as owner trustee under Trust Agreement
_____ dated as of_____, and _____, as loan trustee,
which was recorded by the Federal Aviation Administration on _____
and assigned Conveyance No. _____, as assigned by the following described
instrument:

Instrument	Date of Instrument	FAA Recording Date	FAA Conveyance No.
_____	_____	_____	_____

DESCRIPTION OF LEASE

Lease Agreement _____ dated as of _____ between
_____, as owner trustee under Trust Agreement
_____ dated as of _____, as lessor, and _____, as
lessee, which was recorded by the Federal Aviation Administration on
_____ and assigned Conveyance No. _____, as assigned by the
following described instrument:

Instrument	Date of Instrument	FAA Recording Date	FAA Conveyance No.
_____	_____	_____	_____

APPENDIX H

LEASE MANAGEMENT SERVICES

LIST OF SERVICES

1 Lease Management Services.

   1.1 Invoicing, Collections and Disbursements. Monitor all payments due from the Lessee under the Lease, including, but not limited to, the following services:

     (a) calculate the Lease Rentals, security deposits, maintenance reserves, late payments and any payments in relation to taxes and other payments (including technical, engineering, insurance and other charges);

     (b) invoice the Lessee (and provide all supporting calculations and information as are required under the relevant Lease) or otherwise arrange for all payments due from the Lessee, including rental payments and if applicable, security deposits, maintenance reserves, late payment charges and any payment in relation to taxes and other payments (including technical, engineering, insurance and other charges) to be paid to such account as specified in writing by the Lessor;

     (c) correspond with the Lessee to ensure such payments are made by the Lessee in a timely manner; and

     (d) generally monitor the payment record of the Lessee and provide details thereof in periodic reports.

   1.2 Administration and Monitoring. Administer the Lease and monitor the Lessee's obligations under the Lease including, but not limited to, the following services as applicable:

     (a) make or cause to be made all calculations and determinations as and when required under the Lease Operative Documents and to provide to each of the other parties to the Lease Operative Documents details of any revised, amended or recalculated schedules, figures, sums, amounts or dates;

     (b) undertake any other administrative or management tasks on behalf of the Lessor as the Servicer may agree with the Lessor;

     (c) prepare and keep or cause to be prepared and kept all accounting records of the Lessor;

     (d) administer the Lease according to its terms (including compliance with any conditions subsequent);

     (e) periodically review registration, filing, recording and similar requirements relating to the Lease;

     (f) monitor and review reports received pursuant to the Lease;

(g)   establish a relationship with the Lessee's key points of contact (e.g., management, financial, commercial and technical);

(h)   facilitate the safekeeping by the Lessor of any letters of credit, cash deposits, guarantees, or other credit support held as part of deposits or maintenance reserves and the timely renewal or drawing on or disbursement thereof provided under the Lease and use reasonable efforts to ensure that they are kept valid and current throughout the term of the Lease;

(i)   review from time to time the level of maintenance reserves, security deposits and other amounts that may be adjusted under the Lease;

(j)   if the Servicer becomes aware of any non-compliance by the Lessee under the Lease, it will advise the Lessor of such non-compliance and propose appropriate action to protect the Lessor's interests, including making follow-up telephone calls, sending reminder letters and preparing default notices;

(k)   promptly notify the Lessor if any other event has occurred which Servicer reasonably believes will adversely affect the Lessee's ability to perform any of its material obligations under the Lease;

(l)   review and advise on any request by the Lessee to sub-lease/wet lease the Aircraft and monitor the compliance of the various requirements contained in the Lease concerning any such proposed sublease/wet lease;

(m)   administer and remind the Lessor in advance about any options (such as extension options, early termination options or purchase options) that the Lessee or the Lessor might have under the Lease, and notify the Lessor in relation to any exercise thereof by the Lessee;

(n)   review and advise on any request made by the Lessee in relation to any extension request made by it, including any variations to the lease rentals, maintenance reserves and return conditions;

(o)   at the Lessor's request, use reasonable efforts to make enquiries in the Lessee's domestic market and elsewhere to try to establish, where reasonably possible, the extent of the Lessee's indebtedness to maintenance facilities, trade creditors and any amounts owing on the Aircraft to Eurocontrol/airport authorities; and

(p)   provide all necessary administrative support to complete any documentation and other related matters.

1.3   Maintenance and Technical Services. Administer and monitor the Lessee's performance of its maintenance obligations under the Lease including, but not limited to, the following services as applicable:

(a)   determine the suitability of, and the air authority approval status of, the Lessee's proposed maintenance program and of the proposed maintenance performer(s) for the Aircraft;

(b)   monitor the Lessee's utilization of the Aircraft and its performance of its maintenance obligations under the Lease;

(c)   monitor records keeping, certification of components, life limited parts status, airworthiness directives status, and maintenance status;

(d)   provide monthly utilization reports;

(e)   review workscopes before heavy maintenance/reconfiguration;

(f)   review any request from the Lessee for a contribution from the maintenance reserves or any contribution to airworthiness directives costs and determine the amount (if any) that the Lessor is obliged to contribute pursuant to the provisions of the Lease. Advise the Lessor on any such request and provide the Lessor with evidence (documentary or otherwise) provided by the Lessee in support of such request;

(g)   monitor the heavy maintenance/reconfiguration work, review the relevant invoices;

(h)   monitor the Lessee's draw down of any credit notes for product support relating to the Lease;

(i)   subject to the inspection provisions in the Lease, periodically provide, or arrange for and supervise, a suitably qualified third party to perform an inspection of the Aircraft and an audit of the Aircraft records (which inspection or audit shall be coordinated to the extent possible with a 'C' check or higher level inspection of the Aircraft or at such other time determined by the Servicer) in order to establish:

    (i)   the condition of the Aircraft;

    (ii)   the Lessee's compliance with its maintenance and other relevant Lease obligations;

    (iii)   the accuracy, completeness and accessibility (if there is a default) of the Aircraft records;

    (iv)   verify reported usage of airframe, engines, engine LLPs and major components against the Aircraft records;

    (v)   any remedial action necessary by the Lessee to rectify any problems that are identified;

    (vi)   the Lessor's exposure to potential maintenance costs, and the Servicer will prepare a projection of future exposure based upon the Lessee's maintenance programme and its current monthly utilization of the aircraft;

    (vii)   scan the Aircraft records on to CD-ROM (subject to the Lessee's compliance); and

    (viii)   maintain a record of all material reports received or generated by the Servicer in connection with such inspections.

1.4   <u>Lease Expiry or Termination</u>. Provide any assistance required to monitor Lessee's compliance with the redelivery conditions under the Lease, including, but not limited to, the following services:

(a)   arrange for the appropriate technical inspection or complete the technical inspection of the Aircraft and records to determine whether the Lessee has complied with all required airworthiness directives and mandatory modifications and establish the status of compliance with airframe and engine manufacturer service bulletins;

(b)   review major accident/incident repairs, structural inspections and engine overhauls;

(c)   arrange for the appropriate technical inspection or complete the technical inspection of the Aircraft and records (including supervision of ground tests, engine runs and conducting a test flight) to determine whether the Lessee has satisfied the redelivery conditions specified in the Lease, and make appropriate recommendations as to the correction of any deficiencies in the records;

(d)   negotiate any modifications, repairs, refurbishment, inspections or overhauls to such conditions deemed reasonably necessary or appropriate;

(e)   make recommendations to the Lessor as to whether the redelivery of the Aircraft should be accepted;

(f)   determine the application of any available deposits, maintenance reserves or other payments under the Lease and maintain a record of the satisfaction of such conditions;

(g)   monitor the performance of any maintenance and refurbishment of the Aircraft upon redelivery, including any modifications deemed reasonably necessary or appropriate for the remarketing of the Aircraft;

(h)   use reasonable endeavors to retrieve Aircraft records to the extent available so as to facilitate issuance of certification of airworthiness for export and later use by the next operator; the aircraft; and

(i)   provide assistance to the Lessor with preparation of de-registration and re-registration of

(j)   accept redelivery of the Aircraft and negotiate and record the return acceptance certificate and related materials.

1.5   Insurance. The Lessee will be responsible for the insurances applicable to the Aircraft and the management thereof. However, the Lessor may also request the advice and assistance of the Servicer in relation to the insurances applicable to the Aircraft including, but not limited, to the following services:

(a)   negotiate the insurance provisions of any proposed Lease or other agreement concerning the Aircraft including the structure and substance of the proposed insurance coverage, with such provisions to include such minimum coverage amounts with respect to hull, property and liability insurance as are consistent with market conditions;

(b)   procure that appropriate evidence of insurance exists with respect to the Aircraft in accordance with the terms of the Lease;

(c)   monitor the performance of the obligations of the Lessee relating to insurance under the Lease of the Aircraft;

(d)   settle any claim of damage or loss with respect to the Aircraft;

(e)   advise the Lessor regarding any insurance claim for damage or loss to the Aircraft in excess of the damage notification threshold (if any) in the Lease, including representing the Lessor with respect to the Lessee or other counterparty and its insurance brokers, agent or adjusters and advising the Lessor whether any payment or proposed payment by an insurer or other person with respect to such damage or loss is sufficient and is or shall be properly applied and documented;

(f)   if it considers it necessary or appropriate, engage, on behalf and for the benefit of the Lessor, at the Lessor's expense, an insurance broker (the 'Broker') who shall take directions from the Servicer in respect of the Aircraft and the Servicer shall be entitled to rely reasonably on the actions taken by or recommendations of any such Broker. The Servicer will obtain such advice from the Broker, as it deems appropriate, as to the reasonableness of any insurance arrangements proposed by the Lessee for the Aircraft;

(g)   if at any time the Servicer becomes aware that the Aircraft ceases to be insured for any reason, including as a result of a default by the Lessee or the Aircraft not being leased upon termination of the Lease, the Servicer shall arrange through the Broker at Lessor's expense alternative insurance coverage with such minimum coverage amounts as required by the Lessor; and

(h)   provide such periodic reports regarding insurance matters relating to the Aircraft as the Lessor may reasonably request.

1.6   Information and Custody. Perform and provide the following services:

(a)   keep the Lessor apprised of any significant developments related to the Aircraft, the Lessee or the airline industry (to the extent relevant in the context of this Agreement and the provision of the Services), in each case of which it is aware;

(b)   provide to the Lessor such factual information and data about the Aircraft which the Lessor may reasonably request;

(c)   hold all technical records that relate to the Aircraft that are in the Servicer's possession (and which are not required to be held by the Lessee) in safe custody in accordance with the Standard of Care (it being acknowledged that the Lessor shall at all times hold all original documents relating to the Aircraft); and

(d)   if requested by the Lessor, make and provide to the Lessor such copies of the technical records relating to the Aircraft as the Lessor may from time to time reasonably request, at the Lessor's expense.

2   Remarketing Services. Any marketing services provided by the Servicer under this paragraph 2 must ensure that the Lease remains compliant in all respects with the relevant conditions in the loan agreement pursuant to which the Aircraft being marketed is financed.

2.1   Lease Marketing.

(a)   The Servicer will provide and perform lease marketing services with respect to the Aircraft, including but not limited to:

(i)   negotiating and entering into any commitment for a lease of the Aircraft on behalf of and (pursuant to a power of attorney) in the name of the Lessor; and

(ii)   including within any commitment for a lease of the Aircraft any intermediate lease or leases through any person that the Servicer deems reasonably necessary or appropriate.

(b) The Servicer will commence the negotiation of any lease or leases of Aircraft in a manner consistent with the practices employed by the Servicer with respect to its aircraft operating leasing services business generally and will commence the drafting of, and negotiation with respect to, any leases for Aircraft on the following basis:

    (i) the Servicer will utilize a lease document that includes any requirements set out in the Finance Documents for a future lease, as a starting point in the negotiation of future leases;

    (ii) the Servicer shall, to the extent compatible with market conditions, seek to achieve a rental profile:

        (A) sufficient to pay debt service under the Loan Agreement; and

        (B) consistent with market practice, unless there are mitigating factors such as a long lease term or the proposed lessee is a market leading carrier, provided that, the Servicer is hereby authorised to enter into (in its capacity as attorney of the Lessor under the power of attorney provided pursuant to Section 19 of the Agreement) any extension of a Lease or any amendment or variation of a Lease (unless the Servicer considers that the proposed variation or amendment is material) without obtaining the prior consent of the Lessor or any other person.

(c) The Servicer is not authorised to execute and deliver any binding new lease of the Aircraft on behalf of the Lessor unless the prior written consent of any third party required under the Finance Documents has been obtained.

(d) The Servicer must take such reasonable commercial actions as shall be reasonably necessary or appropriate to deliver the Aircraft pursuant to the terms of the documentation of the Lease or Leases of the Aircraft, including upon an extension of such Leases. The Servicer will attend to all delivery requirements including arranging for:

    (i) the Aircraft to be painted in new livery:

    (ii) the negotiation of the Aircraft documentation package;

    (iii) ferry flights; and

    (iv) all other reasonable steps.

(e) The Servicer will generally provide the lease marketing services set forth in this Section 2 of Schedule 1 through the use of its own marketing staff where it shall deem appropriate and shall utilize third parties to provide such marketing services where it shall deem appropriate, but in any case, the Costs of any third party will be for the sole account of the Servicer.

(f) In performing any marketing activities, the Servicer will always seek to maximize lease revenues and aircraft residual value.

2.2 Sales and Remarketing.

(a) The Servicer will provide and perform sales and remarketing services including, but not limited to:

(i)   arranging the provision of qualified flight crew for ferry flight(s), if required;

(ii)   advising on capital improvements or possible scrapping or other disposition, as may be desirable;

(iii) arranging and supervising the appropriate storage and any required ongoing maintenance of the Aircraft following termination of the Lease and redelivery of the Aircraft thereunder and prior to delivery of the Aircraft to a new lessee or purchaser;

(iv) for Aircraft returning from storage, arranging and supervising a re-activation check, as necessary;

(v)   in respect of a parked Aircraft (where it is intended to re-activate it in due course) arranging appropriate technical review and preservation procedures;

(vi) delivering the Aircraft to a new lessee or purchaser;

(vii) monitoring the market for potential customers and actively and systematically marketing the aircraft to any airline or other relevant person which may be interested in the Aircraft;

(viii) placing advertising in publications as the Servicer considers appropriate;

(ix) contacting potential customers to assess their interest and suitability;

(x)   advising in respect of modification and enhancements to the Aircraft which are likely to improve the marketability and/or returns on the Aircraft having regard to the cost and potential benefits of such modifications and enhancements;

(xi) arranging for inspection of the Aircraft by potential customers;

(xii) assisting in the preparation of LOIs between the Lessor and potential customers, including negotiating and advising on commercial terms;

(xiii) endeavoring to obtain such information as may be required to assist the Lessor determining whether it should approve a potential transaction;

(xiv) upon request of the Lessor, assisting or leading in negotiations on behalf of the Lessor with potential customers on the terms of a purchase;

(xv) coordinating with the Lessor and assisting in the management of the closing process; and

(xvi) providing all necessary administrative support to complete any documentation and related matters.

(b) The Servicer shall provide a quarterly report (delivered not later than ten days after the end of each quarter) during any Remarketing Period, summarizing the marketing strategy to be undertaken on a twelve month rolling basis, to include details of marketing steps to be taken (including commencement of sales campaigns, RFPs to be issued, first round bids to be received, etc.). The report should also highlight any potential conflicts of interest in remarketing the Aircraft.

(c)   Within 30 days of the Servicer's appointment to provide Remarketing Services with respect to the Aircraft, the Servicer will:

    (i)   provide an introductory report on the state of the market for the applicable type of Aircraft (e.g., supply and demand, numbers in storage, generic lease rates and sales prices);

    (ii)   note any regulatory issues that the Servicer considers might impact on marketability;

    (iii)   obtain a desktop valuation from a recognized appraiser for the Aircraft;

    (iv)   provide a commentary on such valuation highlighting any potential discrepancy between the valuation and the price achievable in the market (e.g., market conditions, position in maintenance cycle, configuration, existing Lease terms); and

    (v)   outline the Servicer's proposed strategy for remarketing the Aircraft during the Remarketing Period.

(d)   Once remarketing has commenced, the Servicer shall provide to the Lessor a periodic progress report not less frequently than monthly (to include identification of potential purchasers contacted and advise as to their suitability/seriousness, indicative pricing proposals and other relevant factors).

(e)   At the conclusion of a sale campaign, the Servicer will provide to the Lessor a report summarizing the short-listed bids and making a recommendation based on the terms and conditions of the offers and the identity of the prospective counterparties (advising on risk/reward comparisons between technical/financial factors in the offers, e.g., the likely impact of a technical spend to put the Aircraft in the condition required by a potential purchaser).

(f)   The Servicer shall provide any available information regarding a potential purchaser's source of funds, the location(s) in which the Aircraft will be operated and the identity of the operator of the Aircraft (if other than the purchaser).

3   Default Management Services. Provide any assistance and advice required in relation to any actual or prospective default by the Lessee under the Lease, any related Aircraft repossession and/or any related Lease restructuring, including but not limited to:

(a)   upon becoming aware of any such default, the Servicer will promptly notify the Lessor thereof;

(b)   provide suitable support reasonably expected from a lease manager in the event of any such default, repossession and/or restructuring, which may include any or all of the Lease Management Services;

(c)   prepare a confidential contingency plan for the repossession of the Aircraft;

(d)   advise on best practice in operating lease transaction structuring, including aircraft operation, registration and other issues;

(e) endeavour to obtain such additional information as may be required in the particular circumstances to enable the Lessor to determine whether to approve a particular restructuring; and

(f) provide generic sales and lease marketing advice in relation to the Aircraft, it being agreed that no request for such advice or the giving of such advice shall constitute a request for Remarketing Services or the provision of Remarketing Services, as the case may be, for the purposes of Section 4(b) or otherwise unless explicitly set out in an Activation Notice.

FORM OF CAPE TOWN IDERA

[Insert Date]

To: [Insert Name of Registry Authority]

Re: Irrevocable De-Registration and Export Request Authorisation

> *1 The undersigned is the registered [operator] [owner]* of the [insert the airframe/ helicopter manufacturer name and model number] bearing manufacturers serial number [insert manufacturer's serial number] and registration [number] [mark] [insert registration number/mark] (together with all installed, incorporated or attached accessories, parts and equipment, the 'aircraft').

This instrument is an irrevocable de-registration and export request authorisation issued by the undersigned in favour of [insert name of creditor] ('the authorised party') under the authority of Article XIII of the Protocol to the Convention on International Interests in Mobile Equipment on Matters specific to Aircraft Equipment. In accordance with that Article, the undersigned hereby requests:

(i)   recognition that the authorised party or the person it certifies as its designee is the sole person entitled to:

    (a)   procure the de-registration of the aircraft from the [insert name of aircraft register] maintained by the [insert name of registry authority] for the purposes of Chapter III of the Convention on International Civil Aviation, signed at Chicago, on 7 December 1944, and

    (b)   procure the export and physical transfer of the aircraft from [insert name of country]; and

(ii) confirmation that the authorised party or the person it certifies as its designee may take the action specified in clause (i) above on written demand without the consent of the undersigned and that, upon such demand, the authorities in [insert name of country] shall co-operate with the authorised party with a view to the speedy completion of such action.

The rights in favour of the authorised party established by this instrument may not be revoked by the undersigned without the written consent of the authorised party.

Please acknowledge your agreement to this request and its terms by appropriate notation in the space provided below and lodging this instrument in [insert name of registry authority].

[insert name of operator/owner]

_____

_____

Agreed to and lodged this By: [insert name of signatory] [insert date] Its: [insert title of signatory]

_____

[insert relevant notational details]

_____

* Select the term that reflects the relevant nationality registration criterion.

Note

1   Select the term that reflects the relevant nationality registration criterion.

EVENT OF LOSS/TOTAL LOSS

The Event of Loss/Total Loss of any Aircraft Asset would occur upon:

(a) the destruction of or damage to such property which renders repair of such property uneconomic or which renders such property permanently unfit for normal use;

(b) any damage to such property which results in an insurance settlement with respect to such property on the basis of a total loss or a constructive or compromised total loss of such property;

(c) the theft or disappearance of such property, or the confiscation, condemnation or seizure of, or requisition or other taking of, title to, or use of, such property by any governmental or purported governmental authority (other than a requisition for use by the U.S. Government or any agency or instrumentality thereof), which, in the case of any event referred to in this paragraph (c) other than a requisition or other taking of title to such property shall have continued for a period in excess of consecutive days or shall be continuing upon the expiration of the term, if earlier;

(d) as a result of any law, or an order or other action by an Aviation Authority having jurisdiction over the Aircraft, the use of such property in the normal course of the business of air transportation shall have been prohibited for a period of consecutive days (a 'grounding'), unless the operator, prior to the expiration of such -day period, shall have undertaken and shall diligently be carrying forward steps which are necessary or desirable to permit the normal use of such property by Lessee, but in any event if such use shall have been prohibited for a period of years, [provided that no Event of Loss shall be deemed to have occurred if such prohibition is applicable to all aircraft of the same model and series as the Aircraft and the operator, prior to the expiration of such _____-year period, shall have conformed at least one aircraft of such series in its fleet to the requirements of any such law, order or other action and commenced regular commercial use of the same in such jurisdiction and shall be diligently carrying forward, in a manner which does not discriminate against such property in so conforming such property, steps which are necessary or desirable to permit the normal use of such property by Lessee, but in any event if such use shall have been prohibited as to such property for a period of years];

(e) notwithstanding the foregoing, as a result of any Law, or an order or other action by an Aviation Authority having jurisdiction over the Aircraft, the use of such property in the normal course of the business of air transportation shall have been prohibited at the expiration of the term; and

(f) with respect to any Engine only, any divestiture of title to such Engine.

APPENDIX K

FAA BILL OF SALE

<table>
<tr>
<td colspan="2">UNITED STATES OF AMERICA
U.S. DEPARTMENT OF TRANSPORTATION FEDERAL AVIATION ADMINISTRATION</td>
<td>OMB Control No. 2120-0042
Exp. 04/30/2017</td>
</tr>
<tr>
<td colspan="2" align="center">**AIRCRAFT BILL OF SALE**</td>
<td></td>
</tr>
<tr>
<td colspan="2">FOR AND IN CONSIDERATION OF $ THE
UNDERSIGNED OWNER(S) OF THE FULL   LEGAL
AND BENEFICIAL TITLE OF THE AIRCRAFT
DESCRIBED AS FOLLOWS:</td>
<td></td>
</tr>
<tr>
<td colspan="2">UNITED STATES
REGISTRATION NUMBER   **N**</td>
<td></td>
</tr>
<tr>
<td colspan="2">AIRCRAFT MANUFACTURER & MODEL</td>
<td></td>
</tr>
<tr>
<td colspan="2">AIRCRAFT SERIAL No.</td>
<td></td>
</tr>
<tr>
<td colspan="2">DOES THIS   DAY OF
HEREBY SELL, GRANT, TRANSFER AND
DELIVER ALL RIGHTS, TITLE, AND INTERESTS
IN AND TO SUCH AIRCRAFT UNTO:</td>
<td>Do Not Write In This Block
FOR FAA USE ONLY</td>
</tr>
<tr>
<td rowspan="2">**PURCHASER**</td>
<td>NAME AND ADDRESS
(IF INDIVIDUAL(S), GIVE LAST NAME, FIRST NAME, AND MIDDLE INITIAL.)</td>
<td></td>
</tr>
<tr>
<td>DEALER CERTIFICATE NUMBER</td>
<td></td>
</tr>
<tr>
<td colspan="3">AND TO   EXECUTORS, ADMINISTRATORS, AND ASSIGNS TO HAVE AND TO HOLD
SINGULARLY THE SAID AIRCRAFT FOREVER, AND WARRANTS THE TITLE THEREOF:</td>
</tr>
</table>

IN TESTIMONY WHEREOF   HAVE SET   HAND AND SEAL THIS   DAY OF

	NAME(S) OF SELLER (TYPED OR PRINTED)	SIGNATURE(S) (IN INK) (IF EXECUTED FOR CO-OWNERSHIP, ALL MUSTSIGN.	TITLE (TYPED OR PRINTED)
SELLER			

ACKNOWLEDGMENT (NOT REQUIRED FOR PURPOSES OF FAA RECORDING. HOWEVER, MAY BE REQUIRED BY LOCAL LAW FOR VALIDITY OF THE INSTRUMENT.)

ORIGINAL: TO FAA:

AC Form 8050-2 (01/12) (NSN 0052-00-629-0003)

ACCREDITED INVESTOR – 17 CFR § 230.501

§ 230.501: Definitions and terms used in Regulation D.

As used in Regulation D (§ 230.500 et seq. of this chapter), the following terms shall have the meaning indicated:

(a) **Accredited investor.** Accredited investor shall mean any person who comes within any of the following categories, or who the issuer reasonably believes comes within any of the following categories, at the time of the sale of the securities to that person:

   (1) Any bank as defined in section 3(a)(2) of the Act, or any savings and loan association or other institution as defined in section 3(a)(5)(A) of the Act whether acting in its individual or fiduciary capacity; any broker or dealer registered pursuant to section 15 of the Securities Exchange Act of 1934; any insurance company as defined in section 2(a)(13) of the Act; any investment company registered under the Investment Company Act of 1940 or a business development company as defined in section 2(a)(48) of that Act; any Small Business Investment Company licensed by the U.S. Small Business Administration under section 301(c) or (d) of the Small Business Investment Act of 1958; any plan established and maintained by a state, its political subdivisions, or any agency or instrumentality of a state or its political subdivisions, for the benefit of its employees, if such plan has total assets in excess of $5,000,000; any employee benefit plan within the meaning of the Employee Retirement Income Security Act of 1974 if the investment decision is made by a plan fiduciary, as defined in section 3(21) of such act, which is either a bank, savings and loan association, insurance company, or registered investment adviser, or if the employee benefit plan has total assets in excess of $5,000,000 or, if a self-directed plan, with investment decisions made solely by persons that are accredited investors;

   (2) Any private business development company as defined in section 202(a)(22) of the Investment Advisers Act of 1940;

   (3) Any organization described in section 501(c)(3) of the Internal Revenue Code, corporation, Massachusetts or similar business trust, or partnership, not formed for the specific purpose of acquiring the securities offered, with total assets in excess of $5,000,000;

   (4) Any director, executive officer, or general partner of the issuer of the securities being offered or sold, or any director, executive officer, or general partner of a general partner of that issuer;

(5) Any natural person whose individual net worth, or joint net worth with that person's spouse, exceeds $1,000,000.

    (i) Except as provided in paragraph (a)(5)(ii) of this section, for purposes of calculating net worth under this paragraph (a)(5):

        (A) The person's primary residence shall not be included as an asset;

        (B) Indebtedness that is secured by the person's primary residence, up to the estimated fair market value of the primary residence at the time of the sale of securities, shall not be included as a liability (except that if the amount of such indebtedness outstanding at the time of sale of securities exceeds the amount outstanding 60 days before such time, other than as a result of the acquisition of the primary residence, the amount of such excess shall be included as a liability); and

        (C) Indebtedness that is secured by the person's primary residence in excess of the estimated fair market value of the primary residence at the time of the sale of securities shall be included as a liability;

    (ii) Paragraph (a)(5)(i) of this section will not apply to any calculation of a person's net worth made in connection with a purchase of securities in accordance with a right to purchase such securities, provided that:

        (A) Such right was held by the person on July 20, 2010;

        (B) The person qualified as an accredited investor on the basis of net worth at the time the person acquired such right; and

        (C) The person held securities of the same issuer, other than such right, on July 20, 2010.

(6) Any natural person who had an individual income in excess of $200,000 in each of the two most recent years or joint income with that person's spouse in excess of $300,000 in each of those years and has a reasonable expectation of reaching the same income level in the current year;

(7) Any trust, with total assets in excess of $5,000,000, not formed for the specific purpose of acquiring the securities offered, whose purchase is directed by a sophisticated person as described in § 230.506(b)(2)(ii); and

(8) Any entity in which all of the equity owners are accredited investors.

QUALIFIED INSTITUTIONAL BUYER – 17 CFR
§ 230.144A

(a) Definitions. (1) For purposes of this section, qualified institutional buyer shall mean:

    (i) Any of the following entities, acting for its own account or the accounts of other qualified institutional buyers, that in the aggregate owns and invests on a discretionary basis at least $100 million in securities of issuers that are not affiliated with the entity:

        (A) Any insurance company as defined in section 2(a)(13) of the Act;

NOTE: A purchase by an insurance company for one or more of its separate accounts, as defined by section 2(a)(37) of the Investment Company Act of 1940 (the 'Investment Company Act'), which are neither registered under section 8 of the Investment Company Act nor required to be so registered, shall be deemed to be a purchase for the account of such insurance company.

        (B) Any investment company registered under the Investment Company Act or any business development company as defined in section 2(a)(48) of that Act;

        (C) Any Small Business Investment Company licensed by the U.S. Small Business Administration under section 301(c) or (d) of the Small Business Investment Act of 1958;

        (D) Any plan established and maintained by a state, its political subdivisions, or any agency or instrumentality of a state or its political subdivisions, for the benefit of its employees;

        (E) Any employee benefit plan within the meaning of title I of the Employee Retirement Income Security Act of 1974;

        (F) Any trust fund whose trustee is a bank or trust company and whose participants are exclusively plans of the types identified in paragraph (a)(1)(i) (D) or (E) of this section, except trust funds that include as participants' individual retirement accounts or H.R. 10 plans.

        (G) Any business development company as defined in section 202(a)(22) of the Investment Advisers Act of 1940;

        (H) Any organization described in section 501(c)(3) of the Internal Revenue Code, corporation (other than a bank as defined in section 3(a)(2) of the Act or a savings and loan association or other institution referenced in section 3(a)(5)(A) of the Act or a foreign bank or savings and loan association or equivalent institution), partnership, or Massachusetts or similar business trust; and

(I) Any investment adviser registered under the Investment Advisers Act.

(ii) Any dealer registered pursuant to section 15 of the Exchange Act, acting for its own account or the accounts of other qualified institutional buyers, that in the Qualified Institutional Buyer – 17 CFR § 230.144A aggregate owns and invests on a discretionary basis at least $10 million of securities of issuers that are not affiliated with the dealer, Provided, That securities constituting the whole or a part of an unsold allotment to or subscription by a dealer as a participant in a public offering shall not be deemed to be owned by such dealer;

(iii) Any dealer registered pursuant to section 15 of the Exchange Act acting in a riskless principal transaction on behalf of a qualified institutional buyer;

NOTE: A registered dealer may act as agent, on a non-discretionary basis, in a transaction with a qualified institutional buyer without itself having to be a qualified institutional buyer.

(iv) Any investment company registered under the Investment Company Act, acting for its own account or for the accounts of other qualified institutional buyers, that is part of a family of investment companies which own in the aggregate at least $100 million in securities of issuers, other than issuers that are affiliated with the investment company or are part of such family of investment companies. Family of investment companies means any two or more investment companies registered under the Investment Company Act, except for a unit investment trust whose assets consist solely of shares of one or more registered investment companies, that have the same investment adviser (or, in the case of unit investment trusts, the same depositor), Provided That, for purposes of this section:

(A) Each series of a series company (as defined in Rule 18f-2 under the Investment Company Act [17 CFR 270.18f-2]) shall be deemed to be a separate investment company; and

(B) Investment companies shall be deemed to have the same adviser (or depositor) if their advisers (or depositors) are majority-owned subsidiaries of the same parent, or if one investment company's adviser (or depositor) is a majority-owned subsidiary of the other investment company's adviser (or depositor);

(v) Any entity, all of the equity owners of which are qualified institutional buyers, acting for its own account or the accounts of other qualified institutional buyers; and

(vi) Any bank as defined in section 3(a)(2) of the Act, any savings and loan association or other institution as referenced in section 3(a)(5)(A) of the Act, or any foreign bank or savings and loan association or equivalent institution, acting for its own account or the accounts of other qualified institutional buyers, that in the aggregate owns and invests on a discretionary basis at least $100 million in securities of issuers that are not affiliated with it and that has an audited net worth of at least $25 million as demonstrated in its latest annual financial statements, as of a date not more than 16 months preceding the date of sale under the Rule in the case of a U.S. bank or savings and loan association, and not more than 18 months preceding such date of sale for a foreign bank or savings and loan association or equivalent institution.

(2)   In determining the aggregate amount of securities owned and invested on a discretionary basis by an entity, the following instruments and interests shall be excluded: bank deposit notes and certificates of deposit; loan participations; repurchase agreements; securities owned but subject to a repurchase agreement; and currency, interest rate and commodity swaps.

(3)   The aggregate value of securities owned and invested on a discretionary basis by an entity shall be the cost of such securities, except where the entity reports its securities holdings in its financial statements on the basis of their market value, and no current information with respect to the cost of those securities has been published. In the latter event, the securities may be valued at market for purposes of this section.

(4)   In determining the aggregate amount of securities owned by an entity and invested on a discretionary basis, securities owned by subsidiaries of the entity that are consolidated with the entity in its financial statements prepared in accordance with generally accepted accounting principles may be included if the investments of such subsidiaries are managed under the direction of the entity, except that, unless the entity is a reporting company under section 13 or 15(d) of the Exchange Act, securities owned by such subsidiaries may not be included if the entity itself is a majority-owned subsidiary that would be included in the consolidated financial statements of another enterprise.

(5)   For purposes of this section, riskless principal transaction means a transaction in which a dealer buys a security from any person and makes a simultaneous offsetting sale of such security to a qualified institutional buyer, including another dealer acting as riskless principal for a qualified institutional buyer.

(6)   For purposes of this section, effective conversion premium means the amount, expressed as a percentage of the security's conversion value, by which the price at issuance of a convertible security exceeds its conversion value.

(7)   For purposes of this section, effective exercise premium means the amount, expressed as a percentage of the warrant's exercise value, by which the sum of the price at issuance and the exercise price of a warrant exceeds its exercise value.

INDEX